LIPIDS
of
PATHOGENIC
FUNGI

LIPIDS
of
PATHOGENIC
FUNGI

Edited by

RAJENDRA PRASAD

School of Life Sciences
Jawaharlal Nehru University
New Delhi
India

MAHMOUD A. GHANNOUM

Department of Dermatology
Case Western Reserve University
Cleveland, Ohio
U.S.A.

CRC Press
Taylor & Francis Group
Boca Raton London New York

CRC Press is an imprint of the
Taylor & Francis Group, an **informa** business

First published 1996 by CRC Press
Taylor & Francis Group
6000 Broken Sound Parkway NW, Suite 300
Boca Raton, FL 33487-2742

Reissued 2018 by CRC Press

A Library of Congress record exists under LC control number: 96024683

ISBN 13: 978-1-138-50603-9 (hbk)
ISBN 13: 978-0-203-71152-1 (ebk)

Visit the Taylor & Francis Web site at http://www.taylorandfrancis.com and the CRC Press Web site at http://www.crcpress.com

Acquiring Editor:	Marsha Baker
Cover design:	Denise Craig
PrePress:	Kevin Luong

PREFACE

Lipids of fungi have attracted the attention of a number of scientists because they not only perform storage and architectural function, but are also involved in a number of other vital functions of the cells. With the recent realization that lipids are also involved in the anchoring of glycoprotein and can mediate cellular transduction mechanisms, there are concerted efforts to establish their importance in pathogenicity. A recent increase in various fungal infections, due to *Candida*, *Aspergillus*, *Blastomyces* and *Histoplasma* spp. and Dermatophytes, has attracted an interest to study the biochemistry of responsible fungal pathogens. Lipids provide a site of action of many antifungals, which acts as a barrier for the cellular ingress of various metabolites, in addition to their role in dimorphism, adherence and virulence.

In spite of the importance of lipids in pathogenic fungi, there is hardly any book which exclusively deals with these aspects. There are, however, occasional reviews which have dealt with lipids of some of the pathogenic fungi. In the form of this volume, we have made this attempt at gathering relevant data together into a single comprehensive reference book. At present, when interdisciplinary research has emerged as the only meaningful approach to deal with the complex academic challenges, a book by one or two authors could not satisfactorily cover numerous aspects of a growing field. It is for such reasons that we have compiled a multiauthored book where inputs from various experts in the field have been brought together to present the whole gamut of lipids of pathogenic fungi.

This book does not claim to be the exhaustive account of lipids of all pathogenic fungi but rather deals only with a select few fungi whose infections have become a serious threat to health. The book deals with the lipid composition of various pathogenic fungi and a critical assessment of its metabolism. The emerging role of lipids in pathogenesis is also included in the discussion by several authors. The absence of one or the other aspect of lipids of pathogenic fungi in chapters presented in this volume only reflects the lack of data. It is hoped that this volume will provide needed impetus to conduct research in unexplored areas of pathogenic fungi. One of the chapters is completely devoted to introducing different pathogenic fungi and it should especially be useful to researchers who are new to the field or those who would like to have a brief insight into the clinical manifestations of infections caused by these agents. Two other chapters exclusively deal with antifungals. We thought it would be better if a discussion on the possible antifungals agents, as well as some non-conventional agents, and their targets were also included in the same volume which is devoted to those very organisms against which these drugs are developed. Thus the chapters of the book provide an overview of a growing field. Individually and collectively these chapters should be stimulating to the reader and should serve as springboards for further in-depth studies. It is sincerely hoped that this book will be useful not only to clinicians but also to those who are engaged in the important task of resolving diverse problems associated with fungal pathogenicity.

We would like to thank our contributors for readily agreeing to participate in this book and for their forbearance and patience whilst the final chapters were awaited. Our wives and families have been as usual the main casualties during this project. Our sincere thanks to them for being very supportive of this project and very generous in allowing us the time it required. We remain indebted to our students who took time out to help us by very critically reading the text and in offering valuable suggestions. Thanks are also due to Ms. Marsha Baker, CRC Press, Inc., for her help in the preparation of the camera ready copies. Her assistance was instrumental in shortening the time taken to finalize the volume.

On a personal note, deepest gratitude is due from myself, Mahmoud A. Ghannoum, to the following people without whom, through the tribulations of the past five years, my arrival in and assimilation into the United States would not have been possible: Dr. John E. Bennett, Dr. John E. Edwards, Jr., Dr. Jack Sobel and Dr. Douglas Webb.
My sincere thanks.

R.Prasad
M.A. Ghannoum

CONTRIBUTORS

Charles W. Bacon
Department of Biochemistry,
Emory University,
Atlanta, GA 30322, USA.

Kathleen Barr
Department of Microbiology and Immunology,
Uniformed Services University of the Health Sciences,
Bethesda, MD 20814, USA.

J. Basu
Department of Chemistry,
Bose Institute,
Calcutta-700009, India.

Jacques Bolard
Laboratoire de Physique et Chimie Biomoléculaire,
Université Paris VI,
75252 Paris Cedex 05, France.

H. M. Calvet
Harbor-UCLA Medical Center,
Torrance, CA 90509, USA.

Warren M. Casey
Glaxo Five Moore Drive,
Research Triangle Park, NC 27709, USA.

P. Chakrabarti
Department of Chemistry,
Bose Institute,
Calcutta-700009, India.

Mahmoud A. Ghannoum
Department of Dermatology,
University Hospitals of Cleveland,
Case Western Reserve University,
Cleveland, Ohio 44106, USA

Althea M. Grant
Department of Biochemistry,
Emory University,
Atlanta, GA 30322, USA.

Ashraf S. Ibrahim
Harbor-UCLA Medical Center,
Torrance, CA, 90509, USA.

G. K. Khuller
Department of Biochemistry,
Post Graduate Institute of Medical Education and Research,
Chandigarh-160012, India.

Yasuo Kitajima
Department of Biochemistry,
Gifu University, School of Medicine,
Tsukasamachi-40, Gifu 500, Japan.

Anjni Koul
School of Life Sciences,
Jawaharlal Nehru University,
New Delhi-110067, India.

M. Kundu
Department of Chemistry,
Bose Institute,
Calcutta-700009, India.

D. M. Lösel
Department of Animal and Plant Sciences, University of Sheffield,
Sheffield S10 2TN, UK.

Nandini Manchanda
Department of Biochemistry,
Post Graduate Institute of Medical Education and Research,
Chandigarh-160012, India

Alfred H. Merrill, Jr.
Department of Biochemistry,
Emory University,
Atlanta, GA 30322, USA.

Jeannine Milhaud
Laboratoire de Physique et Chimie Biomoléculaire,
Université Paris VI,
75252 Paris Cedex 05, France.

Pranab K. Mukherjee
School of Life Sciences,
Jawaharlal Nehru University,
New Delhi-110067, India.

Yoshinori Nozawa
Department of Biochemistry,
Gifu University,
School of Medicine, Tsukasamachi-40, Gifu 500, Japan.

Leo W. Parks
Department of Microbiology,
North Carolina State University,
Raleigh, NC 27965, USA.

Rajendra Prasad
School of Life Sciences,
Jawaharlal Nehru University,
New Delhi-110067, India.

Homayoon Sanati
Harbor-UCLA Medical Center,
Torrance, CA, 90509, USA.

M. Sancholle
Laboratoire de Cryptogamie/Phytopathologie,
Université du Littoral,
B.P. 699-62228 Calais Cedex,
France.

Elaine Wang
Department of Biochemistry,
Emory University,
Atlanta, GA 30322, USA.

TABLE OF CONTENTS

Chapter 1

SELECTED HUMAN PATHOGENIC FUNGI

H. M. Calvet and M. A. Ghannoum

CONTENTS

0-8493-4794-7/96/$0.00+$.50

I. INTRODUCTION

In the era of contemporary medicine, fungi are emerging as important opportunistic pathogens. With the employment of modern chemotherapeutic modalities in the 1960's, susceptible hosts were created for several opportunists and with the advent of the AIDS era in the 1980's, a broad range of opportunistic pathogenic fungi are making their appearance on the clinical scene. Journals on infectious diseases frequently report about new pathogens, documenting the expanding role of pathogenic fungi. The role of commensals, such as *Candida* species, is changing as well, with *Candida* now ranking as the third most common causative agent of nosocomial blood stream infection in most hospitals. The overall incidence of *Candida* bloodstream infections has risen greatly over the last decade, ranging from a 75% increase in small hospitals to over 400% increase in some large tertiary care centers.[1]

Along with the increasing importance of fungal infections, there is a steadily increasing understanding of their pathogenesis, clinical manifestations and therapy. Research is going on in many aspects of fungal pathogenesis as well as on epidemiological risk factors for infection, but much remains to be elucidated. The development of azoles has revolutionized the treatment of many fungal infections; yet for others, the treatment of necessity remains amphotericin B or a combination of drugs. For some fungi, a good therapeutic agent has not been identified as yet. Throughout the world, clinicians are faced with the challenging task of not only keeping up with the arrival of new pathogens, but also with the changing recommendations for therapy. An in-depth discussion of all the human pathogenic fungi is beyond the scope of this chapter; therefore, discussion will be limited to the species that represent most important human pathogens.

II. *CANDIDA* SPECIES

Candida spp. are 4-6 μm diameter yeasts that normally colonize the gastrointestinal and lower female genital tract of humans. *Candida albicans* germinates rapidly under physiological conditions and forms a long thin projection called a germ tube, which may be an important virulence factor. Other species of *Candida* do not form germ tubes as readily, allowing for rapid differentiation of *albicans* from non-*albicans* species *in vitro*. *C. albicans* is the most common species causing human disease, but other pathogenic species include *C. tropicalis*, *C. guillermondii*, *C. krusei*, *C. parapsilosis*, *C. lusitaniae* and *C. pseudotropicalis*. *Torulopsis glabrata*, once classified as a *Candida* species, is often included in the group causing candidiasis (Figure 1).

Many components of the immune system are active against *Candida*, maintaining the balance achieved with a colonizing organism. T-lymphocyte function appears to be important in controlling mucosal/cutaneous infection, as

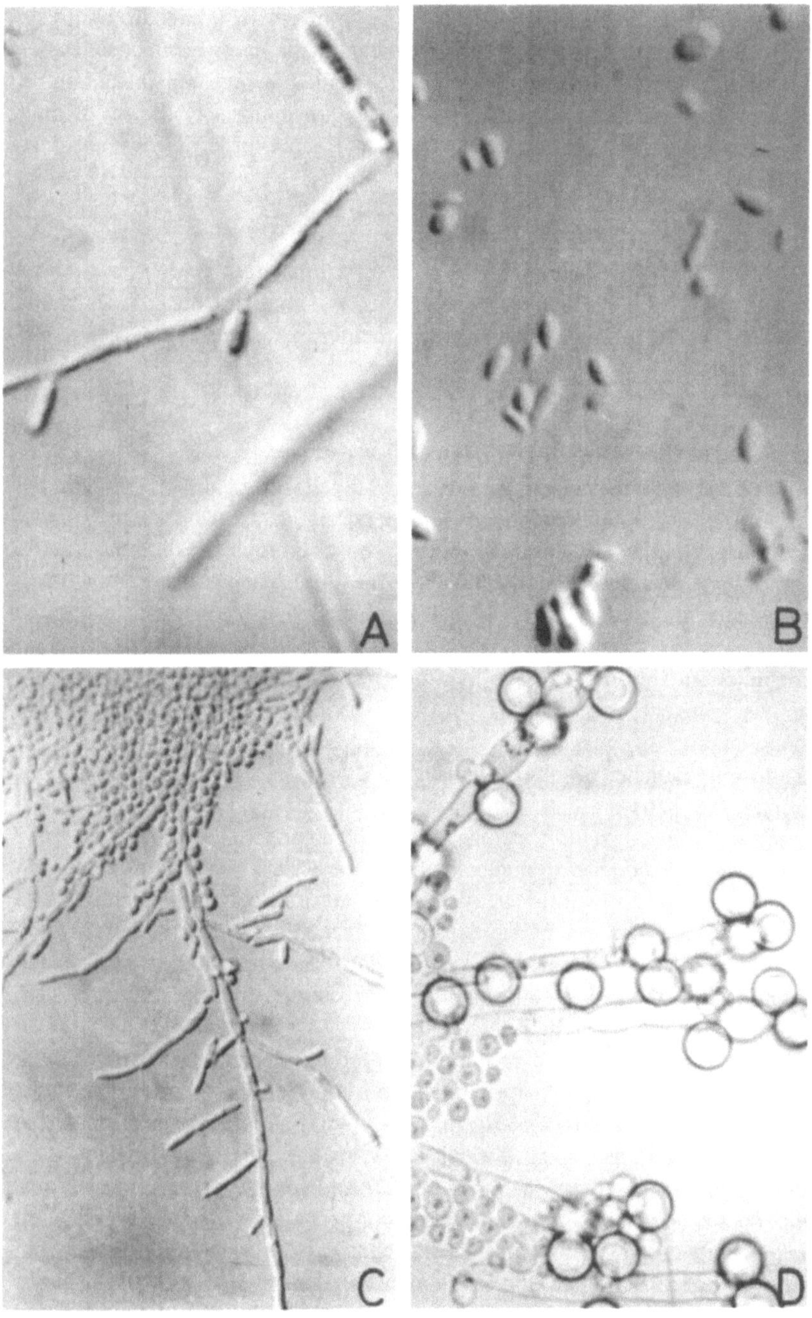

Figure 1. Different species of *Candida*. (a) *C. tropicalis*, (b) *C. lusitaniae*, (c) *C. parapsilosis*, (d) *C. albicans*.

evident by the fact that AIDS patients suffer from severe and recurrent thrush, esophagitis and vaginitis. Similarly, patients with chronic mucocutaneous candidiasis (CMC) suffer from severe mucosal, skin and nail infections. In invasive diseases, neutrophils and monocytes are important because of their phagocytic function, and complement and IgG appear to efficiently opsonize *Candida in vitro.*[2]

A. CLINICAL MANIFESTATIONS

The clinical presentations of *Candida* infections are numerous, and could account for a whole chapter by themselves. A simplified breakdown of the infections would be mucosal/cutaneous infections which may be disseminated, or deep-seated.

1. Mucosal/cutaneous infections

Mucosal and cutaneous infections are relatively common problems, and are the most likely to be encountered by general clinicians. *Candida* may infect any mucosal surface of the body, but the most common sites are the mouth, esophagus and vagina. Oral infection is termed thrush and can present whitish-gray plaques or patches on erythematous mucosa that bleeds easily when the plaques are removed. Thrush can also present only mucosal atrophy or with inflammation and cracking at the margins of the mouth, called angular cheilitis. Patients' complaints of oral irritation may be completely asymptomatic. Esophageal involvement appears like the plaques seen in thrush, and patients usually complain of dysphagia or retrosternal pain or burning. Vaginitis presents vaginal itching and a cheesy discharge; physical examination reveals whitish plaques on erythematous mucosa. As mentioned previously, *Candida* is a normal colonizer of these sites, but various factors can lead to the overgrowth of yeast, leading to localized inflammation and infection. The most common predisposing factor is use of antibiotic, which destroys normal flora and allows overgrowth of *Candida.*[2] Other predisposing factors include diabetes mellitus, use of corticosteroids, or immunosuppression due to chemotherapy, malignancy or HIV.

Cutaneous infections are even more common and can occur in patients with no predisposing factors. *Candida* infection usually occurs in areas with moist or macerated skin, such as intertriginal areas of the axilla, groin, or under the breasts or fat folds. The classical appearance of intertrigo is that of a erythematous plaque with scalloped borders and small round satellite lesions. Obese patients, diabetics and those with poor personal hygiene are predisposed to these infections. Men having sexual contact with women suffering from candidal vaginitis may develop candidal balanitis or inflammation of the penis. In rare cases, *Candida* can cause a more diffuse, macular skin infection, called generalized cutaneous infection, or it can infect the hair follicles. *Candida* rarely causes onychomycosis, but most commonly causes paronychia, an infection starting in the soft tissue just lateral to the nail margin and extending under the nail. *Candida* is also involved in a rare syndrome called CMC, or chronic mucocutaneous candidiasis. The syndrome is characterized by *Candida* infections of the skin, mucous membranes, hair and nails, leading to dysfiguring lesions of the face, scalp and hands. The underlying immune defect causing this syndrome appears to be failure of T-lymphocytes to respond to *Candida* antigens and some other stimuli. Half of the patients with this disorder are allergic to all skin tests, suggesting a wide range of T-lymphocyte defects. Although the patients have normal B-lymphocytes function and antibody titers, they often have problems with recurrent bacterial infections.

2. Disseminated/deep-seated infections

The classification of disseminated or deep-seated candidal infections suffers from a lack of standardization of terminologies. *Candida* infects a number of organ systems, including the eyes, kidneys, skin, liver, brain, spleen and heart by hematogenous seeding during fungemia. Alternatively, organs such as the peritoneum, bladder or kidneys become infected secondary to direct spillage from the gut or introduction of the yeast by instrumentation. The difficulty in developing standardized terminology reflects the difficulty in distinguishing the pathogenesis of these infections. In cases of presumed hematogenous dissemination, often fungemia is not detected. In fact, only 50-60% of the patients with autopsy-proven disseminated candidiasis have *ante-mortem* blood cultures isolate of *Candida* species. The reasons for this low yield are unclear, since *Candida* grows well in the Bactec system (the standard blood culture system used in most hospitals) and even better in Dupont isolators (the lysis-centrifugation system). Nevertheless, for the purpose of differentiation of the disease entities, thought to be associated with hematogenous dissemination, these will be discussed separately from those secondary to instrumentation or gut contamination.

Hematogenously disseminated candidiasis is most commonly a hospital acquired infection, although some patients present community acquired infections, such as intravenous drug abusers or patients with gastrointestinal cancers. Among hospitalized patients, some of the risk factors for fungemia include length of stay in hospital, use of multiple antibiotics, central venous catherization (Hickman, Swan-Ganz, or hemodialysis catheters), use of total parental nutrition, endotracheal incubation, colonization with *Candida* and use of corticosteroids.[3] Burn patients and neutropenic patients also run increased risk due to the lack of normal host defenses. Patients developing fungemia are worse than those without fungemia; Wey et al.[4] showed an overall mortality of 57% in fungemic patients compared to only 19% in case-matched controls, yielding an attributable mortality of approximately 38% for fungemia.

Potential end-organ complications of fungemia are also numerous. Candidal endophthalmitis presents white, fluffy chorioretinal lesions that rapidly progress to involve other structures, often leading to permanent ocular damage or blindness. Infections in the central nervous system (CNS) are rare, but can be present either as meningitis or small parenchymal abscesses. Infection of the cardiovascular system is more common and is serious. *Candida* endocarditis usually occurs in patients with prosthetic valves, underlying valvular disease, prolonged use of central lines, previous endocarditis or intravenous drug use. Pericarditis and myocarditis are rare complications of hematogenous dissemination. Vascular structures can also become infected, most commonly prosthetic vascular grafts. Abdominal abscesses are also seen with hematogenous dissemination, involving liver, spleen, pancreas or kidneys. Subacute or chronic liver and spleen involvement, called hepatosplenic candidiasis or chronic disseminated candidiasis, is seen in patients with prolonged episodes of neutropenia but is rare in immunocompetent patients. This disease occurs when a neutropenic patient on multiple antimicrobials has persistent fever and may have some degree of abdominal pain. The musculoskeletal system can also be involved, with hematogenous seeding leading to osteomyelitis or arthritis. Finally, completing the organ systems, there is skin involvement, with three possible presentations of hematogenously disseminated *Candida.*[2] Nodular erythematous lesions measuring from 0.5-1.0 cm may be seen singly or throughout the whole body, as can lesions resembling ecthyma gangrenosum, which starts as a pustular lesion and then develops central necrosis with surrounding violaceous discoloration. Rarely, it also may be present as purpura fulminans.

Many of the same organ systems may also be involved in localized, or non-hematogenous disease and, in some instances, it is difficult to identify the process which led to the infection. As mentioned previously, *Candida* can infect the eye as a result of contamination during ocular surgery. CNS infection may also occur, usually as a complication of surgery or from infection of a ventriculoperitoneal shunt. In these cases, meningitis is the most frequent manifestation, and is complicated by a high rate of hydrocephalus. Cardiac infections may occur after surgery, most commonly prosthetic valve endocarditis or pericarditis. Visceral abscesses are rarely, if ever, seen in the absence of documented or presumed hematogenously disseminated infection; however, peritonitis or intraperitoneal abscess may often be seen as the sole site of infection. These peritoneal infections are usually either due to perforation of a viscus secondary to cancer, penetrating trauma, inflammatory bowel disease or peptic ulcer disease, or due to infection of a peritoneal dialysis catheter. Peritonitis may secondarily lead to hematogenous dissemination in approximately 25% of cases caused by gut perforation, but is extremely rare in those caused by dialysis catheters. Involvement of the urinary system also differs from the hematogenous picture. *Candida* cystitis is very common in patients with bladder catheters, especially patients with diabetes mellitus or on multiple antimicrobials. Once in the bladder, yeast can also migrate up to the upper urinary tract, causing papillary necrosis, invasion of the calices, and perinephric abscess or fungus ball in the renal pelvis. Obstruction within the urinary system predisposes involvement of the upper tract and secondary hematogenous dissemination. Infection of the musculoskeletal system may occur as a result of trauma, surgery or intra-articular injections. Arthritis secondary to injections is usually due to a non-*albicans* species of *Candida.*[2]

B. DIAGNOSIS

Unlike some other fungal infections such as cryptococcosis and histoplasmosis, there is no reliable method for serodiagnosis of candidal infections. Antigen detection systems, developed to date, are plagued by unacceptably high false negative rates. Latex agglutination assay for the mannan antigen is less than 30% sensitive on initial testing, but testing of repeated samples, or treatment of the sample with protease, may increase the sensitivity up to 80%.[5] Assays for cytoplasmic proteins and heat labile glycoproteins have had similar rates of failure. Antibody detection systems suffer not only from lack of sensitivity, but also from high incidence of false positives secondary to superficial colonization with *Candida.*

As mentioned previously, yield from blood cultures is also unreliable. Positive culture results from other sites are variable, with high incidence of positivity in urine and abdominal cultures, but low positivity in cerebrospinal fluid (CSF). Biopsy of attainable tissue, such as skin lesions, may aid in diagnosis, with histopathology showing yeast forms on methenamine silver stain or tissue gram stain. In many cases, however, the diagnosis is made at the risk for disseminated candidal infection which displays a syndrome consistent with a candidal infection and which does not respond to antimicrobial chemotherapy.

C. TREATMENT

For years, the mainstay of therapy for candidal infection was amphotericin B. Now, with the advent of the triazoles, oral or intravenous azole therapy has presented the clinician with more options. In many cases, the optimal therapy for each type of infection is not known because few controlled clinical trials have been performed. 5-Fluorocytosine (5-FC) is another agent used, often in combination with either an azole or amphotericin B, for certain infections. 5-FC is rarely used alone, however, because of the rapid development of resistance when used as monotherapy. There are differences in the *in vitro* sensitivities to the different agents among the species of *Candida. C. albicans*, which causes the majority of candidal infections, generally displays low minimal inhibitory concentrations (MICs) of both azoles and amphotericin B, but non-*albicans* species of *Candida* are not reliably sensitive to the azoles. In particular, *C. krusei* is known to be inherently resistant to fluconazole. On the other hand, *C. lusitaniae* is known to be resistant to amphotericin B. Therefore, the treatment of candidal infections depends on the inherent sensitivities of the species isolated. The treatments, to be discussed hereafter, are for the most part a reflection of what the medical community in general is practicing, since little data is available to direct therapy.

Cutaneous and mucosal infections are usually treated with topical azoles. A number of different topical ointments are available for cutaneous infection, including nystatin, the imidazoles, clotrimazole, miconazole, econazole, butoconazole and tioconazole, or the triazole terconazole. They appear, for the most part, to be equally efficacious. For mucosal infections involving the mouth, nystatin swish-and-swallow or clotrimazole troches are the first line of therapy, with oral fluconazole being reserved for persistent or very severe cases. *Candida* esophagitis is usually treated with oral fluconazole in mild cases, or intravenous fluconazole or amphotericin B if the patient cannot take oral medications. *Candida* vaginitis, like oral infections, is usually treated initially with topical azole preparations or with oral fluconazole reserved for inadequate responders or for patients likely to be non-compliant with therapy. Chronic mucocutaneous candidiasis is usually treated with long-term oral azole therapy, such as ketoconazole, fluconazole or itraconazole. Various immunostimulants, such as transfer factor, a cell-free leukocyte extract, have been used in this condition, and some have shown to be of benefit but are still at experimental stages.[2]

The treatment of hematogenous infection is a subject of debate. Recently, a controlled, randomized clinical trial was completed comparing amphotericin B and fluconazole for treatment of documented candidemia in non-neutropenic patients. Overall, there was no significant difference in mortality, although there were more early deaths in the amphotericin B arm. Looking at efficacy, there was a trend of better efficacy in the amphotericin B arm that did not reach statistical significance in either the primary or the secondary analyses.[6] Most of the isolates were *C. albicans*; therefore, no conclusions could be drawn about the treatment of fungemia due to non-*albicans* species. It remains to be demonstrated that higher doses of fluconazole would prove to be more efficacious than amphotericin B, or that the addition of 5-FC to either therapy would be of benefit. The decision to use a particular agent, therefore, depends on the clinical situation and the preference of the clinician since the superiority of either agent is still arguable.

The treatment of other deep-seated infections is less well studied. Therapy of ocular infections often involves intravitreal, intravenous and/or topical amphotericin B, and vitrectomy may be needed in certain cases. Treatment of cardiovascular infections involves surgical removal of the infected focus, such as the valve or vascular graft. Medical therapy, following surgery, usually consists of high doses of amphotericin B for an unspecified period of time. Intra-abdominal infections, such as liver, spleen, pancreas or renal abscesses, are more difficult to treat since amphotericin B does not predictably penetrate these organs. In hepatosplenic candidiasis in the neutropenic host, amphotericin B was the traditional drug of choice, but the response rates were poor (only 60-65% responding to long term therapy). A recent report has demonstrated efficacy of fluconazole in this disease, with 14 of 16 treated patients responding to fluconazole, but no randomized studies comparing amphotericin B to fluconazole have been performed.[7] Amphotericin B penetrates into the peritoneum; therefore, either intravenous or intraperitoneal therapy may be used to treat candidal peritonitis. Fluconazole is also known to reach therapeutic levels in peritoneal fluid, making it a less toxic, more easily administered alternative for susceptible *Candida* species. Amphotericin B does not penetrate into urine, so oral fluconazole would be the drug of choice for renal infections from the ascending route. For isolated candidal cystitis, either amphotericin B bladder washings, or oral fluconazole or 5-FC are usually adequate; however, if the patient has a chronic indwelling urinary catheter, incidence of relapse with any of these therapies is high.

III. *ASPERGILLUS* SPECIES

Aspergillus species are found throughout the world with ubiquitous distribution in the environment. *Aspergillus* forms hyphae 2-4 μm wide which are septated and branched at an acute angle to the main stem (Figure 2). The fungus grows well on many kinds of organic material, and *Aspergillus* spores are found throughout the hospitals, in air samples, dust and environment cultures. Exposure to the spores through the respiratory tract is a common occurrence, yet *Aspergillus* is not a common cause of human disease except in the severely immunocompromised population.[8] Among the many species of *Aspergillus* causing invasive disease, *A. fumigatus*, *A. flavus* and *A. niger* are the most common species. Over twelve species have been reported to cause disease in humans. The main host defenses are the phagocytic cells, primarily neutrophils. Complement aids in hyphal damage, but is not a necessary factor, and antibody does not appear to play a significant role. Disseminated *Aspergillus* infections, therefore, tend to occur predominantly in neutropenic hosts, especially those with hematologic malignancies or prolonged neutropenia due to chemotherapy. Invasive aspergillosis can also occur in transplant or on immunosuppressive therapy patients. It is seen with increasing frequency in the AIDS population, usually in patients with very low T-lymphocyte count and other multiple opportunistic infections.

A. CLINICAL MANIFESTATIONS

The presentation of disease differs among the immunocompetent and immunocompromised patient. Spores are introduced into the respiratory tract via inhalation, and can produce disease of the lung or sinuses. There are several well described pulmonary processes that affect immunocompetent patients. Allergic bronchopulmonary aspergillosis is usually seen in patients suffering from asthma or reactive airway disease. It causes fleeting pulmonary infiltrates which are due to transient bronchial plugging with thick mucus that contains *Aspergillus* hyphae. Repeated plugging may lead to bronchiectasis, but some patients experience no permanent damage. In other patients with

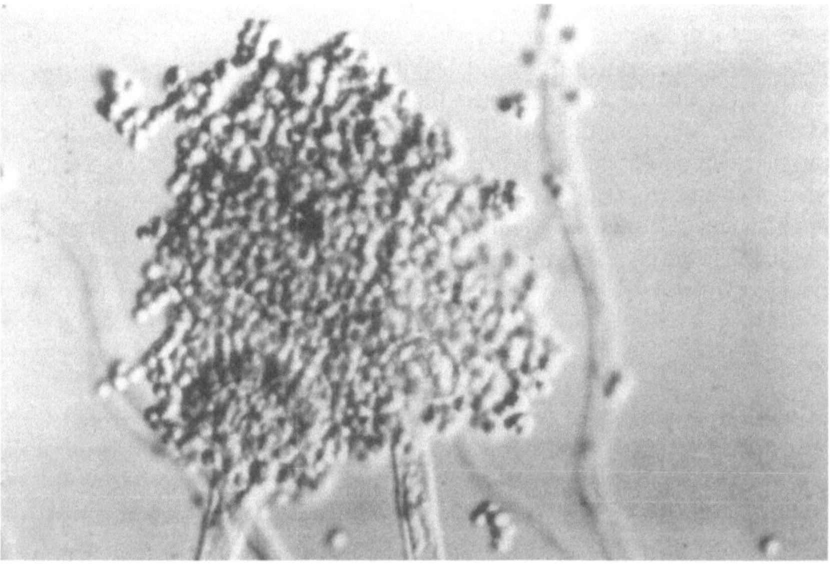

Figure 2. *Aspergillus fumigatus.*

chronic underlying lung conditions, such as tuberculosis, chronic bronchitis or bronchiectasis, asthma, or sarcodiosis, *Aspergillus* species can colonize structural defects, such as cysts, cavities or ectatic bronchi, and grow to form a mass of hyphae which is visible on chest X-ray. These masses of hyphae, or "fungus balls", are called aspergillomas, and in general are benign. In rare cases, however, the fungus will invade adjacent structures, causing fatal hemoptysis, bronchopleural fistula, fungal osteomyelitis, or lung abscess. In the sinuses, *Aspergillus* can form a benign hyphal ball or can slowly invade, eventually involving bony structures or organs abutting the sinuses, such as the eyes or brain. The fungus tends to be angiotropic, invading blood vessels and causing ischemic necrosis of surrounding tissues.

Other diseases in the immunocompetent host include ocular infections, with introduction of the spores through ocular trauma or surgery. Corneal trauma can lead to keratitis which can progress to invade deeper structures, and introduction of the spores during surgery can lead to endophthalmitis. Introduction of the spores during surgery can also complicate cardiac and orthopedic surgery, leading to prosthetic valve endocarditis or infection of prosthetic joints.[8] In the external ear canal, *Aspergillus* can occasionally be seen growing in the debris found in patients with chronic otitis externa, but this seldom represents actual infection. Rarely, *Aspergillus* may cause brain abscess, usually in users of intravenous drugs.

In neutropenic patients, *Aspergillus* can cause a rapidly progressive and invasive pneumonia, initially heralded by dense pulmonic infiltrates and high fever. With prolonged neutropenia, the fungus will then disseminate to other organs, ultimately causing death. Recovery of neutrophil counts, prior to dissemination, leads to cavitation of the pneumonitis, which may be complicated by significant hemoptysis or expectoration of necrotic lung.[8] Invasive pulmonary aspergillosis may also occur in patients with normal neutrophil counts but abnormal neutrophil function, such as patients with organ transplants or those receiving very high doses of corticosteroids.

In the AIDS population, *Aspergillus* can cause an indolent but progressive invasion of lung parenchyma which may present as a cavitation, an ill-defined infiltrate, or pleural-based or nodular lesion.[9] It tends to affect end-stage patients with very low CD4 lymphocyte counts, usually less than 30, and neutropenia is seen in about half of the cases. Other possible associated factors include use of corticosteroids, cytotoxic chemotherapy or marijuana, and presence of other pre-existing pulmonary infections such as *Pneumocystis carinii* pneumonia (PCP) or cytomegalovirus (CMV) pneumonia. Patients usually have symptoms for several weeks to months prior to diagnosis; most common symptoms are fever, cough, or chest pain due to invasion of local sensory neurons. Other sites of infection have been reported in AIDS patients, including the brain in approximately 10% of reported cases. Fifteen percent of cases have been documented to have two or more sites of infection.[10]

B. DIAGNOSIS

In general, aspergillosis is diagnosed through a combination of culture, demonstration of the organisms in tissues and clinical presentation consistent with disease. The problem with culture alone is that the organism may be difficult to isolate from blood and when isolated from sputum or the sinuses, it may represent colonization instead of actual infection. Therefore, a high index of suspicion based on clinical presentation is first needed, followed by biopsy of the appropriate tissue with histological confirmation of tissue invasion by the organism or isolation of the fungus from a normally sterile site. This approach is often not feasible with the neutropenic patient with invasive pulmonary aspergillosis, but recent advances in serodiagnosis may help in solving that problem. A recently developed monoclonal antibody EIA against the glycomannan of *Aspergillus*, which is transiently present in the serum of patients with disseminated disease, has shown high sensitivity and specificity for diagnosing aspergillosis in neutropenic patients.[5] The glycomannan is also present in urine for a prolonged time, and assaying urine samples may lead to a higher yield.

For the non-invasive pulmonary diseases, antigen-antibody precipitins, such as double immunodiffusion or counter-immune electrophoresis (CIE), are helpful in diagnosis.[5] Often patients with allergic bronchopulmonary aspergillosis have to make peripheral eosinophilia which prompts the clinician towards the diagnosis. Precipitins will confirm the diagnosis in approximately 70% of the cases. The test is even more helpful in patients with aspergilloma, yielding a 90% positive result. More sensitive assays are being developed.

C. TREATMENT

Aspergillus species are resistant to most azole antifungals, except itraconazole, making amphotericin B the drug of choice for invasive disease, even though varying resistance to amphotericin B also occurs. Neutropenic patients, with suspected or documented aspergillosis, are initially prescribed high doses of amphotericin B, and therapy is continued for a long time. 5-FC is occasionally used in combination with amphotericin B, but no prospective studies have been done as yet to document efficacy. Despite appropriate therapy, mortality is high in neutropenic patients unless the neutropenia resolves. Surgery plays a prominent role in the treatment of some *Aspergillus* infections, especially in sinus disease, endocarditis and brain abscess. Asymptomatic aspergillomas may be followed without therapy because of the low incidence of invasion, but may require surgical removal, if significant hemoptysis occurs. Allergic bronchopulmonary aspergillosis, in general, does not require antifungal therapy, but tends to respond to systemic glucocorticoids.

The role of itraconazole in treatment of *Aspergillus* infections is not well defined, but is mainly used in immunocompetent patients with mild sinus disease or in AIDS patients who are not tolerant to amphotericin B. A study comparing itraconazole with amphotericin B for invasive pulmonary aspergillosis is currently underway. In AIDS patients, outcome is generally poor no matter what therapy is used, with deaths occurring on an average of 2-4 months after diagnosis, either from aspergillosis or other AIDS-related conditions.[10]

IV. *HISTOPLASMA CAPSULATUM*

Histoplasma capsulatum is a dimorphic fungus displaying a yeast form measuring 2×3 μm and a mycelial form of septate branching hyphae with lateral or terminal spores. The organism is found throughout the Ohio River Valley area of the United States, as well as in many parts of Central and South America and scattered areas of other tropical and subtropical zones. The fungus is easily found in soil contaminated with avian excrement, with blackbirds, pigeons, chickens and bats as the main culprits. The spores are the infectious form; their 2-6 mm diameter size makes them amenable to aerosolization and inhalation with eventual lodging in the alveoli or terminal bronchioles. Within the lung parenchyma, the spores germinate in 2-3 days and then are phagocytosed by lung macrophages. Other macrophages are recruited to the area which then phagocytose the germinating fungus, causing a small parenchymal infiltrate to develop. These infected macrophages then migrate to regional lymph nodes, causing infiltrates at new sites and stimulating host defense mechanisms. An inflammatory response follows, causing caseation necrosis and eventual calcification of involved nodes, leaving lifelong evidence of the exposure to the fungus.[11] However, killing by macrophages is not complete and some yeast forms may persist within the tissues for many years, leading to the possibility of reactivation of disease if the host were to develop defects in cell mediated immunity. Exact mechanism of immunity is unknown, but T-cell function appears to be critical, and presence of specific antibodies, either through previous infection or experimental immunization, appears to be protective. Humoral immunity appears to wane with time.[11]

A. CLINICAL MANIFESTATIONS

Depending on the size of the innoculum and the immunity of the host, presentation of infection ranges from asymptomatic to fulminant disease with imminent death. The most common scenario, occurring in 50-99% of the infections, is that of inhalation of a small to moderate amount of organisms leading to mild, non-specific symptoms rarely prompting the individual to seek medical attention. These patients usually experience fever, headache, chills, cough and they may also have various degrees of chest pain, myalgia and fatigue. Symptoms in these mild cases generally last less than five days, and fewer than 25% of these patients will have radiographic evidence of disease.[11] Heavier inoculations will lead to more severe symptoms lasting a week or longer, and leading to small infiltrates and enlarged hilar and mediastinal lymph nodes on chest X-ray. These infiltrates eventually resolve and may calcify, like the involved lymph nodes, leading to self-limited disease in about 60% of the patients with heavy exposure.

Approximately 40% of patients with symptomatic primary infection present with syndromes other than self-limited pulmonary disease.[12] Disseminated disease, discussed later, develops in about 10%. Arthritis and erythrema nodosum may be the dominant symptoms affecting 5% of patients, and another 10% present with pericarditis; both syndromes are thought to be mediated through immune mechanisms. Chronic pulmonary disease occurs in another 10% of symptomatic primary infections and presents in two fashions. The first is characterized by nodular or alveolar type infiltrate usually in the central lung fields, and histopathology shows organisms in fluid-filled lung spaces. The second is characterized by the presence of thick walled cavities, which probably represent infection of pre-existing bullae, and organisms can be demonstrated in the necrotic wall of the cavity. Cavities are usually located subapically, leading to confusion with tuberculous cavities, and enlarge down toward the base of the lung. Pulmonary histoplasmosis may also lead to other local complications, such as tracheal, bronchial or esophageal obstruction secondary to enlarged lymph nodes, acute pericarditis secondary to adjacent affected lymph nodes, pleural effusion or mediastinal masses. Broncholithiasis may occur secondary to extravasation of lymph nodes into the bronchial tree, and sinus tracts may form secondary to caseous lymph nodes. A rare, and difficult to treat, complication is fibrosing mediastinitis, which leads to entrapment of great vessels and other mediastinal structures.

Disseminated disease complicates approximately one in 2000 acute infections in immunocompetent persons,[12] with a higher incidence in infants and immunocompromised patients. Disseminated disease can involve any organ system with significant portion of lymphoid tissue, such as the liver, spleen, bone marrow and gastrointestinal tract. Less commonly, there may be involvement of the adrenals or the meninges. In mild cases, there are focal accumulation of infected macrophages, which is the most common pathogenesis of adrenal, meningeal and intestinal submucosal disease. Immunocompetent patients may present with an acute, subacute or chronic time course. Acute presentation is more common in infants, and is characterized by high fever, hepatosplenomegaly, gastrointestinal symptoms, anemia, leukopenia and occasionally thrombocytopenia. Death, if left untreated, may occur within several weeks. The subacute presentation develops over the course of months, and is characterized by focal areas of involvement, such as intestinal ulceration, adrenal destruction, meningitis or endocarditis. Fever, hematologic abnormalities and hepatosplenomegaly are present to a lesser degree than those in the acute form, and is fatal within one year if untreated, unless local complications such as perforated colon or adrenal insufficiency causes an early demise. Chronic presentation occurs in adults only and is characterized by months to years of mild symptoms, such as asthenia and weight loss. Fever, hepatosplenomegaly and hematologic abnormalities are usually absent or present only to a mild degree. Focal lesions similar to those seen in the subacute form are also seen in the chronic disease, as are skin lesions or oral ulcerations. Eventual outcome is not well known, but patients have been known to survive with symptoms lasting longer than ten years prior to diagnosis.[11]

Disseminated histoplasmosis also affects a significant number of AIDS patients, with an incidence of 5-27% in endemic areas.[13] It has been reported in patients from non-endemic areas as well, or in patients who have resided for a long time outside an endemic area, suggesting that it is most commonly a relapse of previous infection. AIDS patients may occasionally present with fulminant disease causing a sepsis syndrome, or more commonly with a subacute febrile illness characterized by weeks or months of weight loss, fever, hepatosplenomegaly and lymphadenopathy. Skin and mucous membrane lesions are also relatively common. Similar to immunocompetent patients, anemia, leukopenia and thrombocytopenia may be prominent laboratory findings. Reports of localized disease, such as colitis, are also in the literature. In general, the patients presenting with sepsis syndrome do very poorly, whereas others, if diagnosed in an early stage, respond to treatment but require life-long suppression therapy.

B. DIAGNOSIS

Pulmonary disease may be diagnosed by isolation of yeast from the sputum, but yeast is found in the sputum of less than 10% of patients with acute disease and only 60% of patients with chronic disease. Therefore, the diagnosis in immunocompetent patients with the pulmonary form of the disease is usually made through serological tests, but these are not 100% sensitive. The three main forms of serodiagnosis are the radioimmunoassay (RIA) for IgG and IgM antibodies, immunodiffusion (ID) of the H and M glycoproteins and complement fixation (CF) of yeast and mycelial phase antigens. The CF test for the yeast antigen is positive in up to 75% of patients with acute pulmonary disease by 6 weeks post-exposure, 93% of patients with chronic pulmonary histoplasmosis, but only 50-85% of patients with disseminated disease.[14] RIA shows a greater sensitivity, turning positive earlier than the CF in acute disease, but is not a widely available test. Immunodiffusion is less sensitive than CF or RIA but is very specific; therefore, it is useful to aid the confirmation of the diagnosis since the CF test has cross-reactivity with *Blastomyces dermatitidis*. CF titers may remain elevated for years after exposure. So positive titers do not indicate acute disease. Skin testing is not useful for diagnosis because of the high incidence of positive tests in endemic areas and the significant number of false negatives in patients with disease. Skin testing may also boost the CF titer to the mycelial phase histoplasmin antigen.

Due to varying degrees of immunosuppression,[13] serology is not very useful in disseminated disease. The diagnosis in these cases is made by isolation of the organism from blood culture or other sterile site and demonstration of the organism in tissue histopathology, or detection of the *H. capsulatum* polysaccharide antigen (HPA) in blood, urine or cerebrospinal fluid. Cultures often take several weeks to turn positive, so biopsy of appropriate tissues or assay for the HPA can yield a more immediate answer. HPA is positive in 80-90% of disseminated cases, but in only 20-25% of localized infections.[12] Levels of HPA drop with therapy and a subsequent rise may indicate relapse. HPA is the best test for diagnosis in AIDS patients, and periodic determination of urinary antigen levels is indicated to follow response.

C. TREATMENT

Treatment differs between the immunocompetent and immunocompromised host. Mild acute pulmonary disease usually does not require antifungal therapy, nor do the immunologic complications of disease, such as erythema nodosum or pericarditis. In immunocompetent patients, ketoconazole is effective for the chronic pulmonary disease, severe acute pulmonary disease and mild cases of non-meningeal disseminated disease. Itraconazole has also been shown to be effective for both chronic pulmonary disease and disseminated disease.[15] Fluconazole is being investigated as another possible agent, but the *in vitro* activity of fluconazole against *H. capsulatum* is lower than that of ketoconazole or itraconazole. Suggested duration of oral therapy is 2-3 months for acute pulmonary disease, 12 months for chronic pulmonary disease and at least 6 months for disseminated disease.[15] Actual duration of therapy should depend on the patient's clinical and serological response to therapy. Amphotericin B is used in more severe disseminated/acute pulmonary cases, or in patients with chronic pulmonary disease who are not likely to comply with long term oral azole therapy. Optimal duration of therapy with amphotericin B is not known, but relapse rates of 25% have been reported in patients receiving less than 30 mg/kg. Adjunctive corticosteroids may be of benefit in addition to amphotericin B therapy in patients with acute pulmonary disease and hypoxia.[5] Unfortunately, no treatment is useful for the rare but fatal complication of fibrosing mediastinitis.

In the AIDS or immunosuppressed population, induction therapy with at least 500 mg of amphotericin B is the current standard, followed by maintenance therapy. Ketoconazole was ineffective as an agent for induction, but some studies have shown itraconazole to have a 85% response rate to initial induction therapy. Failure to use maintenance therapy leads to relapse rates in up to 80% of patients,[13] and various agents have been investigated for use in maintenance therapy. Among the azoles, ketoconazole is not effective for maintenance, failing in 40-50% of the cases. Wheat et al.[16] found itraconazole to be very effective, with 95% of the patients being disease free during 97 patient-years of follow-up. Maintenance therapy with weekly amphotericin B also works reasonably well, with a 10-20% rate of relapse. The superiority of any agent for maintenance is not known since no comparative trials have been performed, but the effectiveness, ease of administration and low toxicity of itraconazole have made it the current drug of choice.

V. *CRYPTOCOCCUS NEOFORMANS*

Cryptococcosis is caused by an encapsulated yeast measuring approximately 4-6 μm in diameter (Figure 3). The capsule varies in size depending on growth conditions, viz., carbon dioxide concentration, nutrient availability and temperature. In general, the fungus tends to be sparsely encapsulated in nature and highly encapsulated in host tissues. There are four different serotypes of the organism, classified by the capsular polysaccharide: A, B, C and D. A and D are grouped together as var. *neoformans* and B and C as var. *gattii*. These two varieties differ in geographical distribution, biochemical characteristics and DNA content. It is unclear if these varieties cause different clinical manifestations, but most of the clinical isolates in the U.S. are serotypes A or D. *Cryptococcus* enjoys a world-wide distribution, found in soil and especially in pigeon droppings, possibly because of the organism's ability to utilize creatinine as a nitrogen source.[17]

Host immunity to *Cryptococcus* has not been completely delineated. It is clear that functional host defenses are important since many people are exposed and develop a positive skin test, but very few show signs of disease. Most, but not all, cases of disease are seen in patients with altered cellular immunity, such as patients with AIDS, lymphomas, sarcodiosis, or patients

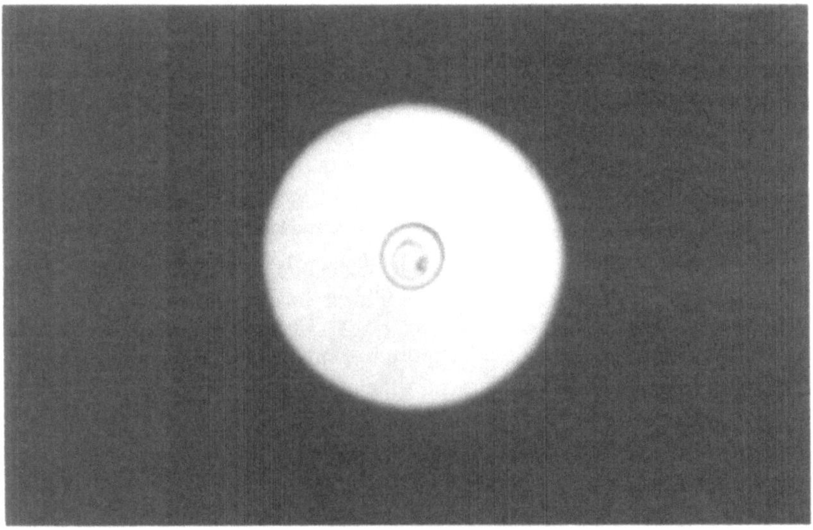

Figure 3. *Cryptococcus neoformans.*

taking large doses of corticosteroids or other immunosuppressive agents. There is also a possible predilection in patients with diabetes mellitus, but this association has not been proven. Cells shown to be important in the defense against cryptococcosis include neutrophils, monocytes, macrophages and natural killer cells. Complement and antibodies may also play a role, but what amount they contribute to host defense against the organism is not clear.

A. CLINICAL MANIFESTATIONS

Pathogenesis of cryptococcosis appears to be similar to histoplasmosis since it initially starts by inhalation of fungus particles. However, unlike histoplasmosis, there are few cases of symptomatic primary infection, and most patients with positive skin tests will have no radiographic evidence of prior disease. Isolated pulmonary disease may occur which, if left untreated, may spontaneously resolve, remain stable or worsen. Findings on radiography are that of areas of focal pneumonitis, which only occasionally cavitate. Lymphadenopathy is unusual. Dissemination may occur at any time, even when the pulmonary process is resolving.[17]

The most common organ affected by dissemination is the CNS, but patients may also present with skin, bone or eye involvement. Meningitis or meningoencephalitis are the most common presentations of CNS disease and are also the most common cause of death from cryptococcosis. Symptoms may be present for a long time before the patient seeks medical attention, and generally consist of headache, dizziness or lack of coordination. As the infection progresses, sleepiness and visual changes develop, and if untreated coma eventually supervenes. Waxing and waning symptoms, combined with a paucity of physical findings, may lead the clinician away from considering a

CNS infection. Most patients are afebrile, have no nuchal rigidity, and focal neurological or physical findings such as cranial nerve palsies or papilledema are present in only 20-33% of patients. Certain findings in AIDS patients portend a grave prognosis, including poor mental status on presentation, high titer of antigen in cerebrospinal fluid (CSF), low numbers of leukocytes in CSF and presence of positive cultures from extraneural sites.[18] Acute mortality is around 30%, but is much higher in patients with these poor prognostic signs. Long term sequelae include cranial nerve palsies and hydrocephalus.

Skin lesions occur in only 10% of patients, which usually consists of a painless papule that slowly enlarges then ulcerates, draining serous fluid with numerous yeasts.[17] In AIDS patients, skin lesions may appear like umbilicated papules, mimicking molluscum contagiosum. Bony lesions are seen in 5-10% of patients, with the most common sites being the vertebra, pelvis, skull and ribs. X-rays reveal well-circumscribed lytic lesions without periosteal elevation, and histopathology reveals a "cold abscess", or collection of organisms in an inflammatory exudate with a relative paucity of surrounding inflammation. These collections can extend beyond the involved bone, similar to tuberculous abscesses. The most serious complication of the bone involvement is spinal cord compression secondary to vertebral disease. Eye involvement may include chorioretinitis, optic atrophy or papilledema. Blindness is the most serious outcome and is usually permanent. Rarely, cryptococcosis may also involve other organs, such as the adrenals, heart, prostate, muscles or kidneys.

B. DIAGNOSIS

Culture is very useful for detecting the yeast in all body fluids and tissues. Culture of CSF is usually positive in meningitis, and the blood may be positive as well in approximately 10%. *Cryptococcus* can be directly visualized in CSF by examining the fluid mixed with India ink; the yeast is clearly visible surrounded by a clear zone created by the presence of the capsule. India ink may also be useful in detecting the yeast in urine, sputum or autopsy material. Yeast and the capsules will also be visible in histopathologic sections stained with mucicarmine, whereas only the yeast will be detected by other fungal stains.

Cryptococcal disease is also relatively easy to diagnose due to the wide availability of the latex agglutination (LA) assay for the polysaccharide capsule antigen. LA can detect the antigen in serum or CSF, although often the antigen is present in other body fluids as well. False negatives are more common in immunocompetent hosts, since they generally have a lower organism burden than the immunocompromised hosts, but they also may occur secondary to the prozone phenomenon or to infection with acapsular variants.[5] False positives may occur in patients with *Capnocytophaga* sepsis, certain malignancies, or infection with *Trichosporum beigelii.*[5] There is an enzyme immunoassay for the capsular polysaccharide which is highly sensitive and specific, but not as widely available as the latex agglutination assay.

C. TREATMENT

For cryptococcal infections the optimal treatment is a controversial issue. In a small study of AIDS patients with cryptococcal meningitis it was shown that the combination of amphotericin B and 5-FC was superior to fluconazole alone.[18] A larger study done by the AIDS Clinical Trial Group (ACTG) and the Mycosis Study Group (MSG) showed no statistical difference in response rates between amphotericin B alone and fluconazole in the treatment of cryptococcal meningitis in AIDS patients, but there was a trend to better response with amphotericin B.[18] Despite therapy, early mortality ranges from 10-40%. Retrospective studies have shown that the relapse rate in AIDS patients with cryptococcal meningitis treated only with induction therapy is 50-60%, making maintenance therapy the standard of care. Fluconazole, 200 mg/day, has been shown to be superior to weekly infusions of 1 mg/kg amphotericin B for maintenance therapy,[18] but many investigators argue that the dose of amphotericin B used was too low. Both therapies, however, have an unacceptable rate of relapse, underscoring the difficulties in treating this illness in the immunocompromised.

In the immunocompetent host, therapy consists of a period of treatment with amphotericin B with 5-FC, which has been shown to be superior to amphotericin B alone.[19] The duration of therapy is not clear, but most patients are treated with at least one gram of amphotericin B. Whether maintenance therapy is needed is not known since the problem has not been studied, but previous literature suggests that the relapse rate is much lower than that in AIDS patients. Many clinicians treat, for a period of time, with oral fluconazole following the course of amphotericin B. Patients with isolated extraneural disease are treated in a similar fashion, although some patients with isolated pulmonary disease may resolve without therapy. It is clear that, regardless of therapeutic decisions, patients should be followed for an extended period of time and intermittently assessed for evidence of relapse.

VI. DERMATOPHYTES

Dermatophytes are a group of fungi that cause infection of superficial keratinized tissues, such as epidermis, hair and nails. As a rule, the infection is limited to these structures, and deep invasion is rarely seen. There are three genera of fungi included in this group: *Trichophyton*, *Microsporum* and *Epidermophyton* (Figure 4). There are forty species within these genera, eleven of which are not considered to be human pathogens. The fungi are described as anthropophilic, zoophilic or geophilic depending on the source of transmission to man, i.e., man to man, animal to man, or soil to man, respectively. The anthropophilic dermatophytes listed in Table 1, have varied geographic predilections,[20,21] as do the zoophilic dermatophytes, listed in Table 2. The geophilic dermatophytes are soil inhabitants and generally infect animals, but

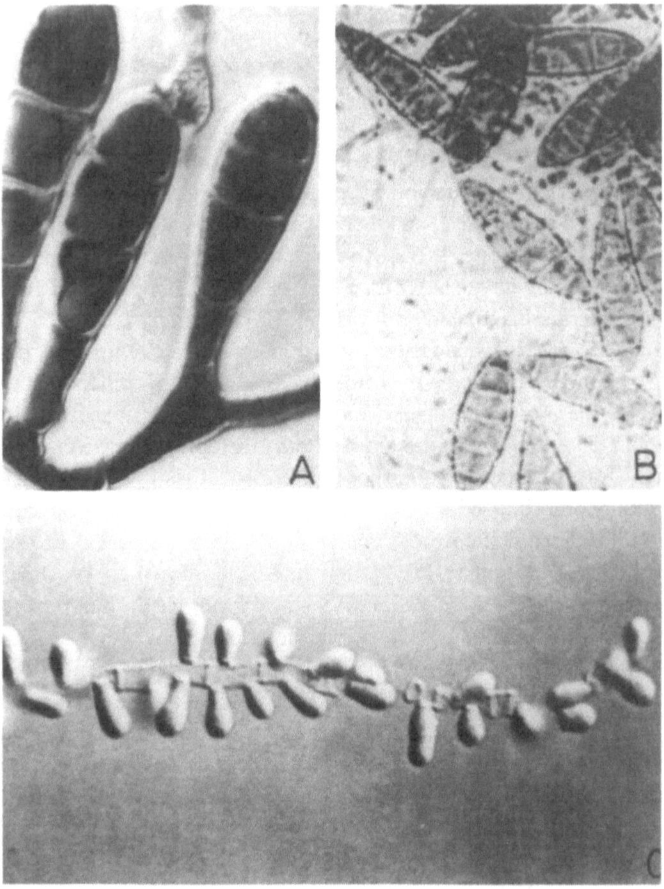

Figure 4. Some common dermatophytes. A. *Epidermophyton floccosum*, B. *Microsporum gypseum*, C. *Trichophyton rubrum*.

some species have been reported to cause human disease, including *T. ajelloi*, *T. terrestre*, *M. fulvum*, *M. gypseum*, and *M. cookei*. The infectious forms of these fungi are the arthrospores, which can survive for long periods of time away from the host.[22] The exact mode of transmission is not known, but *in vitro* studies have demonstrated invasion of local structures after adherence of arthrospores to keratinocytes. The severity of disease depends on the host immunity, tropism of the organism and possibly host genetic factors. T-lymphocyte mediated immunity appears to be important, since there is a well demonstrated delayed-type hypersensitivity response to these fungi leading to increased epidermopoiesis.[20] Local immunity may also play a role, but has not been well characterized. Because of the superficial site of infection, humoral immunity and neutrophils do not contribute much to host defenses against

Table 1
Anthropophilic Dermatophytes

Species	Location	Dermatophytoses
T. concentricum	South Pacific, South America	Tinea imbricata
T. gourvilli	Central Africa	Tinea capitis
	world-wide	Tinea pedis, all others
	Portugal, Sardinia, Africa	Tinea barbae, pedis
T. rubrum	world-wide	Tinea pedis, cruris
T. schoenleinii	North Africa	Tinea corporis, capitis
T. soudanese	West and Central Africa	Tinea capitis
T. tonsurans	North and Central America	Tinea capitis
T. violaceum	Europe, Africa, Asia, S. America, Middle East	Tinea capitis
T. yaoundei	Central Africa	Tinea capitis
M. audouinii	North and Central America, West Africa	Tinea corporis, capitis
M. ferrugineum	Central Africa, Far East	Tinea capitis
E. floccosum	---	Tinea cruris, pedis, corporis

Table 2
Zoophilic Dermatophytes

SPECIES	HOST	LOCATION
T. erinacei	hedgehogs	Europe, New Zealand
T. equinum	horses	world-wide
T. mentagrophytes - mentagrophytes	rodents	world-wide
T. quinckeanum	mice	world-wide
T. simii	monkeys	India and Far East
T. verrucosum	cattle, horses	world-wide
M. canis	cats, dogs	world-wide
M. canis var. *distortum*	cats, dogs	Australia, New Zealand
M. gallinae	poultry	world-wide
M. nanum	pigs	world-wide
M. persicolor	voles	Europe

these organisms.

A. CLINICAL MANIFESTATIONS

The medical terminology of dermatophytoses is not reflective of the species of fungi involved, but of the site of involvement. The term “tinea” is derived from a Latin word meaning “gnawing worm”, and is modified by the Latin names for the various parts of the body affected. Hence, tinea capitis, tinea corporis, tinea pedis, tinea cruris and tinea unguium represent dermatophytosis of the scalp, trunk, foot, groin and nails, respectively. Moist areas favor replication of the fungus, leading to the classical sites of the interdigital areas of the feet and the intertriginous areas of the groin for tinea pedis and tinea cruris. The lesions of tinea pedis are characterized by scaling and fissuring, and cause itching and burning in affected areas. Small vesiculations may also be seen, and the infection often spreads and is less well demarcated than those caused by the geophilic or zoophilic dermatophytes. In general, the lesions are initially round, then more irregularly shaped, with raised, erythematous borders and central pallor with fine scaling. Occasionally, on the lower extremities, tinea corporis may present with a more nodular, follicular appearance with overlying scales. A variant of tinea corporis caused by *Trichophyton concentricum*, called tinea imbricata, appears as concentric rings with prominent scaling. In tinea capitis, the fungi first invade the hairshaft in the follicle, then spread to the scalp once the infected portion grows out, causing localized scalp infection. Infected hairs appear dull and discolored, and easily break off. Occasionally, the scalp may appear very inflamed, covered with an exudative crust. Onychomycosis, or tinea unguium, begins with infection along the lateral margins or leading edge of the nail then spreads throughout the nail, eventually involving the bed. Infection leads to thickening and discoloration of the nail, along with accumulation of subungual debris. A rare manifestation of dermatophytosis is mycetoma, or invasion of the subcutaneous tissues, which causes thickening of the skin, formation of granules of clumped hyphae and possibly sinus formation with extrusion of granules.

B. DIAGNOSIS

Diagnosis is made by obtaining samples of infected materials, such as skin scrapings, hair or nail clippings, then culturing for fungus. Observation of hyphae in specimens after digestion of the keratinocytes with potassium hydroxide will confirm a fungal infection, but will not aid in speciation. Fluorescence of involved hair under a Wood's lamp may also help in identification, with yellow-green fluorescence indicating infection with *Microsporum audouinii*, *M. canis* or *M. ferrugineum*, and bluish-white fluorescence indicating infection with *T. schoenleinii*.[20] Serological tests are not used on a clinical basis.

C. TREATMENT

Treatment depends on the immunological competency of the host and the extent and location of the disease. Mild, localized disease is generally treated with topical antifungal agents, ointments for dry areas and powders for moist areas. The most commonly used agents are the azoles, such as miconazole, clotrimazole, econazole, tioconazole, or ketoconazole, but other effective agents include cyclopiroxamine and naftifine.[22] Treatment is usually for 2-4 weeks. Tinea capitis or widespread tinea corporis in the immunocompetent host, or any extensive disease in the immunocompromised patient, should be treated systemically. Most experiences to date have been with griseofulvin, which appears to be the most effective, with ketoconazole being a secondary agent. Treatment courses of 6-12 weeks are often required for these infections and, despite adequate treatment, relapse is not uncommon. Nail infections are extremely difficult to treat, but some investigators have reported success with a combination of oral griseofulvin and topical azole for prolonged periods of time.

VII. *FUSARIUM* SPECIES

Fusarium spp. are soil saprophytes with a broad geographic distribution. They are major pathogens of plants, but play a relatively minor role in human mycoses, with only 32 cases of reported disseminated disease by 1990.[23] Species reported to cause human disease include *F. solani*, *F. oxysporum*, *F. moniliforme*, *F. proliferatum* and *F. chlamytosporum*. Invasive *Fusarium* infections are termed hyalohyphomycoses because of the glassy, hyaline appearance of the light colored hyphae in tissue sections. Other invasive fungi with similar appearance are also grouped as hyalohyphomycoses, including *Geotrichum*, *Penicillium* and *Pseudallescheria*, among others. The pathogenesis of the disease is not known, but it is thought that the fungus gains access through the paranasal sinuses, nails, gastrointestinal tract, or intravenous line and then disseminates. Neutrophil function is critical to host defense, with the vast majority of cases occurring in patients with hematologic malignancies and prolonged neutropenia. Other modes of host defense are not well defined.

A. CLINICAL MANIFESTATIONS

Fusaria can cause disease either through intoxication or infection. *Fusaria* produces mycotoxins, which can cause aplastic anemia and myopathy when ingested. This form of disease, termed "alimentary toxic aleukia", was first reported in Russia in 1913 after a large number of people ingested grain contaminated with *Fusarium*. The disease is characterized by fever, gastrointestinal symptoms such as nausea, vomiting and diarrhea, neurological disturbances, stomatitis, dermatitis and bone marrow suppression.[23] Occasionally, some of these toxic effects may be seen in patients infected with the organism, although *in vivo* production of toxins by infecting organisms has not been demonstrated.

Infection with *Fusarium* ranges from mild skin infection to fulminant disseminated infection. *F. oxysporum* is the second most common cause of onychomycosis, but deep skin infections are rare. Ocular infections, such as corneal ulcer or endophthalmitis, may occur as a result of trauma or surgery. Other focal infections, such as osteomyelitis, may follow trauma.[24] Isolated sinus disease has also been reported. In the neutropenic host, *Fusarium* can cause rapidly disseminated disease, with mortality reaching 70%. These patients present with refractory fever on broad spectrum antibiotics and even fungal prophylaxis. Approximately 80% develop skin lesions, which may be erythematous macules, palpable or non-palpable purpura, or flaccid pustules.[23] The lesions eventually necrose, developing central eschars. Many sites of involvement may be evident, including sinuses, brain, lung, abdominal organs and musculoskeletal system. The infection becomes fatal if it remains undiagnosed. The recovery of bone marrow function is thought to be a critical factor in the recovery of patients diagnosed in time. There are also reports of patients with fungemia, but no signs of tissue invasion or organ involvement. All these patients had indwelling catheters and relatively short episodes of neutropenia, and they responded well to catheter removal and antifungal therapy.[24]

B. DIAGNOSIS

Diagnosis is made by isolation of the fungus from blood or an otherwise sterile body site. Unlike *Aspergillus* and other newer opportunistic fungal pathogens, *Fusarium* is relatively easy to isolate from blood cultures, with 50% or more yielding the pathogen.[25] Skin biopsy gives an even higher yield, either through visualization of the fungus in the biopsy specimen or culture of the fungus from the specimen. *Fusarium* infection can be suspected histopathologically, but the appearance of the hyphae in tissue is similar to the other hyalohyphomycoses, so culture confirmation is needed. There are no serologic diagnostic methods available to date.

C. TREATMENT

In vitro susceptibility testing of *Fusarium* has demonstrated resistance to amphotericin B, miconazole, ketoconazole and 5-FC, but has shown synergy between amphotericin B and rifampicin. Merz et al.[26] had good response with high doses of amphotericin B (1.5 mg/kg/day) and 5-FC, achieving a cure in 5/5 cases treated. Recovery of bone marrow function appears to be as important as prompt diagnosis and initiation of therapy. Among the 11 successfully treated cases, at least nine resolved neutropenia during treatment, compared to only two of the 11 unsuccessful cases whose bone marrow status was commented upon.

VIII. REFERENCES

1. **Pfaller, M.A. and Wenzel, R.P.,** Impact of the changing epidemiology of fungal infections in the 1990's, *Eur. J. Clin. Microbiol. Infect. Dis.,* 11(4), 287, 1992.
2. **Edwards, J.E., Jr.,** *Candida* species, in *Principles and Practice of Infectious Diseases*, 3rd ed., Mandell, G.L., Douglas, R.G. and Bennett, J.E., Eds., Churchill Livingstone, Inc, New York, 1990, chap. 235.
3. **Wey, S.B., Mori, M., Pfaller, M.A., Woolson, R.F. and Wenzel, R.P.,** Risk factors for hospital-acquired candidemia, *Arch. Int. Med.*, 149, 2349, 1989.

4. **Wey, S.B., Mori, M., Pfaller, M., Woolson, R.F. and Wenzel, R.P.**, Hospital acquired candidemia, *Arch. Int. Med.*, 148, 2542, 1988.
5. **deRepentigny, L.**, Serodiagnosis of candidiasis, aspergillosis and cryptococcosis, *Clin. Infect. Dis.*, 14 (Suppl. 1), S11, 1992.
6. **Rex, J.H., Bennett, J.E., Sugar, A.M., Pappas, P.G., van der Horst, C.M., Edwards, J.E., Washburn, R.G., Scheld, W.M., Karchmer, A.W., Dine, A.P., Levenstein, M.J. and Webb, C.D.**, A randomized trial comparing fluconazole with amphotericin B for the treatment of candidemia in patients without neutropenia, *NEJM*, 331(20), 1325, 1994.
7. **Bodey, G.P.**, Azole antifungal agents, *Clin. Infect. Dis.*, 14 (Suppl. 1), S161, 1992.
8. **Bennett, J.E.**, *Aspergillus* species, in *Principle and Practice of Infectious Diseases*, 3rd ed., Mandell, G.L., Douglas, R.G. and Bennett, J.E., Eds., Churchill Livingstone, Inc, New York, 1990, chap. 236.
9. **Denning, D.W., Follansbee, S.E., Scolaro, M., Norris, S., Edelstein, H. and Stevens, D.A.**, Pulmonary aspergillosis in the acquired immunodeficiency syndrome, *NEJM*, 324 (10), 654, 1991.
10. **Khoo, S.H. and Denning, D.W.**, Invasive aspergillosis in patients with AIDS, *Clin. Infect. Dis.*, 19 (Suppl. 1), S41, 1994.
11. **Loyd, J.E., Des Prez, R.M. and Goodwin, R.A., Jr.**, *Histoplasma capsulatum*, in *Principles and Practice of Infectious Diseases*, 3rd ed., Mandell, G.L., Douglas, R.G. and Bennett, J.E., Eds., Churchill Livingstone, Inc, New York, 1990, chap. 242.
12. **Wheat, J.E.**, Histoplasmosis in Indianapolis, *Clin. Infect. Dis.*, 14 (Suppl. 1), S91, 1992.
13. **Sarosi, G.A. and Johnson, P.C.**, Disseminated histoplasmosis in patients infected with human immunodeficiency virus, *Clin. Infect. Dis.*, 14 (Suppl. 1), S60, 1992.
14. **Davies, S.F.**, Serodiagnosis of histoplasmosis, *Seminars in Respiratory Infections*, 1 (1), 9, 1986.
15. **Wheat, J.E.**, Histoplasmosis: recognition and treatment, *Clin. Infect. Dis.*, 19 (Suppl. 1), S19, 1994.
    Wheat, J.E., Hafner, R., Wulfsohn, M., Spencer, P., Squires, K., Powderly, W.G., Wong, B., Rinaldi, M., Saag, M., Hamill, R., Murphy, R., Connolly-Strinfield, P., Briggs, N. and Owens, S., Prevention of relapse of histoplasmosis with itraconazole in patients with the acquired immunodeficiency syndrome, *Ann. Int. Med.*, 118 (8), 610, 1993.
16. **Diamond, R.D.**, *Cryptococcus neoformans*, in *Principles and Practice of Infectious Diseases*, 3rd ed., Mandell, G.L., Douglas, R.G. and Bennett, J.E., Eds., Churchill Livingstone, Inc, New York, 1990, chap. 241.
17. **Powderly, W.G.**, Therapy for cryptococcal meningitis in patients with AIDS, *Clin. Infect. Dis.*, 14 (Suppl. 1), S54, 1992.

18. **Bennett, J.E., Dismukes, W.E., Duma, R.J., Medoff, G., Sande, M.A., Gallis, H., Leonard, J., Fields, B.T., Bradshaw, M., Haywood, H., McGee, Z.A., Cate, T.R., Cobbs, C.G., Warner, J.F. and Alling, D.W.,** A comparison of amphotericin B alone and combined with flucytosine in the treatment of cryptococcal meningitis, *NEJM*, 301, 126, 1979.
19. **Kwon-Chung, K.J. and Bennett, J.E.,** Dermatophytes, in *Medical Mycology*, Kown-Chung, K.J. and Bennett, J.E., Eds., Lea and Febiger, Philadelphia, 1992, chap. 6.
20. **Elewski, B.E.,** Superficial mycoses, dermatophytoses and selected dermatomycoses, in *Cutaneous Fungal Infections*, Elewski, B.E., Ed., Igaku-Shoin, New York, 1990, chap. 2.
21. **Hay, R.J.,** Dermatophytosis and other superficial mycoses, in *Principle and Practice of Infectious Diseases*, 3rd ed., Mandell, G.L., Douglas, R.G. and Bennett, J.E., Eds., Churchill Livingstone, Inc, New York, 1990, chap. 245.
22. **Helm, T.N., Longworth, D.L., Hall, G.S., Bolwell, B.J., Fernandez, B. and Tomecki, J.,** Case report and review of resolved fusariosis, *J. Amer. Acad. Dermat.*, 23, 393, 1990.
23. **Anaissie, E., Nelson, P., Beremand, M., Kontoyiannis, D. and Rinaldi, M.,** *Fusarium*-caused hyalohyphomycosis: an overview, *Curr. Top. Med. Mycol.*, 4, 231, 1992.
24. **Barrios, N.J., Kirkpatrick, D.V., Murciano, A., Stine, K., Van Dyke, R.B. and Humbert, J.R.,** Successful treatment of disseminated *Fusarium* infection in an immunocompromised child, *The American J. Ped. Haemat./Oncol.*, 12 (3), 319, 1990.
25. **Merz, W.G., Karp, J.E., Hoagland, M., Jett-Goheen, M., Junkins, J.M. and Hood, A.F.,** Diagnosis and successful treatment of fusariosis in the compromised host, *J. Infect. Dis.*, 158 (5), 1046, 1988.

Chapter 2

FUNGAL LIPIDS

D. M. Lösel and M. Sancholle

CONTENTS

0-8493-4794-7/96/$0.00+$.50

I. INTRODUCTION

Since pathogens are encountered in most groups of fungi, their lipid composition is likely to reflect basic patterns of fungal chemotaxonomy, although modified by interaction, biotrophic or saprotrophic. Fungi pathogenic on mammals and other homoiotherms have to survive and grow at temperatures higher than most normal environmental conditions. At the same time, they are protected from effects of fluctuations in ambient temperature, water potential, nutrient supply or toxic substances which affect free-living fungi and often induce adaptive changes in their lipid composition. Similarly, pathogenesis of poikilothermic animals and plant tissues may ensure more uniform growth conditions than the saprotrophic state and ones in which the host tissue provides some degree of environmental regulation.

In an earlier review of the lipids of pathogenic fungi, Chopra and Khuller[1] summarized information on the lipid composition of selected human pathogens. They made comparisons between species pathogenic to insects and plants and discussed topics such as chemotaxonomy, metabolism, effects of environmental factors, roles of lipids and enzymes of lipid metabolism in pathogenesis, virulence and allergenicity. With increased investigation in the intervening period, apparent differences in lipid composition between pathogens and non-pathogens have become less clear, while the range of fungi reported to exhibit pathogenicity has increased. It now includes species previously regarded as harmless saprophytes as well as some plant pathogens.[2]

The diversity and versatility of fungi in lipid synthesis are still being explored to only a limited extent. The structures of acyl and terpenoid lipids of fungi and other microorganisms have been comprehensively reviewed.[3,4] Extensive data on the composition and distribution of these and on related fungal metabolites had previously been compiled by Turner,[5] Turner and Aldrich[6] and Weete,[7,8] and physiological aspects were discussed.[9] More recently, detailed accounts of the lipids in yeasts[10] and filamentous fungi[11] are found in *Microbial Lipids*, edited by Ratledge and Wilkinson.[12,13] The influence of environmental factors on fungal lipid composition has been reviewed by Rose[14] and will be considered later in this book.[15] The present chapter brings together some selected observations on lipids of the major groups of fungi, with particular emphasis on species known or suspected to be pathogenic to man and animals. Since pathogenesis of plants is generally more amenable to physiological study and experimentation than infections of animal tissues, occasional reference will be made to lipids of plant pathogens, the interactions of which with host tissue may throw light on aspects which also merit study in mammalian systems.

II. TOTAL LIPID CONTENT

A. LIPID ACCUMULATION

The amount of lipid present in fungal cultures ranges from < 5% to over 25% of biomass, varying with stage of development, culture conditions and species. A study of 12 species of *Mortierella* found that all of them were active producers of lipid which could account for 40-60% of biomass.[16] Oleaginous fungi, defined as capable of accumulating more than 20% of their dry weight as lipid, include species of *Lipomyces* in which the lipid content may reach 40%. Conversion of such lipid reserves into new biomass has been demonstrated in eight strains of the oleaginous yeasts, *Candida curvata, Lipomyces starkeyi, Rhodosporidium toruloides* and *Trichosporum cutaneum,* transferred to carbon starvation conditions after accumulation of up to 34% of biomass as lipid.[17] During this lipid utilization, the activities of ATP: citrate lyase (EC 4.1.3.8) and malate dehydrogenase (EC 1.1.1.42) decreased, while those of carnitine acetyltransferase (EC 2.3.1.7) and isocitrate lyase (EC 4.1.3.1) increased.[18] The oleaginicity of yeasts and filamentous fungi can be correlated with ATP-citrate lyase activity[19] and has been reviewed by Ratledge.[20]

The formation of lipid drops is a common feature of vegetative growth of saprophytic fungi and generally increases greatly during the formation of resting and reproductive structures. *Aspergillus foetidus* and *Fusarium oxysporum* accumulated fatty inclusions, when growing on sucrose, tridecane and emulsol as carbon sources, maximum amounts being found after 48 h in the former and 72 h in the latter.[21] Both neutral lipid content and cephalosporin C production by *Paecilomyces persicinus* were maximal after 72 h of incubation, at which stage the polar lipid fraction was relatively low.[22] Similarly, the lipid content of *A. versicolor* decreased when the production of sterigmatocystin began.[23] Although most of this lipid is cytoplasmic in location, significant amounts have been recorded from cell walls.[11] Walls of both yeast and mycelial forms of *Blastomyces dermatidis* contained relatively high proportions of saturated fatty acids of carbon chain lengths 20, 22, and 24.[24] Lipid droplets within resting sporangia of the potato wart pathogen *Synchytrium endobioticum,* with palmitic, oleic and nonadecanoic acids as major fatty acids, differed in composition from the sporangium wall, where the major fatty acids, stearic, oleic, arachidic ($C_{20:0}$) and arachidonic ($C_{20:4}$) acids, were accompanied by significant amounts of wax esters with branched chains.[25]

B. LIPID CONTENT OF PATHOGENS

Histological examination of infected animal or plant tissue frequently reveals lipid bodies within the fungal pathogen.[9] Electron micrographs showing lipid bodies in cells of filamentous and yeast-like forms as well as spores of *B. dermatitidis, Paracoccidioides brasiliensis,*

T. cutaneum and *C. albicans* are included in Garrison's review of the ultrastructural cytology of fungi pathogenic on human and animal tissues.[26] Such inclusions often account for a substantial proportion of intracellular volume in mature cells. Yeast cells and blastospores of the Dutch elm disease fungus *Ceratocystis ulmi,* grown on proline as a nitrogen source, contained lipid storage bodies.[27] There was some evidence that the enriched unsaturated fatty acid content of the pathogen, when grown in susceptible elm trees, correlated well with the degree of pathogenicity.

The greater amount of lipid and higher proportions of oleic acid and sterols, in arthrospores of a mouse-virulent strain of *Coccidioides immitis* than in a non-virulent strain, led to suggestions that lipid content could be significant in the initiation of infection.[28] Ghannoum and co-workers presented evidence that individual phospholipids, sterols and sterol esters from *C. albicans* and *C. tropicalis* but not from the weakly adherent *C. pseudotropicalis* blocked adherence of these organisms to buccal epithelial cells *in vitro,* whereas triacylglycerols and free fatty acids had no effect.[29] *Scopulariopsis brevicaulis*, a soil saprophyte which causes human and animal onchomycosis, was found to resemble other pathogens in its high lipid content (34% of biomass after 12 d).[30] Lipids from the dermatophytes *Fonsecaea pedrosoi, F. compactum, Cladosporium carrionii* and *Phialophora verrucosum*, causal organisms of chromoblastomycosis, have been reported to induce granulomatous reactions in mice injected with charcoal particles coated with the lipid extracts.[31] Subsequent investigations have, however, provided no clear evidence that pathogens have a higher lipid content than saprophytic fungi, among which a wide range of lipogenicity and lipophilicity exists.[18,32] On the other hand, as discussed later, some fungi pathogenic to insects and plants are normally dependent on lipids of host tissues, particularly sterols, for their development.[9]

Attempts have also been made to correlate the dimorphism in pathogenic fungi, the alteration from mycelial to yeast-like morphology, with their lipid composition and, particularly, with wall lipids. However, although higher lipid contents were recorded in cell walls of hyphal than of yeast-form cells of *B. dermatitidis, C. albicans* and *Mucor rouxii,* the opposite was true of *Histoplasma capsulatum* and *Paracoccidioides brasiliensis.*[7] In a comparative study of yeast and mycelial forms of *C. albicans,* Ghannoum et al. found the total lipid content of the former to be consistently lower than that of mycelium. Polar lipid was predominant in both forms after 12 h growth, whereas after 96 h, there were substantial amounts of apolar lipid, mainly sterol esters and triacylglycerols, reaching higher levels in the mycelium than in the yeast form. More free sterol was present in yeast cells, while the mycelium contained more sterol glycoside and esterified sterol glycoside.[33] Investigation of differences in lipid composition in the dimorphic human pathogens *F. pedrosoi, Cladosporium carrionii and Sporothrix schenckii* led Dennettiere et al.[34] to conclude that the lipid metabolism of these species was temperature-dependent and that the observed changes in fatty acid composition may not be directly related to dimorphism. Dimorphism of *F. pedrosoi* at 37°C was accompanied by changes in amounts of oleic and linoleic acids but similar changes occurred in *Cladosporium carrionii* which was not dimorphic at this temperature. When dimorphism of the latter was induced at 23°C at pH 2.5, the fatty acid composition was identical to that of the mycelial phase at 23°C and pH 5.5. No differences in fatty acid composition were detected between mycelial and yeast forms of *S. schenckii*.

III. COMPOSITION OF FUNGAL LIPIDS

The lipid content of fungal cells comprises mainly of membrane components and storage lipids. The former largely determine the structure and permeability properties of the membrane system which delimits organelles and compartmentalises metabolic processes, as well as providing sites of insertion or attachment of proteins. Fungal membrane lipids include phosphoacylglycerols, glycosylacylglycerols and sphingolipids, polar molecules which orientate themselves in a definite manner at the interface of aqueous and lipid phases to form membranes, the stability of which may be further modified by the presence of sterols. Storage lipids, principally triacyl- and diacylglycerols, are synthesized within the endoplasmic reticulum and accumulate in lipid bodies and lipid droplets. Triacylglycerols have also been found in pure cell wall fractions and it has been suggested that they may be derived from lipid bodies excreted through the plasmalemma. Whether free fatty acids or storage compounds, such as sterol esters of fatty acids, are also present in such lipid reserves remains controversial.

A. ACYL LIPIDS

1. Fatty Acids

Fatty acids normally occur in bound form, as components of acyl glycerols and phosphoacyl glycerols, glycolipids, sphingolipids, sterol esters and other complex lipids. Reports of substantial amounts of free fatty acids are generally attributed to lipase activity during extraction although evidence for their accumulation at certain developmental stages of some species of fungi has been presented.[10,11] Most of the available data on fungal lipids refer to total fatty acids, thereby obscuring the differences between individual membrane lipid classes and reserve lipids at successive stages of the cell cycle.

a) Long-chain Fatty Acids

The fatty acids of fungi are predominantly of carbon chain lengths C_{16} and C_{18}, with varying degrees of unsaturation, accompanied by smaller amounts of even-numbered carbon chains from 14 to 24 and small amounts of odd-numbered $C_{15\text{-}19}$ fatty acids, sometimes associated with particular substrates. Among yeasts, however, $C_{18:0}$ generally accounts for less than 10% of total fatty acids and may even be absent, although it reaches levels similar to that of $C_{16:0}$ in *Schizosaccharomyces cerevisiae* and around 30% of total fatty acids in *Cryptococcus albidus.*[10] Oleic and linoleic acids are the most widely distributed unsaturated fatty acids while the occurrence of trienoic and other polyenoic fatty acids differs greatly within and among different fungal classes. The degree of saturation of fatty acids varies in growth and reproductive stages, e.g. in *Entomophthora obscura*, a pathogen of aphids, the degree of unsaturation was low during vegetative growth, decreasing with age of mycelium, but increased as spores matured, when the (uncommon) $C_{19:1}$ content increased at the expense of $C_{19:0}$.[35] The commonly occurring trienoic fatty acid in fungi is α-linolenic acid (9,12,15-octadecatrienoate) but γ-linolenic acid (6,9,12-octadecatrienoate), first noted in fungi by Shaw,[36,37] is the normal 18:3 component of Chytridiomycetes, Oomycetes and Zygomycotina, where it is accompanied in some genera by α-linolenic acid.

Alteration in the degree of unsaturation, carbon chain length and positional distribution of the fatty acids of acyl lipids in response to fluctuating environmental conditions may contribute to the maintenance of membrane function and stability which is generally important in poikilotherms but it has not been studied extensively in fungi. When *Taphrina deformans* was grown at 5°C, instead of its optimum growth temperature (18°C), there was no difference in growth rate but the unsaturation of the fatty acids in the plasma membrane increased strongly.[38] Suutari et al.[39] detected no qualitative changes in fatty acid composition in a series of yeasts, within the range of temperatures (10° to 35°C) permitting growth, but various adjustments were observed in the amounts and proportions of fatty acids. Throughout this range, both the lipid content and the amount of palmitoleic acid in *Saccharomyces cerevisiae* increased with decreasing temperature but the degree of unsaturation was virtually unaltered, whereas *Rhodotorula toruloides* showed increase in lipid content with an increase in temperature, accompanied by lowering of the degree of unsaturation. *C. utilis* and *Lipomyces starkeyi* exhibited a biphasic response, depending on the temperature range, with the ratio of C_{16}/C_{18} fatty acids and the fatty acid content decreasing with increase in temperature below the critical range of 20° - 26°C. Above this temperature, the degree of unsaturation decreased and the fatty acid content increased with temperature, as in *R. toruloides*. The fatty acid content of *L. starkeyi* showed a substantial increase below 20°C, at the expense of palmitic and oleic acids. These responses were compared with earlier observations that shortening of fatty acid chain length and insertion of the first double bond more strongly affected membrane fluidity than did the insertion of subsequent double bonds.[40,41]

b) Less Common Fatty Acids

Very long-chain fatty acids (>C_{20}) are widely distributed among the Mastigomycotina and Zygomycotina but seldom reported from Ascomycotina and Basidiomycotina. They are, however, found in mycelium and yeast cells of the pathogenic *B. dermatitidis* and in some oleaginous yeasts, e.g., the basidiogenous genera *Rhodotorula* and *Rhodosporidium* but less frequently in ascosporogenous yeasts. In *S. cerevisiae* ATCC 9896, long-chain fatty acids (C_{20-26}), including 2-hydroxy-$C_{26:0}$,[42] accounted for about 15% of the plasmalemma fatty acids which were also present in the other membrane systems isolated. In both yeasts and filamentous fungi, such long-chain fatty acid components appear to be largely associated with sphingolipids.

Short and medium-chain fatty acids are widespread in yeasts[10] and filamentous fungi[11] but tend to get lost during normal preparation procedures and are seldom specifically investigated. Substantial amounts have been found in *S. cerevisiae,*[43] particularly during microaerobic or anaerobic growth or under conditions of glucose repression.[44,45] From skin tests on intradermally sensitised guinea pigs, Hellgren and Vincent[46] obtained evidence that medium-chain length (particularly C_{10}-C_{12}) fatty acids, previously identified from the mycelium of dermatophytes, could elicit inflammatory reactions in mammalian skin. The physiological activity of short-chain fatty acids reported among volatile metabolites of fungi has been discussed.[47]

The branched-chain and hydroxy fatty acids detected in normally small amounts amongst all the lipids of some fungi are probably the main components of sphingolipids or nitrogen-free glycolipids, although over 40% of the ergot oil present in sclerotia of *Claviceps purpurea* consists of D-12-hydroxy-*cis*-octadec-9-enoic acid (ricinoleic acid),[48] a derivative of linoleic acid.[49] Ricinoleic acid, which is absent from vegetative mycelium but can account for 40% of the ergot oil in the resting sclerotia,[50] is synthesized in parallel with the medically important ergot alkaloids, during the development of sclerotial plectenchyma. Substantial amounts of biosynthetically related dihydroxy octadecanoic acids have been detected, together with smaller proportions of ricinoleic acid, in infections of other host plants by different *Claviceps* species.[49,50] Several even-numbered and odd-numbered branched-chain fatty acids form a significant proportion of the lipid content in species of *Conidiobolus*, showing an inverse relationship with oleic acid during development.[51-53]

Compounds closely related to fatty acids also occur, e.g., long-chain fatty aldehydes and, among secondary metabolites, the large class of polyketides,[4,5] derived from acetate. A series of long-chain fatty aldehydes, previously not known in fungi, was isolated by Stoessl from cultures of *Cercospora arachidicola* growing on potato dextrose agar and shown to consist predominantly of penta- and hexadecanal and heptadec-8-enal, with small or trace amounts of tetra-, hexa- and nonadecenal, tetra-, penta-2-, and

Table 1
Summary of Distribution of Fungal Fatty Acids

FATTY ACIDS	OCCURRENCE
Medium-chain fatty acids	Sporadically recorded in filamentous fungi,[11] and dermatophytes.[40]
$C_{10:0-12:0}$	Absent or low in yeasts but substantial in *Saccharomyces cerevisiae*, particularly during oxygen limitation[10]
$C_{16:0}$, $C_{18:1}$, $C_{18:2}$	Major fatty acids in most of the species [8,10,11]
$C_{14:0}$, $C_{16:1}$, $C_{18:0}$	Minor fatty acids in most of the species[8,10,11]
γ - $C_{18:3}$	Minor fatty acid of Chytridiomycetes, Oomycetes, Zygomycotina [11,37]
α - $C_{18:3}$	Minor fatty acid of most Ascomycotina, Basidiomycotina[37] and yeasts but absent or very low in *Saccharomyces cerevisiae*.[10]
C_{12-26} hydroxy fatty acids	Mainly in sphingolipids and glycolipids of yeasts[10] and filamentous fungi [9,127,128]
Branched-chain, odd- and even-numbered fatty acids	*Conidiobolus* spp.[51,53]
D-12-hydroxy-*cis*-octadec-9-enoic (ricinoleic) and other OH-$C_{18:0}$ fatty acids	*Claviceps purpurea*,[40] *Claviceps* spp.[49]
threo-9,10-dihydroxystearic acid	*Claviceps* spp.[49] and spores of some rusts[11]
cis-9,10-epoxyoctadecanoic acid	Major fatty acid in rust spores.[8,11] Low in axenic mycelium of *Puccinia graminis*.[72]
$C_{19:4}$	*Mortierella alpina*[57]
$C_{20:3}$	*Dactylaria ampulliforme*,[67] *Mortierella alpina*[57]
$C_{20:4}$ (arachidonic acid)	Chytridiomycetes, Oomycetes and some Zygomycotina[11,57]
$C_{20:5}$	*Mortierella alpina* on medium containing 1-hexadecene,[58] *Phytophthora infestans*[60]
$C_{22:0}$	Low levels sporadically reported[10,11]
$C_{22:2-6}$	Some Chytridiomycetes and Oomycetes[11]

penta-dec-6-enal as well as heptadec-8,11-dienal, together with a mono- and a di-unsaturated C_{16} aldehyde.[54]

c) *Polyunsaturated Fatty Acids*

In recent years, polyunsaturated fatty acid (PUFA) production by fungi has been extensively investigated in the context of the biosynthesis of the essential fatty acids, γ-linolenic acid and arachidonic acid (*cis*-5,8,11,14-eicosatetraenoic acid). Biosynthesis of tri-, tetra-, penta- and hexa-enoic acids by fungi is of particular interest in relation to their importance as dietary factors in human nutrition and also as precursors of the medically important eicosanoid hormones, prostaglandins, thromboxanes and leucotrienes.[55,56]

Among arachidonic acid synthesising strains of *Mortierella* spp. investigated by Shimizu et al.,[57] a soil isolate, *M. alpina* 1S-4, produced 95% of the total mycelial fatty acids as odd-chain fatty acids, distributed in both neutral and polar lipid fractions, mainly 5,8,11,14-*cis*-nonadecatetraenoic acid. The biosynthesis of this was presumed to mimic that of arachidonic acid. On 5% N-heptadecane and 1% yeast extract as substrate, $C_{19:4}$ accounted for up to 11.2% of total fatty acids of mycelial lipid (44.4 mg g^{-1} mycelial dry wt., a yield of 0.68 mg ml^{-1} culture broth). Decreased synthesis of $C_{19:4}$ and accumulation of $C_{19:3}$ resulted from the addition of sesamin, a specific inhibitor of Δ^5-desaturation. *Mortierella* species incapable of C_{20} PUFA synthesis accumulated C_{17} fatty acids but not C_{19} PUFA when grown on odd-chain fatty substrates. The same team also demonstrated the production of a novel ω-1-eicosapentaenoic acid, *cis*-5,8,11,14,19-eicosapentaenoic acid (20:5, ω1), by *M. alpina* 1S-4 grown on a medium containing 4% 1-hexadecene.[58] The fatty acid composition and unsaturation of *Mortierella* species has also been shown to be regulated by the carbon: nitrogen ratio in the culture medium; elevated carbon content (C:N = 40:1) results in cultures with increased proportions of saturated C_{12}-C_{16} and monenoic C_{18} fatty acids and others with a higher content of polyenoic fatty acids in their triacylglycerols.[59]

Arachidonic and eicosapentaenoic acids have been identified as the active components of a lipoglycoprotein complex from *Phytophthora infestans* eliciting production of the phytoalexin rishitin which is implicated in hypersensitive resistance reactions and necrotization in potato tissue.[60] Degradation of these PUFA, in the presence of hydrogen peroxide, superoxide and radical-generating systems, was attributed to superoxide-mediated formation of hydroxyl radicals by Merzlyak and colleagues,[61] who discussed the role of lipid peroxidation as a mechanism of fungicidal activity of oxygen radicals and their possible involvement in the production of $C_{20:4}$ metabolites eliciting the host defence reactions. A highly active 5-lipoxygenase in potato tissue has been shown by Castoria et al. to convert arachidonic acid to an unstable intermediate, 5-S-hydroperoxy-eicosatetraenoic, capable of inducing phytoalexin accumulation at much lower concentrations than that required for arachidonic acid.[62]

d) Application of Fatty Acid Analysis in Fungal Chemotaxonomy

A short summary of current information on the distribution of fatty acids in fungi is given in Table 1. Despite the lack of uniformity in culture conditions, from which most fatty acid data have been derived, and the limited number of records in which individual lipid classes have been analysed separately, a succession of reviewers have sought correlations between taxonomic groups of fungi and their fatty acid composition.[8,11,37,52] The chemotaxonomic use of microbial lipids has been discussed by Lechevalier and Lechavalier.[63] Possibly on account of the diversity in morphological characters for distinguishing genera and species of fungi, chemotaxonomy of filamentous fungi has received less attention than that of yeasts and bacteria.[8,63] Fatty acid profiles are employed in identifying bacterial pathogens of man, animals and plants. The application of this approach to fungi would be of particular value for identifying non-sporulating cultures and 'difficult' genera but requires data from a larger number of species, cultured under strictly defined conditions. The pioneering review of Shaw,[37] who noticed the association of γ-linolenic acid with species of Zygomycotina and commented on other correlations, as well as subsequent intensive studies of specific genera, e.g. *Entomophthora,*[35,64,65] *Conidiobolus*[51,53] and of a range of arbuscular mycorrhizal fungi,[66] have demonstrated the chemotaxonomic potential of fatty acid analysis of fungi.

As might be expected, the morphologically and biologically diverse classes of Siphomycetes (Mastigomycotina and Zygomycotina) differ widely in composition of lipids as well as in their cell wall components, hyphal structure and motility of reproductive cells. Medium-chain length (C_{12} to C_{14}) and odd numbered fatty acids as well as significant proportions of polyenoic fatty acids of carbon chain lengths from 20 to 24 are more common in these divisions of the Fungi than that among Septomycetes (Ascomycotina and Basidiomycotina). γ-linolenic acid, often accompanied by α-linolenic acid, is widespread in Chytridiomycetes and Oomycetes (Mastigomycotina) as well as in the Zygomycotina, which differ from the Mastigomycotina in lacking motile stages. Thus, the fatty acid composition of the Siphomycetes, which include some human and animal pathogens, offers a number of chemotaxonomically useful markers.[8,11]

Throughout the Septomycetes, α-linolenic acid is generally the only trienoic fatty acid recorded, apart from a single report of the occurrence of γ-linolenic acid, accompanied by $C_{20:3}$, in *Dactylaria ampulliforme.*[67] Most human and animal pathogens belong to the Ascomycotina, the largest division of fungi, or are assigned to the Deuteromycotina ("Fungi Imperfecti"), since they are generally encountered as anamorphs of which the teleomorphs (sexual stages) are presumed to be Ascomycotina. Currently available fatty acid data from these divisions of the fungi offer disappointingly few clear taxonomic separations. Oleic and linoleic acids are generally the predominant fatty acids. α-linolenic acid is of widespread occurrence, normally in smaller amounts than other C_{18} fatty acids, but is absent from many yeasts and Septomycetes. However, it does reach substantial levels in some plant pathogens, particularly rust fungi, apparently reflecting the fatty acid composition of their host tissues.[11] The occurrence of substantial amounts of ricinoleic acid in sclerotia of *Claviceps purpurea* and its production in smaller amounts accompanied by related dihydroxyoctadecanoic acids, mentioned above,[49,50] offers an unusually clear chemotaxonomic marker. The fatty acid content of 36 strains from 13 species of the genus *Monascus*, a traditional source of natural pigments used in various oriental fermented foods, showed very less qualitative variation, $C_{16:0}$, $C_{16:1}$, $C_{18:0}$, $C_{18:1}$ and $C_{18:2}$ being the major components. From quantitative analysis, however, it was concluded that classification of strains and species using fatty acid analysis should be possible and that the *Monascus* species differed clearly from the strains of *A. oryzae, A. niger, Penicillium notatum* and *Paecilomyces variotii* as studied by the same workers.[68]

Among Basidiomycotina, fatty acid records are most commonly derived from the readily-collected fruit-bodies of macrofungi of the Hymenomycetes and Gasteromycetes and may differ from the composition of vegetative mycelium, which has been studied in relatively few species. Oleic and linoleic acids are the predominant components in both classes. Several species have shown a higher degree of unsaturation in lipids of fruit-bodies than in mycelium.[69,70] Hymenomycetes which have been studied intensively in mycelial stages include the cultivated mushroom *Agaricus bisporus.*[71] The lipid composition of the smut and rust fungi, biotrophic phytopathogenic microfungi included in the Basidiomycotina, has been studied in relation to their interactions with host tissue.[8,9,72] The smut fungi are readily grown saprophytically and appear to resemble Hymenomycetes in their fatty acid composition, whereas the rust fungi, which are more difficult to culture as mycelium, have been mainly studied in spore stages and show lipid characteristics distinct from other Basidiomycetes.[72] Although pathogenicity to humans and animals has not usually been associated with Basidiomycotina, instances of infection have been recorded, particularly in compromised patients, e.g. *Coprinus cinereus* and *Schizophyllum commune* (Hymenomycetes) and two genera of smut fungi, *Ustilago* and *Filobasidiella* spp.[2] The presence of *cis*-9,10-epoxy-octadecanoic acid as a major fatty acid is characteristic of rust spores, accompanied in some species by *threo*-8,10-dihydroxystearic acid, which is also found in some *Claviceps* species.[8,11,72,131] A rapid decrease of the former was noted by Sock,[72] during germination of *Puccinia graminis* urediniospores to a low level in axenic mycelium with a concomitant increase in α-linolenic acid.

e) Peroxidation of Fatty Acids

Products of lipid peroxidation are of interest in the physiology of fungal-infected plants and have been demonstrated to accumulate at sites of both fungal penetration and insect attack. Oxylipins, a term recently introduced to denote products of fatty acid oxidation, have been shown to exert a wide range of physiological effects in plant tissues,[73,74] the best known being jasmonic acid, a cyclopentanone derived from peroxidation products of linolenic acid. The involvement of peroxidation products of PUFA from *Phytophthora infestans* in the hypersensitive response of potato tissue has already been discussed (III.A.1.c).

The key role of lipid peroxidation in the biosynthesis of aflatoxins by fungi such as *A. flavus* and *A. parasiticus* growing on wheat, maize and sunflower seeds, both *in vitro* and *in vivo,* has been demonstrated by Fanelli and Fabbri,[75] who found that the antioxidants, butylated hydroxytoluene and butylated hydroxyanisoline, reduced aflatoxin biosynthesis by aflatoxigenic *Aspergilli*, for which the seed surface lipid was an important carbon source for growth. The stimulating effect of lipoperoxidation in the production of aflatoxins and their congeners by different strains of these species was obtained by adding epoxides, hydroperoxides and carbon tetrachloride to culture media but could be inhibited in the presence of cysteamine, a free radical scavenger.[76] Ascorbic acid in the presence of ferrous ions can also induce lipid peroxidation and aflatoxin synthesis.[77]

2. Acylglycerols

The predominant cell lipids are acylglycerols, in which fatty acids are attached to the carbon skeleton of glycerol, or phosphoacylglycerol in which the *sn*-1 and -2 positions of glycerol are acylated and another moiety, e.g., choline, ethanolamine or inositol, is linked via phosphate to the *sn*-3 position. Substantial amounts of other acyl lipids, particularly glycolipids and sphingolipids, may also occur in fungal cells but have been studied in only a limited range of species.

That the proportions of different lipid classes, as well as their individual composition, differ in successive stages of the life cycle of yeasts and filamentous fungi has been demonstrated in studies of a wide range of fungi, as discussed previously.[8,9,10,11] A representative pattern is illustrated in the yeast *Dipodascus uninucleata*,[78] where lipid content decreased rapidly during spore germination from 55 mg g^{-1} to 3.3 mg g^{-1} dry wt. then increased gradually during hyphal development to 12.0 mg g^{-1} dry wt. at the stage of ascospore liberation. The developmental cycle was accompanied by changing proportions of storage and membrane lipids.

3. Phosphoacylglycerols

a) Phospholipid Classes

As with so many aspects of fungal lipid metabolism, much more information is available concerning the composition, sub-cellular location and functioning of glycerophospholipids in yeasts than that in filamentous fungi. Phosphatidylcholine (PC) and phosphatidyl ethanolamine (PE) are the major phospholipid classes throughout the fungi, generally accompanied by smaller amounts (3-10%) of phosphatidyl inositol (PI), phosphatidyl serine (PS) and diphosphatidylglycerol (DPG, 2-5%). Phosphatidylglycerol (PG), which has been reported sporadically in yeasts and filamentous fungi,[10,11] comprised 2.6 to 5.6% of the phospholipids of azole-sensitive and -resistant strains of *C. albicans* studied by Hitchcock and co-workers.[79] In an azole-sensitive strain and in an azole- and polyene-resistant mutant of *C. albicans*, examined by this group, phosphatidylglycerol decreased from exponential to stationary phase, while diphosphatidylglycerol increased substantially.[80] Like the accompanying increases in PC relative to PE, this was considered likely to reflect biosynthetic conversion of precursor to product as culture growth slows.

Other phospholipids, phosphatidic acid, lysophosphatidyl choline and lysophosphatidyl ethanolamine, occurring in normally much smaller amounts and varying with the stage of development of fungal cultures, may represent intermediates of lipid synthesis or breakdown or these could result from phospholipase activity during extraction. Experiments by Hitchcock and colleagues demonstrated that the omission of preliminary extraction by boiling in isopropanol resulted in loss of 39% (by weight) of phospholipids (without change in triacylglycerol or sterol ester content) of *C. albicans*, compared with 14% loss of phospholipid during routine extraction.[79] It was suggested that an apparent increase in phospholipase activity in the stationary phase of *C. albicans* from a very low level in exponential phase cultures might be due to a decrease in efficacy of the inactivation step.[80]

b) Biosynthesis

Investigation by Mago and Khuller of the incorporation of [^{14}C]-acetate and [^{32}P]-phosphate into the mainly microsomal phospholipid fraction of *C. albicans* revealed that PS was synthesized at the highest rate, followed by PC, PE and PI.[81] From the incorporation of labelled serine, ethanolamine and choline, it was concluded that serine was a biosynthetic precursor of PC, PE and PS; ethanolamine is a precursor of PC and PE, and choline of PC only. Phospholipid biosynthesis of PC in *C. albicans* has been shown by Klig et al. to be regulated by exogenous choline, thus differing from *S. cerevisiae* in which the biosynthesis of PC via methylation of PE appears to be regulated in response to inositol and choline levels and is not suppressed by the provision of choline alone.[82]

The phospholipid composition of individual fungi can be modified by nutritional and other factors affecting biosynthetic pathways. Exogenous phosphate, up to a certain level, increased the phosphorus content of mycelium and its lipid fraction in *F. oxysporum* but higher concentrations had no major effect. The most susceptible compounds were phosphatidic acid and phosphatidyl inositol, while phosphatidyl choline was least affected.[83] Radiolabeling showed higher specific activities for phosphatidic acid, phosphatidylglycerol-3-phosphate and cardiolipin. In *Pythium irregulare*, a carbon/phosphorus ratio (C/P) of 30 was found to be the most suitable for mycelial growth; C/P ratios of 10 and 20 resulted in maximum synthesis of total lipids and phospholipids, respectively, while the highest concentration of phosphorus (C/P ratio 5) was toxic, resulting in poor growth and reduced phospholipid synthesis.[84] At higher concentration of phosphate, there was relatively increased synthesis of PC and PE. Addition of choline to cultures of *F. graminearum* increased the hyphal extension rate and production of sparingly branched mycelium but had no significant effect on phospholipid composition.[85] A species of *Fusarium* had the highest content of PC and PS when grown in ethanol-containing medium whereas PE and PG were higher in molasses-containing medium.[86] The extent to which enzymes controlling phospholipid synthesis in *Cunninghamella japonica* can be regulated by additions of cycloheximide and actinomycin D has been studied with a view to determine whether the composition of phospholipids and neutral lipids may be controlled by using inhibitors of RNA and protein synthesis. Correlation between growth phases of the fungus and changes in certain characteristics of membrane lipids were detected.[87]

c) Functioning

The general structural roles of the major phospholipids in membranes of fungal cells[8,10,11] corresponded to those established for other eukaryotes. As discussed above (III. A. 1.a), the remodelling of membrane phospholipids by changes in chain length, saturation or positions of fatty acids can facilitate the maintenance of membrane structure and permeability in fluctuating environmental conditions. However, this has not been studied in detail in fungi.

Changes in the proportions of phospholipid classes during fungal growth, in some cases has been correlated with modifications of the physico-chemical properties of membranes. In *C. japonica*, PE is predominant in the lipids of the early active period of growth, which are of low microviscosity, high oxidative activity and of a high degree of unsaturation. As growth activity declines, mycelial lipid content increases with changing proportions of neutral and polar lipids, while antioxidative activity and the degree of unsaturation decreases.[88] In the case of *Dipodascopsis uninucleata*, cited above,[78] PUFA were more abundant in phospholipids than in other lipid fractions, at all stages of the life cycle, increasing during germination but decreasing during vegetative growth and differentiation, when the content of oleic acid increased in inverse ratio to PUFA. Both [^{3}H] arachidonic acid and [1-^{14}C] oleic acid provided separately were rapidly incorporated into phospholipids. The former was taken up more rapidly during the growth phase, decreasing during differentiation, when the incorporation of the latter increased, a significant proportion appearing in triacylglycerols.

Evidence has been accumulating in recent years that phosphatidyl inositol, located in the plasma membrane, may have a role in signalling systems in fungi,[89-91] analogous to that demonstrated in certain mammalian and plant tissues.[92] Both synthesis and phosphorylation of phosphatidylinositol are likely to take place at the plasma membrane of *S. cerevisiae*, since enzymes of phosphoinositide synthesis have been found to be associated with post-Golgi secretory vesicles destined for the plasma membrane.[93]

4. Nitrogen-free Glycolipids

Although sporadically recorded for many years,[8,9,11] fungal glycolipids have been overlooked in general analyses of fungal lipids. They have, however, attracted much more attention during the past decade, particularly in relation to the biotechnological importance of acylated sugars and sugar

Table 2
Examples of Nitrogen-Free Glycolipids from Fungi

GLYCO-LIPID CLASS	SOURCE	COMPOSITION
Hydroxy-acid glycosides	*Aspergillus niger*[96]	2-glucosyl-*trans*-octadec-3-enoic acid
	Ustilago zeae[98]	dihydroxyhexadecanoic acid β-cellobiose
	Ustilago nuda[99]	trihydroxyhexadecanoic acid β-cellobiose
	Rhodotorula, Cryptococcus, Candida[10]	extracellular hydroxy fatty acid sophorosides
Glycosyl diacyl-glycerols	*Saccharomyces cerevisiae*,[100] *Hansenula anomala*[101]	monogalactosyldiacylglycerols
	Blastocladiella emersonii,[102] *Choanephora cucurbitarum*[103]	mono- and digalactosyldiacylglycerols
	Aspergillus niger,[104] *Erysiphe pisi*[105]	diglucosyl- and digalactosyldiacylglycerols
Acylated sugars and sugar alcohols	*Claviceps purpurea*[107]	acyltrehalose
	Aureobasidium pullulans[108]	acyltrehalose and acyl mannitol
	Agaricus bisporus[106]	mono- and poly-acylglucoses, acyltrehalose
	Ustilago zeae[109,110]	tetra-acyl-β-mannosyl-erythritol; C_{12-18} fatty acids, with C_{16} predominant.
Sterol glycosides	*Aspergillus niger*[96]	
	Choanephora cucurbitarum[103]	
Polyprenol glycosides	*Aspergillus fumigatus*[112]	hexahydropolyprenol phosphates; 18-20 isoprene units per molecule.
	Neurospora crassa[113]	mannosyl-1-phospho-polyisoprenol mannosyl phosphoryldolichol
	Trichoderma reesei[114]	

alcohols as biosurfactants of wide application (see reviews by Tulloch[94] and Kitamoto[95]). A diversity in nitrogen-free glycolipids has been reported in cultures of various Ascomycotina and Basidiomycotina. Some representatives are listed in Table 2.

a) Hydroxyacid Glycosides

Simple acylated sugars have been occasionally identified, mainly sophorolipids in which 2-O-β-D-glucopyranose (sophorose) is attached by a β-glycosidic link to hydroxy fatty acids, e.g., 2-glucosyloxy-*trans*-octadec-3-enoic acid from *A. niger*[96] and the ustilagic acids, di- and tri-hydroxy-hexadecanoic acid β-cellobiosides, in the smut fungi *Ustilago zeae*[97,98] and *U. nuda*,[99] respectively. Sophorosides of hydroxy fatty acids also accumulate extracellularly in cultures of *Rhodotorula, Cryptococcus* and *Candida.*[98]

b) Glycosyl Diacylglycerols

Glucosyl and galactosyl glycolipids have been reported from a limited number of fungi of various affinities. Monogalactosyldiacylglycerols in *S. cerevisiae*[100] and *Hansenula anomala*,[101] mono- and digalactosyldiacylglycerols in *Blastocladiella emersonii* chitosomes[102] and in *Choanephora cucurbitarum*,[103] and diglucosyldiacylglycerol plus digalactosyldiacylglycerol in *A. niger*[104] and *Erysiphe graminis*[105] can comprise of ~20% of the lipid content of fungal hyphae. Besides being structural components of membranes, there is an evidence of the involvement of these molecules in cell wall polysaccharide synthesis.[9]

c) Acylated Sugars and Sugar Alcohols

Earlier reports of this group of fungal lipids include mono- and polyacylglucoses from *Agaricus bisporus*,[106] acyltrehalose from *Claviceps purpurea*[107] and *Aureobasidium pullulans*[108] and a tetra-acyl-β-mannosyl-erythritol from *U. zeae.*[109,110] Industrial application of fungal mannosyl-erythritol lipids and related metabolites as biosurfactants has resulted in a rapid increase in the information concerning their composition and properties in recent years.[94,95,111]

d) Sterol Glycosides

(see Section C.2.b)

e) Polyprenol-containing Glycosides

Polyprenol-containing glycosides have been implicated in sugar transport across membranes of yeasts and filamentous fungi, e.g., *A. fumigatus.*[112] They have also been implicated in glycoprotein synthesis in *Neurospora crassa*[113] and in *Trichoderma reesei*, where mannosylphosphoryl dolichol was identified as an intermediate.[114] The involvement of polyprenol glycosides in transfer of sugars, acetylglucosamine and oligosaccharides has been discussed by Brennan.[9]

5. Sphingolipids

A wide diversity of sphingolipids, particularly glycosphingolipids in which sugar residues are conjugated with the primary hydroxyl group of the *N*-acylated sphingosine base, have been identified from the relatively small number of fungal species so far investigated. Long-chain hydroxy fatty acids from yeasts and filamentous fungi are generally associated with sphingolipids of the cell wall and plasma membrane, or secreted extracellularly. Fungal sphingolipids, examples of which are given in Table 3, are briefly discussed in this section. More detailed accounts of their structure and distribution in yeasts and filamentous fungi are available in previous reviews.[3,8-11]

a) Ceramides

Ceramides consisting of long-chain bases of the sphingosine type (C_{14-22}) esterified to fatty acids of composition similar to the acylglycerol fractions have been described from a number of yeasts and filamentous fungi.[8,9,10,11] Commonly occurring bases are: (2S,3R)-2-amino-octadecane-1,3-diol

Table 3
Examples of Fungal Sphingolipids

SPHINGOLIPID CLASS	SOURCE	COMPOSITION
Ceramides	Various yeasts[9,10] and filamentous fungi[96,119,123-125]	Long-chain sphingosine-type bases[a] esterified to fatty acids and hydroxy fatty acids
Glycosyl ceramides (cerebrosides)	*Sporothrix schenkii*,[116] *Agaricus bisporus* and *Clitocybe tabescens*[123]	Monoglucosyl ceramides
	Candida albicans[118]	Glucosyl cerebroside: 9-methyl-C_{18}-sphinga-4,8-dienine with 2-OH $C_{18:0}$ as the predominant fatty acid.
	Lentinus edodes[119]	9-methyl-4,8-sphingadienine, with glucose and 2-hydroxy fatty acids
Glycophospho-sphingolipids	*Ophiostoma ulmi*[122]	Inositol phosphoceramides
	Saccharomyces cerevisiae[121,122]	PI-containing sphingolipids
	Histoplasma capsulatum[127,128]	Ceramide-PI-dimannoside-galactoside
	Aspergillus niger[106]	Ceramide-PI-dimannoside-trigalactoside

[a](2S, 3R)-2-amino-octadecane-1,3-diol, (2S, 3R, 4E)-2-amino-4-octadec-4-ene-1,3-diol, (2S, 3S, 4R)-2-amino-octadecane-1,3,4-triol, or (2S, 3S, 4R, 8E)-2-amino-octadec-8-ene-1,3,4-triol.

(Sphi-nganine), (2S, 3R, 4E)-2-amino-4-octadec-4-ene-1,3-diol (Sphingosine), (2S, 3S, 4R)-2-amino-octadecane-1,3,4-triol (Phytosphingosine) and (2S, 3S, 4R, 8E)-2-amino-octadec-8-ene-1,3,4-triol (Dehydrophytosphingosine).[3] The fatty acid components generally include 2-hydroxy fatty acids which accounted for up to 22% of the total fatty acids in several species of Basidiomycotina.[115,121]

b) Glycosyl Ceramides

Glycosyl ceramides with an attached sugar moiety are produced by a number of fungi and include cerebrosides and glycosphingolipids similar to those associated with the nervous system, e.g. the glucocerebroside in the dimorphic human pathogen *S. schenkii.*[116] The carcinogenic effects of grain and other materials contaminated with the mycotoxic fungus *F. moniliforme* on mammals have been attributed to inhibition of sphingolipid biosynthesis by fumonisins, a family of compounds with structural similarities to the long-chain (sphingoid) bases of sphingolipids.[117] The associated ccumulation of sphinganine suggested that fumonisins interfered with the conversion of [^{14}C]-sphinganine to N-acyl-[^{14}C]-sphinganine, a step that precedes introduction of the 4,5-*trans* double bond of sphingosine. It was postulated that disruption of the de novo pathway of sphingolipid synthesis may be a critical event in the diseases associated with the consumption of fumonisins.

Matsubara and co-workers attempted to use cerebrosides for diagnosis of candidiasis. They isolated from the dimorphic human pathogen, *C. albicans*, a cerebroside consisting of glucose linked in β-configuration to a ceramide, the predominant long-chain base of which was identified as 9-methyl-C_{18}-sphinga-4,8-dienene, with the main fatty acid 2-hydroxystearic acid (62%).[118] This compound which is widely distributed in fungi has been reported to be similar in structure and fungal fruiting-inducing activity to cerebrosides in *Schizophyllum commune, Penicillium funiculosum* and *F. amygdali* and several edible fungi, e.g., *Pleurotus ostreatus* and *Lentinus edodes.* Some cerebrosides produced by the species of *Lentinus* stimulate fruiting of other Hymenomycetes such as *S. commune.*[119] A glycosphingolipid from the smut fungus *U. violacea,* which infects *Silene dioica* and produces black smut spores in place of pollen in the anthers of stamens, was identified as a host-specific recognition factor.[120]

c) Glycophosphosphingolipids

Fungal glycophosphosphingolipids in which sugars and oligosaccharides are attached via phosphoinositol to a ceramide have been a focus of interest in recent years in relation to their cell surface role and possible implication in signalling system. PI containing sphingolipids are associated with the plasmamembrane. They can account for about 30% of total phospholipid in *S. cerevisiae,*[121] where the role of glycophosphatidylinositol in anchorage of glycoproteins has been investigated.[122] Evidence implicating an inositol phosphatide pathway in the dimorphism of the plant pathogen *Ophiostoma ulmi* has been presented by Brunton and Gadd[89] but this aspect has not been investigated in dimorphic pathogens of man and animals.

Information on fungal glycolipids and sphingolipids has expanded during the last 10-15 years and some progress has been made in understanding of the function of these compounds in cell biology. Relatively few homologous mycoglycolipids have been fully characterized. They have been identified in most classes of yeasts and filamentous fungi. Among the latter ones, sphingolipids have been studied in greater detail in some of the Hymenomycetes, fruit-bodies of which have offered large amounts of material for analysis.[123-125] Although sphingolipids are associated with the plasma membrane and cell wall development in yeasts and bacteria,[8,9] Manocha and co-workers were unable to find any correlation between virulence and cererobroside content in different strains of *Paracoccidioides brasiliensis.*[126] A series of novel glycolipids containing inositols, isolated from the yeast form of *Histoplasma capsulatum*, were shown to react with antibodies from sera of patients with histoplasmosis.[127] All contained equimolar amounts of hydroxysphinganine (phytosphingosine) and hydroxy or non-hydroxy 24:0 fatty acid and yielded inositol phosphate on hydrolysis. Strong ammonolysis of some resulted in dimannosyl inositol or galactosyldimannosylinositols which were virtually absent from the mycelium and, in one case, a novel galactofuranose glycosphingolipid.[128]

B. HYDROCARBONS

Although rarely investigated, hydrocarbons have been reported among surface lipids of a diversity of fungal structures, mainly in higher fungi. Aliphatic hydrocarbons, in amounts up to 12 mg g^{-1} d.w. of cell wall, occur with other lipids in spores of various fungi, e.g. *Botrytis fabae, Alternaria tenuis, Rhizopus stolonifer,*[129,130] rusts[131,132] and smuts[133,134] but appear to be absent from spores of powdery mildews (*Erysiphe* and *Sphaerotheca* spp.), *Nectria galligena, Penicillium expansum* and *Mucor rouxii.*[129,135,136] The amount and composition of these fungal hydrocarbons varies with species and growth conditions, generally including a homologous series of even- or odd-numbered alkanes with chain lengths in the range C_{15} to C_{36}. In addition, branched-chain compounds have been found in surface lipids of *Aureobasidium pullulans*[108] and in some smut spores.[133] Since hydrocarbons have also been detected in wall material of vegetative mycelium,[137] sclerotia of *Sclerotinia sclerotiorum*[108] and the woody, bracket fruit-bodies of *Fomes ignarius*, growing on *Populus tremuloides* (poplar),[138] it is very likely that further investigation may show such compounds to be more widely distributed than realised at present.

Short-chain hydrocarbons have been detected among physiologically active volatile metabolites of some Ascomycotina and a wide range of Basidiomycotina. These have included ethylene, triacetylene, polyacetylenes and trimethylethylene.[11] The finding of squalene, the C_{30} isoprenoid hydrocarbon on the biosynthetic pathway leading to sterols, in *R. arrhizus*[139] as well as in yeast cells and chitosomes of *M. rouxii,*[71] has been attributed to oxygen deficiency.

C. ISOPRENOIDS

The production of a very wide range of frcc and bound sterols, α-carotenes and a diversity of isoprenoid secondary metabolites has been recorded from fungi.[4,5,7,8,10,11]

1. Sterols

a) Sterol Content and Composition

Apart from certain Oomycetes and some specialized pathogens of animals and plants which are dependent on substrate sterols, most fungi can synthesise sterols. Plant pathogens in the genera and *Phytophthora* are, however, generally capable of vegetative growth on sterol-free media, although they lack membrane sterols.[140] For morphogenesis of reproductive structures, however, these fungi require substrate sterols, e.g., sitosterol or cholesterol, which are normally supplied by host plant tissues. Both in yeasts and filamentous fungi, there is evidence for the functional significance of sterols in stabilizing membrane structure and modifying permeability[10,141] as well as of "sparking" functions at very low levels of sterols and related terpenoids (e.g., the steroid hormones, oogoniols and antheridiols,

Table 4
Summary of Distribution of Fungal Sterols[10,11]

OCCURRENCE	MAJOR STEROLS
MASTIGOMYCOTINA	
Chytridiomycetes	Cholesterol, campesterol, β-sitosterol
Hyphochytriomycetes	Campesterol, ergost-5-enol, β-sitosterol, stigmasterol
Oomycetes	
Saprolegniales	Cholesterol, desmosterol, fucosterol, 24-methylene-cholesterol
Peronosporales	Sterols generally absent unless supplied in medium
Zoophagus insidians	Ergosterol
Pythium and *Phytophthora* spp.	No sterols present, unless obtained from substrate
ZYGOMYCOTINA	
Zygomycetes	Generally ergosterol, sometimes with β-sitosterol, 5-dihydroergosterol, episterol or brassicasterol
Glomus spp.	Cholesterol, campesterol
ASCOMYCOTINA	
Hemiascomycetes	
Ascosporogenous yeasts	Ergosterol and zymosterol ± 24(28)-dehydroergosterol or lanosterol
Protomyces, *Taphrina deformans*	Brassicasterol
Euascomycetes	
Plectomycetes	Episterol, brassicasterol, 22-dihydroergosterol
Loculoascomycetes, Pyrenomycetes	Mainly ergosterol
Discomycetes	Brassicasterol, ergosterol
Lichens	Ergosterol, episterol, lichesterol
DEUTEROMYCOTINA	Mainly ergosterol
BASIDIOMYCOTINA	
Hymenomycetes	
Agaricales	Ergosterol, 5-dihydroergosterol
Polyporales	Ergosterol with 22-dihydroergosterol, 5-dihydro-ergosterol, or fungisterol, plus low levels of cholesterol and lanosterol
Gasteromycetes	Ergosterol
Ustilaginomycetes	Ergosterol with fungisterol ± C_{28}-dienol
Urediniomycetes	Stigmast-7-enol ± stigmasta-5,7-dienol or stigmasta-7,24 (28)-dienol, with some cholesterol and fungisterol

triggering the development of sexual structures in *Achlya*[142]) in morphogenesis and reproduction.[143-145]

Changes in amounts of free and esterified sterols have been correlated with conservation of sterols during alternating phases of the growth cycle. Parks et al.,[143] after observing increased esterification of sterols upon entry into stationary phase and rapid hydrolysis to free ergosterol when transferred to fresh medium, suggested that there was a critical and transient requirement for ergosterol in the actively growing fungal cells and that esterification conserves sterols for availability in active growth phases. Studies on mutants defective in sterol synthesis often show "an accommodating physiological or genetic adjustment" of phospholipid content, allowing growth without sterol.[143] The strict requirement of sterols by some members of the Pythiaceae, for example, has been thrown in doubt by the apparent induction of oospores on defined media supplemented with phospholipids. Detailed examination by Kerwin and Duddles suggested that only traces of sterols were necessary for sparking and critical domain roles, acting synergistically with phospholipids containing unsaturated fatty acid moieties to induce oospore formation.[146] In *Pythium ultimum*, enrichment of the phospholipid fraction of cell lipid with unsaturated fatty acids, taken up from exogenous lipid, had promoted oospore induction while enhanced levels of unsaturated fatty acids in the neutral lipids had increased oospore viability.

Although ergosterol ([22E]-ergosta-5,7,22-trienol), frequently referred to as the fungal sterol, accounts for a high proportion of the sterol content of many Zygomycetes, Ascomycotina and Basidiomycotina, a very wide range of sterols have been reported in fungi (Table 4) and some striking chemotaxonomic associations are indicated.[7,8,10,11] Since published information on sterol composition (like that on fatty acids) is beset with variation reflecting culture conditions, developmental phases[147] or extraction procedures, its chemotaxonomic application still requires data from a larger number of species, grown in defined conditions and at standardised stages of development. This is at present nearer to achievement with yeasts than for filamentous fungi. Sterol responses to temperature have been examined in a range of *Mucor* species[148] and in *Neurospora* mutants.[149] Sterol biosynthesis is particularly sensitive to oxygen availability, although the obligately anaerobic rumen organism *Neocallimastix frontalis* (Chytridiomycetes) appears to be an exception.[150] Oxygen limitation may result in the accumulation of biosynthetic precursors of the normal principal sterols and of bound sterols.[10,151]

b) Sterols in Responses of Fungal Pathogens to Antibiotics

Certain features of the sterol composition of membranes of pathogens contribute to their difference in the affinity for antibiotics. The effects of triazoles and other sterol-inhibiting fungicides on fungal cells have been discussed in this context.[152,153] Antibiotic resistant cultures of *Cryptococcus neoformans*[154] and *Candida* spp.[155] have been found to have decreased amounts of ergosterol in which the presence of double bonds at positions 5-6 and 7-8 has been correlated with polyene sensitivity, whereas the presence of less planar sterols, e.g., fecosterol, which are unsaturated at position 8-9 or have, in addition, methyl group substitutions at positions 4 and 14, appears to confer greater resistance[156] (for more details on sterols of *Candida* and *Cryptococcus* see chapters 5 and 7).

Interactions between phospholipids and sterols have also been implicated in the antibiotic sensitivity of *C. albicans.* More non-esterified sterol was present in azole-resistant strains, in which phospholipid/sterol ratios were approximately half of those in azole-sensitive strains as detected by Hitchcock and co-workers,[79] who also drew attention to other evidence that resistance to triazoles in this species was due to membrane impermeability rather than the insensitivity to ergosterol synthesis. The same team demonstrated that, in a laboratory mutant of *C. albicans* exhibiting both azole and polyene resistance, ergosterol, which was the only non-esterified sterol and the major esterified sterol in other strains, was replaced by methylated sterols, mainly lanosterol, 24-methylene-24,25-dihydrolanosterol and 4-methylergostadiene-3-ol, and thus lacked a key target site for azoles.[80]

Among plant pathogenic fungi, the special features of Oomycetes capable of vegetative growth without synthesising sterols, discussed above, attracted early interest,[140,153] on account of the resulting resistance of important pathogens, e.g., *Phytophthora* and *Pythium* spp. to polyene antibiotics such as nystatin. *Taphrina deformans* (Ascomycotina), which grows biotrophically in leaves of peach and related species, has been the subject of intensive study in relation to the action of sterol biosynthesis inhibitors.[157-160] Weete and co-workers have discussed the effects of sterol biosynthesis inhibitors on the fatty acid metabolism of this species.[161]

c) Chemotaxonomic Potential of Sterols

As with their fatty acid composition and features of structure and life cycle, the Siphomycetes ("lower" fungi) are more distinctly grouped with respect to sterol composition than the higher fungi, offering a greater diversity of chemotaxonomic characters (Table 4). Records of fungal sterols are, however, beset with variations in quality of analyses. The presence of such compounds as fucosterol, 24-methylene cholesterol, cholesterol or desmosterol in various Oomycetes supports a phylogenetic relationship with algae and higher plants.[11] These sterols are generally absent from the Zygomycotina in which ergosterol is a major component, accompanied by 22-dihydroergosterol, although cholesterol in varying amounts and sitosterol have been reported in some *Mucor* species. On the other hand, *Glomus mosseae* and *G. caledonicus*, obligately biotrophic mycorrhizal fungi which are generally assigned to the Zygomycotina, lacked ergosterol and 22-dihydroergosterol, although like some *Mucor* spp. they produced stigmast-5-enol, and resembled Oomycetes in having cholesterol and campesterol as major sterols.[162]

Among the Septomycetes, or "higher fungi", the very large divisions of Ascomycotina and Deuteromycotina (*Fungi imperfecti*, mainly anamorphs of Ascomycotina) are inadequately studied, although detailed information on sterol composition is available for certain species.[10,11] Sterol synthesis has been intensively studied using mutant strains of *S. cerevisiae.* Yeast strains which are relatively insensitive to anaerobic conditions have been reported to synthesize 24(28)-dehydroergosterol rather than ergosterol.[163]

Ergosterol is the predominant sterol in Ascomycotina, accompanied by smaller amounts of cholesterol, fungisterol, 5-dihydroergosterol or a wide variety of other sterols.[8,10,11] More extensive investigation of their sterol composition may provide a valuable chemotaxonomic tool for this largest division of fungi in which morphological features alone have not provided a satisfactory separation of classes. Since the preponderance of fungal pathogens can be assigned to the Ascomycotina, such chemotaxonomic study is likely to be of diagnostic importance.

The classes of Basidiomycotina are morphologically more clearly defined than those of the Ascomycotina and from such studies they appear to have some distinct features of sterol composition.[11] The relatively large fruit-bodies of the Hymenomycetes and Gasteromycetes have been more frequently analysed than their vegetative mycelium. In these, ergosterol is the predominant sterol, accompanied in the Agaricaceae by substantial proportions of 22-dihydroergosterol but, in the Polyporaceae, by 5-dihydroergosterol, fungisterol and smaller amounts of cholesterol and lanosterol.

Smut fungi and rusts, which are plant pathogenic microfungi of the Basidiomycotina, differ markedly in sterol composition.[11] Smut fungi, which include species of *Ustilago* and *Filobasidiella*, implicated in pathogenesis of animals, resemble the Hymenomycetes, whereas rust fungi, which have not been found to be pathogenic to animals, lack ergosterol but have stigmasta-7,24 (28)-dienol as major sterol, with smaller amounts of stigmasta-5,7-dienol, the synthesis of which has been demonstrated in germinating urediniospores of *Uromyces phaseoli*.[164] Like their different fatty acid composition, this could contribute to discussions on the vexed question whether smuts and rusts should be considered as distinct classes, Ustilaginomycetes and Urediniomycetes, or as the orders Ustilaginales and Uredinales.

2. Combined Sterols

a) Sterol Esters

The role and location of esterified sterol fractions in fungi have become clearer in recent years.[143] In yeasts there is evidence for the occurrence of sterol esters in low density, lipid-enriched cytoplasmic vesicles.[165] The amounts of sterol esters found in lipid extracts of fungal cultures and fungus infected plant tissue change with developmental phases and environmental factors,[10,143,166] sometimes varying in inverse proportion to free sterols. Since this fraction characteristically includes a variety of biosynthetic intermediates of sterols, it is likely that sterol esters provide a reserve of membrane sterol precursors.[167]

b) Glycosyl Sterols

Although the steryl glycosides have been detected in the various fungi, e.g., *A. niger*, their functions have not been extensively investigated. Conversion of sterols into water-soluble form, e.g., a mannosyl-sterol complex from *S. cerevisiae*, may facilitate their accumulation in cytosol or vacuole at certain phases of the life cycle. Sterol glycosides have been shown to increase during biotrophic infection of *Choanephora cucurbitarum* by the mycopathogen *Piptocephalis virginianae.*[168]

3. Other Isoprenoids

a) Carotenoids

Carotenoids are widely distributed in fungi,[8,11,169,170] commonly associated with reproductive structures, photoreceptive cells and in mitochondrial membranes.[171] Their significance at all these sites may be partly due to antioxidant activity which offers protection against photooxidation of unsaturated fatty acids of membrane lipids.

b) Other Terpenoids

A very large number of terpenoids of varying complexity have been identified and described in fungi.[4,5] Although largely classified as secondary metabolites, many of these compounds have important physiological effects on other organisms and include plant growth regulators as well as a wide range of antibiotics, mycotoxins and carcinogens of major importance in medicine.[2,11]

IV. CONCLUSION

Within the scope of a short chapter we have attempted to provide an orientating survey of lipid components of pathogenic and non-pathogenic fungi, their significance and the variation encountered. Comprehensive treatment of the structures and composition of fungal lipids as well as their physiological roles have been cited earlier. With the exception of the Oomycetes and other Mastigomycotina which are plant-like in structural and biochemical aspects, some features of lipid metabolism of fungi, such as pathways of fatty acid, sphingolipid and sterol biosynthesis, lie closer to those of the animal kingdom than to those of plants, with which this distinct kingdom of organisms has traditionally been linked. Some emphasis has been laid here on the variability of lipid composition encountered even within genera and species of fungi, to the extent that it reflects the versatility of fungi in responding to environmental and nutritional constraints and consequently the caution which has to be observed in attempting chemotaxonomic generalizations from non-standardized conditions. Nevertheless, diagnosis of pathogens isolated from tissues may, ultimately, be facilitated by the inclusion of lipid among other taxonomic characteristics. In any case, a knowledge of the role of lipid components of fungal cells in their interaction with host tissue can be expected to contribute to a better understanding of the nature of their pathogenicity and to allow more precise targeting of antibiotics and fungicides which exert their effects on synthesis of membrane lipid components or may be accumulated in lipid bodies. Experimentation has been easier with plant pathogens but some of the findings of physiological plant pathology may lead to insights into fungal-animal interactions.

ACKNOWLEDGMENTS

It is a pleasure to record our thanks to Dr. Yolande Dalpé, Agriculture Canada, Centre for Agricultural Land Resources, Ottawa, for substantial assistance in the literature search for this Chapter.

V. REFERENCES

1. **Chopra, A. and Khuller, G.K.,** Lipids of pathogenic fungi, *Prog. Lipid Res.*, 22, 189, 1983.
2. **Howard, D.H. and Howard, L.F.** (Eds.), *Fungi Pathogenic for Humans and Animals*, Part A, *Biology*, Marcel Dekker, New York and Basel, 1983.
3. **Ratledge, C. and Wilkinson, S.G.,** Fatty acids, related and derived lipids, in *Microbial Lipids*, Vol. 1, Ratledge, C. and Wilkinson, S.G., Eds., Academic Press, London, 1988, chap. 2.
4. **Ratledge, C. and Wilkinson, S.G.,** Terpenoid lipids, in *Microbial Lipids*, Vol. 1, Ratledge, C. and Wilkinson S.G., Eds., Academic Press, London, 1988, chap. 3.
5. **Turner, W.B.,** *Fungal Metabolites*, Academic Press, London, 1971.
6. **Turner, W.B. and Aldrich, D.C.,** *Fungal Metabolites* II, Academic Press, London, 1983.
7. **Weete, J.D.,** *Fungal Lipid Biochemistry*, Plenum Press, New York, 1974.
8. **Weete, J.D.,** *Lipid Biochemisrty of Fungi and Other Organisms*, Plenum Press, New York, 1981.
9. **Brennan, P.J. and Losel, D.M.,** Physiology of fungi: selected topics, *Adv. Microbial Physiol.*, 17, 48, 1978.
10. **Rattray, J.B.M.,** Yeasts, in *Microbial Lipids*, Vol. 1., Ratledge, C. and Wilkinson, S.G., Eds., Academic Press, London, 1988, chap. 9.
11. **Lösel, D.M.,** Fungal lipids, in *Microbial Lipids*, Vol. 1, Ratledge, C. and Wilkinson, S.G., Eds., Academic Press, London, 1988, chap. 10.

12. **Ratledge, C. and Wilkinson, S.G.**, Eds., *Microbial Lipids*, Vol. 1, Academic Press, London, 1988.
13. **Ratledge, C. and Wilkinson, S.G.**, Eds., *Microbial Lipids*, Vol. 2, Academic Press, London, 1989.
14. **Rose, A.H.**, Influence of the environment on microbial lipid composition. In *Microbial Lipids*, Vol. 2, Ratledge, C. and Wilkinson, S.G., Eds., Academic Press, London, 1989, chap. 16.
15. **Prasad, R.**, Lipids in the structure and function of yeast membrane, *Advances in Lipid Research*, 21, 187, 1985.
16. **Radzhabova, A.A., Allakhverdiev, A.D. and Gumbatova, R.I.**, Regulation of lipid synthesis and unsaturation in *Mortierella* fungi, *Microbiologiya*, 59, 982, 1990.
17. **Holdsworth, J.E. and Ratledge, C.**, Lipid turnover in oleaginous yeasts, *J. Gen. Microbiol.*, 134, 339, 1988.
18. **Holdsworth, J.E., Veenhuuis, M. and Ratledge, C.**, Enzyme activities in oleaginous yeasts accumulating and utilizing exogenous or endogenous lipids, *J. Gen. Microbiol.*, 134, 2907, 1988.
19. **Botham, P.A. and Ratledge, C.**, A biochemical explanation for lipid accumulation in *Candida* 107 and other oleaginous microorganisms, *J. Gen. Microbiol.*, 114, 361, 1979.
20. **Ratledge, C.**, Biotechnology of oils and fats, in *Microbial Lipids*, Vol. 2, Ratledge, C. and Wilkinson, S.G., Eds., Academic Press, London, 1989, chap. 22.
21. **Scherbakov, M.A.**, Fatty inclusions in mycelia of *Aspergillus foeitdus Naka* Thom and Raper and *Fusarium oxysporum* Schlecht, Synd. et Hans, *Ukr. Bot. Zh.*, 47, 96, 1990.
22. **Papacharilaou, E. and Pisano, M.A.**, Changes in the lipid composition of *Paecilomyces persicinus* during growth and cephalosporin C production, *Appl. Env. Microbiol.*, 48, 1084, 1984.
23. **Chevant, L., Le Bars, J. and Sancholle, M.**, Metabolism lipidique chez L'*Aspergillus versicolor* (Vuill.) Tiraboschi. Relation avec la biogene de la sterigmatocystine, *Mycopathologia*, 60, 151, 1977.
24. **Massoudnia, A. and Scheer, E.R.**, The influence of medium on the chemical composition of *Blastomyces dermatitidis*, *Current Microbiol.*, 7, 25, 1982.
25. **Bal, A.K., Dey, A.C. and Hampson, M.C.**, Resting sporangium of *Synchytrium endobioticum*: its structure and composition of the lipids and fatty acids, *Arch. Microbiol.*, 140, 178, 1985.
26. **Garrison, R.G.**, Ultrastructural cytology of pathogenic fungi, in *Fungi Pathogenic for Humans and Animals*, Howard, D.H. and Howard, L.F., Eds., Part A, *Biology*, Marcel Dekker, New York and Basel, 229,1983.
27. **Kulkarni, R.K. and Nickerson, K.W.**, Pleomorphism in *Ceratocystis ulmi* refractile lipid bodies, *Can. J. Bot.*, 61, 153, 1983.

28. **Anderes, E.A., Finley, A.A. and Walsh, H.A.,** The lipids of an auxotrophic avirulent mutant of *Coccidioides immitis*, *Sabouraudia*, 11, 127, 1973.
29. **Ghannoum, M.A., Burns, G.R., Abu Elteen, K. and Radwan, S.S.,** Experimental evidence for the role of lipids in adherence of *Candida* spp. to human buccal epithelial cells, *Infect. Immun.*, 54, 189, 1986.
30. **Fonvieille, J.L. and Sancholle, M.,** *Scopulariopsis brevicaulis* (Bainier): study of the lipid content in relation to growth, in *Biogenesis and Function of Plant Lipids*, Mazliak, P., Beneviste, P. and Douce, R., Eds., Elsevier/North Holland Biomedical Press, Amsterdam, 235, 1980.
31. **Silva, C.L. and Ekizlerian, S.M.,** Granulomatous reactions induced by lipids extracted from *Fonsecaea pedrosoi*, *Fonsecaea compactum*, *Cladosporium carrionii* and *Phialophora verrucosum*, *J. Gen. Microbiol.*, 131, 187, 1985.
32. **Ratledge, C.,** Microbial lipids: commercial realities or academic curiosities, in *Applied Single Cell Oils*, Kyle, D.J. and Ratledge, C., Eds., AOCS: Champaign, III, 1, 1992.
33. **Ghannoum, M.A., Janini, G., Khamnis, L. and Radwan, S.S.,** Dimorphism associated variations in the lipid composition of *Candida albicans*, *J. Gen. Microbiol.*, 132, 2367, 1986.
34. **Dennettiere, B., Ibrahim-Granet, O. and De Bievre, C.,** Temperature phospholipid variations and dimorphism in *Fonsecaea pedrosoi*, *Cladosporium carrionii* and *Sporothrix schenckii*, *J. Mycol. Med.*, 1, 302, 1991.
35. **Latgé, J.P. and de Bievre, C.,** Lipid composition of *Entomophthora obscura*, *J. Gen. Microbiol.*, 121, 151, 1980.
36. **Shaw, R.,** The occurrence of γ-linolenic acid in fungi, *Biochim. Biophys. Acta*, 98, 230, 1965.
37. **Shaw, R.,** The polyunsaturated fatty acids of microorganisms, *Adv. Lipid Res.*, 9, 107, 1966.
38. **Sancholle, M., Weete, J.D. and Rushing, A.,** Sterols and fatty acids in the plasma membrane of *Taphrina deformans* cultured at low temperature and with propiconazole, in *Biological Role of Plant Lipids*, Biacs, P.A., Gruiz, K. and Kremmer, T., Eds., Plenum Publishing Corp., Amsterdam, 413, 1988.
39. **Suutari, M., Liukkonen, K. and Lassko, S.,** Temperature adaptation in yeasts: the role of fatty acids, *J. Gen. Microbiol.*, 136, 1469, 1990.
40. **Davis, P.J., Fleming, B.D., Coolbear, K.P. and Keough, K.M.W.,** Gel to liquid-crystalline phase transitions of two pairs of positional isomers and unsaturated mixed isomer phosphatidylcholines, *Biochemistry*, 20, 3633, 1981.
41. **Coolbear, K.P., Berde, C.B. and Keough, K.M.W.,** Gel to liquid-crystalline phase transitions of aqueous dispersions of polyunsaturated mixed-acid phosphatidylcholines, *Biochemistry*, 22, 1466, 1983.

42. **Nurminen, T. and Suomalainen, H.,** Occurrence of long-chain fatty acids and glycolipids in the cell envelops fractions of bakers' yeast, *Biochem. J.*, 125, 963, 1971.
43. **Taylor, G.T.,** The influence of some non-ionic detergents on the synthesis and secretion of medium chain length fatty acids by *Saccharomyces, FEMS Microbiol. Lett.*, 6, 103, 1979.
44. **Wilson, K. and McLeod, B.J.,** The influence of conditions of growth on the endogenous metabolism of *Saccharomyces cerevisiae*: effect on protein, carbohydrate, sterol and fatty acid content and on viability, *Antonie van Leeuwenhoek*, 42, 397, 1976.
45. **Astin, A.M., Haslam, J.M. and Woods, R.A.,** The manipulation of cellular cytochrome and lipid composition in a heme mutant of *Saccharomyces cerevisiae, Biochem. J.*, 166, 275, 1977.
46. **Hellgren, L. and Vincent, J.,** Contact sensitizing properties of some fatty acids in dermatophytes, *Sabouraudia*, 14, 243, 1976.
47. **Lösel, D.M.,** Functions of lipids: specialized roles in fungi and algae, in *Microbial Lipids*, Vol. 2, Ratledge, C. and Wilkinson, S.G., Eds., Academic Press, London, 1989, chap. 18.
48. **Mantle, P.G., Morris, L.J. and Hall, S.W.,** Fatty acid compositions of sphacelial and sclerotial growth forms of *Claviceps purpurea* in relation to the production of ergoline alkaloids, *Trans. Brit. Mycol. Soc.*, 53, 441, 1969.
49. **Morris, L.J.,** Fatty acid composition of *Claviceps* species. Occurrence of (+)-threo-9,10-dihydroxystearic acid, *Lipids*, 3, 260, 1968.
50. **Mantle, P.G.,** Fatty acid composition of triglyceride oils from sphacelial, sclerotial and stromatal tissues of *Claviceps purpurea and Claviceps sulcata, Trans. Brit. Mycol. Soc.*, 59, 325, 1972.
51. **Tyrrell, D. and Weatherston, J.,** The fatty acid composition of some Entomophthoraceae. IV The occurrence of branched-chain fatty acids in *Conidiobolus* species, *Can. J. Microbiol.*, 22, 1058, 1976.
52. **Brennan, P.J., Griffin, P.F.S., Lösel, D.M. and Tyrrell, D.,** The lipids of fungi, *Prog. Chem. Fats Other Lipids*, 14, 51, 1974.
53. **Tyrrell, D.,** The fatty acid composition of some Entomophthoraceae. II The occurences of branched-chain fatty acids in *Conidiobolus danaesporus* Drechsl, *Lipids*, 3, 368, 1968.
54. **Stoessl, A.,** Metabolites of *Cercospora arachidicola*. The first report of long-chain fatty aldehydes from a fungus, *Can. J. Microbiol.*, 31, 129, 1985.
55. **Radwan, S.S.,** Sources of C20-polyunsaturated fatty acids for biotechnological use, *Appl. Microbiol. Biotechnol.*, 35, 441, 1991.
56. **Sancholle, M. and Lösel, D.M.,** Lipids in fungal biotechnology, in *The Mycota II*, Kück, U. and Esser, K., Eds., Springer Verlag, Heidelberg, 1995, chap. 20.
57. **Shimizu, S., Akimoto, K., Kawashima, H., Shinmen, Y., Jareonkitmongkol, S. and Yamada, H.,** Stimulatory effect of peanut

oil on the production of dihomo-γ-linolenic acid by filamentous fungi, *Agri. Biol. Chem.*, 53, 1437, 1989.
58. **Shimizu, S., Jareonkitmongkol, S., Kawashima, H., Akimoto, K. and Yamada, H.**, Production of a novel ω-1-eicosapentaenoic acid by *Mortierella alpina* 1S-4 grown on hexadecene, *Arch. Microbiol.*, 156, 163, 1991.
59. **Radzhabova, A.A., Allakhverdiev, A.D. and Gumbatova, R.I.**, Regulation of lipid synthesis and unsaturation in *Mortierella* fungi, *Mikrobiologiya*, 59, 982, 1990.
60. **Ozeretskovskaya, D.L, Avdyushko, S.A., Chalova, L.I., Chalenko, G.I. and Yurganova, L.A.**, Arachidonic and eicosapentaenoic acids as the active principle of a biogenic elicitor. A lipoglycoprotein complex isolated from the causative agent of the late blight of potato, *Dokl. Akad. Nauk. SSSR*, 292, 738, 1987.
61. **Merzlyak, M.N., Reshetnikova, I.V., Chivkunova, O.B., Ivanova D.G. and Maximova, N.I.**, Hydrogen peroxide-dependent and superoxide-dependent fatty acid breakdown in *Phytophthora infestans* zoospores, *Plant. Sci. (Limerick)*, 72, 207, 1990.
62. **Castoria, R., Fanelli, C., Fabbri, A.A. and Passi, S.**, Metabolism of arachidonic acid involved in its eliciting activity in potato tuber, *Physiological and Molecular Plant Pathology*, 41, 127, 1992.
63. **Lechevalier, H. and Lechevalier, M.P.**, Chemotaxonomic use of lipids - an overview, in *Microbial Lipids*, Vol. 1, Ratledge, C. and Wilkinson, S.G., Eds., Academic Press, 1988, Chap. 12.
64. **Tyrrell, D.**, The fatty acid composition of 17 *Entomophthora* isolates, *Can. J. Microbiol*, 13, 755, 1967.
65. **Mumma, R.O. and Bruszewski, T.E.**, The fatty acids of *Entomophthora coronata*, *Lipids*, 5, 915, 1970.
66. **Sancholle, M. and Dalpé, Y.**, Taxonomic relevance of fatty acids of arbuscular mycorrhizal fungi and related species, *Mycotaxon*, 49, 187, 1993.
67. **Sumner, J.L. and Evans, J.C.**, The fatty acid composition of *Dactylaria* and *Scolecobasidium*, *Can. J. Microbiol.*, 17, 7, 1971.
68. **Nishikawa, J., Sato, Y., Kashimura, J. and Iizuka, H.**, Cellular fatty acids composition of the genus Monascus, *J. Basic Microbiol.*, 29, 369, 1989.
69. **Holtz, R.B. and Schisler, L.C.**, Lipid metabolism of *Agaricus bisporus* (Lang) Sing.: analysis of spore and myecelial lipids, *Lipids*, 6, 176, 1971.
70. **Sumner, J.L.**, The fatty acid composition of Basidiomycetes, *N.Z.J. Bot.*, 11, 435, 1973.
71. **Weete, J.D., Furter, R., Hänseler, E. and Rast, D.M.**, Cellular and chitosomal lipids of *Agaricus bisporus* and *Mucor rouxii*, *Can. J. Microbiol.*, 31, 1120, 1985.

72. **Sock, J.**, Über das Lipidspektrum und den Lipidgehalt phytopathogene Pilze unter besondere Berücksichtigung keimender Uredosporen sowie axenisch wachsenden Mycels des Weizenschwarzrostes-*Puccinia graminis* f.sp. tritici, Rasse 21, Dr. phil. Dissertation, Georg-August-Universitat, Gottingen, 1994.
73. **Grechkin, A.N.**, A physiological product of plant lipoxygenase pathway, in *11th International Meeting on Plant Lipids*, Kader, J.C. and Mazliak, P., Eds., Paris, 1995.
74. **Grechkin, A.N., Kuramshin, K.A., Latypov, S.K., Safonova, Y.Y. and Gafarova, T.E.**, Hydroperoxides and α-ketols. Novel products of the plant lipoxygenase pathway, *Eur. J. Biochem.*, 1991.
75. **Fanelli, C. and Fabbri, A.A.**, Relationship between lipids and aflatoxin biosynthesis, *Mycopathologia*, 107, 115, 1989.
76. **Fanelli, C., Fabbri, A.A., Panfili, G., Castoria, R., De Luca, C. and Passi, S.**, Aflatoxin congenor biosynthesis induced by lipoperoxidation, *Exp. Mycol.*, 13, 61, 1989.
77. **Patel, U.D., Bapat, S.R. and Dave, P.J.**, Induction of aflatoxin biosynthesis in *Aspergillus parasiticus* by ascorbic acid mediated lipid peroxidation, *Current Microbiol.*, 20, 159, 1990.
78. **Kock, J.L.F. and Ratledge, C.**, Changes in lipid composition and arachidonic acid turnover during the life cycle of the yeast *Dipodascus uninucleata*, *J. Gen. Microbiol.*, 139, 459, 1993.
79. **Hitchcock, C.A., Barret-Bee, K. and Russell, N.J.**, The lipid composition of azole-sensitive and azole-resistant strains of *Candida albicans*, *J. Gen. Microbiol.*, 132, 2421, 1986.
80. **Hitchcock, C.A., Barret-Bee, K. and Russell, N.J.**, The lipid composition and permeability to azole of an azole- and polyene-resistant mutant of *Candida albicans*, *J. Med. Vet. Mycol.*, 25, 29, 1987.
81. **Mago, N. and Khuller, G.K.**, Lipids of *Candida albicans*: subcellular distribution and biosynthesis, *J. Gen. Microbiol.*, 136, 993, 1990.
82. **Klig, L.S., Friedli, L. and Schmid, E.**, Phospholipid biosynthesis in *Candida albicans*, regulation by the precursors inositol and choline, *J. Bacteriol.*, 172, 4407, 1990.
83. **Arneja, J.S.**, Effect of phosphate on lipid metabolism in *Fusarium oxysporum*, *Acta Mycol.*, 23, 149, 1987.
84. **Raheja, R.K., Paul, Y. and Sharma, B.N.**, Effect of inorganic phosphate in the medium on phospholipids of *Pythium irregulare*, *J. Res. Punjab Agric. Univ.*, 20, 160, 1983.
85. **Wiebe, M.G., Robson, G.D. and Trinci, A.P.J.**, Effect of choline on the morphology, growth and phospholipid composition of *Fusarium graminearum*, *J. Gen. Microbiol.*, 135, 2155, 1989.
86. **Ievleva, N.R., Bragintseva, L.M. and Tentsova, A.I.**, The influence of the sources of carbon in the medium on the composition of phospholipids in a fungus of the genus *Fusarium*, *Farmatisya* (Mosc.),

33, 16, 1984.
87. **Mihailova, M.V., Feofilova, E.P., Rozantsev, E.G., Arkhipova, G.V. and Hrckova, E.A.,** The effect of cycloheximide and actinomycin D on the content of phospholipids and neutral lipids during the growth of *Cunninghamella japonica*, *Mikrobiologiya*, 53, 816, 1984.
88. **Kuznetsova, L.S., Polotebnova, M.V., Feofilova, E.P., Grigoryan, G.L. and Rozantsev, E.G.,** Changes in lipid physico-chemical properties during growth of the mucor fungus *Cunninghamella japonica*, *Biol. Nauki. (Mosc)*, V.0 (2), 73, 1987.
89. **Brunton, A.H. and Gadd, G.M.,** Evidence for an inositol signal pathway in the yeast-mycelium transition of *Ophiostoma ulmi*, the dutch elm disease fungus, *Mycol. Res.*, 95, 484, 1991.
90. **Robson, G.D., Trinci, A.P.J., Wiebe, M.G. and Best, L.C.,** Phosphatidylinositol 4,5-bisphophate (PIP_2) is present in *Fusarium graminearum*, *Mycol. Res.*, 95, 1082, 1991.
91. **Lösel, D.M., Meilleur, E.T. and Abood, J.K.,** Role of membrane lipids in powdery mildew infection of cucumber, *J. Plant Physiol.*, 143, 575, 1994.
92. **Morse, M.J., Slatter, R.L., Crain, R.C. and Cote, G.C.,** Signal transduction and phosphatidylinositol turnover in plants, *Physiol. Plant.*, 76, 118, 1989.
93. **Kinney, A.J. and Carman, G.M.,** Enzymes of phosphoinositide synthesis in secretory vesicles destined for the plasma membrane in *Saccharomyces cerevisiae*, *J. Bacteriol.*, 172, 4115, 1990.
94. **Tolloch, A.P.,** Glycosides of hydroxy fatty acids, in *Handbook of Lipid Research*, vol. 6, Kates, M., Ed., Plenum Press, New York, 463, 1990.
95. **Kitamoto, D.,** Production of surfactants by microorganisms, *J. Japan Oil Chemists Soc.*, 41, 839, 1992.
96. **Laine, R.A., Griffin, P.F.S., Sweeley, C.C. and Brennan, P.J.,** Monoglucosyl oxyoctadecenoic acid - a glycolipid from *Aspergillus niger*, *Biochemistry*, 11, 2267, 1972.
97. **Lemieux, R.U.,** Biochemistry of the Ustilaginales VIII. The structures and configurations of the ustilic acids, *Can. J. Chem.*, 31, 396, 1953.
98. **Stodola, F.H., Deinema, M.H. and Spencer, J.F.T.,** Extracellular lipids of yeasts, *Bacteriol. Rev.*, 31, 194, 1967.
99. **Bhattacharjee, S.S., Haskins, R.H. and Gorin, P.A.J.,** Location of acyl groups on partially acylated glycolipids from strains of *Ustilago* (smut fungi), *Carbohydr. Res.*, 13, 235, 1970.
100. **Baraud, J., Maurice, A. and Napias, C.,** Composition and distribution of lipids in *Saccharomyces cerevisiae* cells, *Bull. Soc. Chim. Biol.*, 52, 421, 1970.
101. **Ng, K.H. and Lanéele, M.A.,** Lipids of the yeast *Hansenula anomela*, *Biochemie*, 59, 97, 1977.
102. **Mills, G.L. and Cantino, E.C.,** Lipid composition of the zoospores of *Blastocladiella emersonii*, *J. Bacteriol.*, 118, 192, 1974.

103. **Deven, J.M. and Manocha, J.S.,** Effects of various cultural conditions on the fatty acid and lipid composition of *Choanephora cucurbitarum*, *Can. J. Microbiol.*, 22, 443, 1976.
104. **Hackett, J.A. and Brennan, P.J.,** Reaction of fungal ceramides containing α-hydroxy acids with the periodate-Schiff reagents, *J. Chromatog.*, 117, 436, 1976.
105. **Sloss, R.I.,** Lipid metabolism in *Pisum sativum* infected by *Erysiphe pisi*. Ph.D. Thesis, Imperial College, University of London, 1976.
106. **Byrne, P.F.S. and Brennan, P.J.,** The lipids of *Agaricus bisporus*, *J. Gen. Microbiol.*, 89, 245, 1975.
107. **Cooke, R.C. and Mitchell, D.T.,** Carbohydrate physiology of sclerotia of *Claviceps purpurea* during dormancy and germination, *Trans. Brit. Mycol. Soc.*, 54, 93, 1970.
108. **Merdinger, E., Kohn, P. and McClain, R.C.,** Composition of lipids in extracts of *Pullularia pullulans*, *Can. J. Microbiol.*, 14, 1021, 1968.
109. **Bootthroyd, B., Thorn, J.A. and Haskins, R.H.,** Biochemistry of the Ustilaginales XII. Characterization of extracellular glycolipids produced by *Ustilago* spp., *Can. J. Biochem. Physiol.*, 34, 10, 1956.
110. **Fluharty, A.L. and O'Brien, J.S.,** The mannose and erythritol-containing glycolipid from *Ustilago maydis*, *Biochemistry*, 8, 2627, 1969.
111. **Frautz, B., Lang, S. and Wagner, F.,** Formation of cellulose lipids by growing and resting cells of *Ustilago maydis*, *Biotechnol. Lett.*, 8, 757, 1986.
112. **Stone, K.J., Butterworth, A.H.W. and Hemming, F.W.,** Characterization of the hexahydropolyprenols of *Aspergillus fumigatus fresenius*, *Biochem. J.*, 102, 443, 1967.
113. **Gold, M.H. and Hahn, H.J.,** Role of a mannosyl lipid intermediate in the synthesis of *Neurospora crassa* glycoproteins, *Biochemistry*, 15, 1808, 1976.
114. **Kruszewska, J., Messner, R., Kubicek, C.P. and Palamarczyk, G.,** O-glycosylation of proteins by membrane fractions of *Trichoderma reesei* QM 9414, *J. Gen. Microbiol.*, 135, 301, 1989.
115. **Ratledge, C.,** Microbial oils and fats: an assessment of their commercial potential, in *Progress in Industrial Microbiology*, Bull, M.J., Ed., 16, 119, 1982.
116. **Cardoso, D.B.S., Anguluster, J., Travassos, L.R. and Alviano, C.S.,** Isolation and characterization of a glucocerebroside monoglucosyl ceramide from *Sporothrix schenkii*, *Microbiol. Lett.*, 43, 279, 1987.
117. **Wang, E., Norred, W.P., Bacon, C.W., Riley, R.T. and Merrill, A.H., Jr.,** Inhibition of sphingolipid biosynthesis by fumonisins. Implications for diseases associated with *Fusarium moniliforme*, *J. Biol. Chem.*, 226, 14486, 1991.

118. **Matsubara, T., Hayashi, A., Banno, Y. and Morita, T.**, Cerebroside of the dimorphic human pathogen, *Candida albicans*, *Chem. Phys. Lipids*, 43, 1, 1987.
119. **Kawai, G.**, Molecular species of cerebrosides in fruiting bodies of *Lentinus edodes* and their biological activity, *Biochim. Biophys. Acta*, 1001, 185, 1989.
120. **Kokontis, J. and Ruddat, M.**, Glycosphingolipid recognition factor from *Ustilago violacea*, *Plant Physiol.*, 75 (Suppl. 1), 167, 1984.
121. **Patton, J.L. and Lester, R.L.**, The phosphoinositol sphingolipids of *Saccharomyces cerevisiae* are highly localized in the plasma membrane, *J. Bacteriol.*, 173, 3101, 1991.
122. **Conzelmann, A., Puoti, A., Lester, R.L. and Desponds, C.**, Two different types of lipid moieties are present in glycophosphoinositol anchored membrane proteins of *Saccharomyces cerevisiae*, *EMBO J.*, 11, 457, 1992.
123. **Prostenik, M. and Cosovic, C.**, Lipids of higher fungi II. The nature of cerebrins and cerebrosides from the mushroom *Clitocybe tabescens*, *Chem. Phys. Lipids*, 13, 117, 1974.
124. **Ondrusek, V. and Prostenik, M.**, Sphingolipids of fruit-bodies of the Basidiomycete *Lactarius deliciosus*, *Exp. Mycol.*, 2, 156, 1978.
125. **Cosovic, C. and Prostenik, M.**, Lipids of higher fungi V. The occurrence of long-chain erythro-2,3-dihydroxy fatty acids in *Polyporus officinalis*, *Chem. Phys. Lipids*, 23, 349, 1979.
126. **Manocha, M.S., San-Blas, G. and Centeno, J.L.**, Lipid composition of *Paracoccidioides brasiliensis*: possible correlation with virulence of different strains, *J. Gen. Microbiol.*, 117, 147, 1980.
127. **Barr, K. and Lester, R.L.**, Occurence of novel antigen phosphoinositol-containing sphingolipids in the pathogenic yeast *Histoplasma capsulatum*, *Biochemistry*, 23, 5581, 1984.
128. **Barr, K., Laine, R.A. and Lester, R.L.**, Carbohydrate structures of three novel phosphoinositol-containing sphingolipids from the yeast *Histoplasma capsulatum*, *Biochemistry*, 23, 5589, 1984.
129. **Fisher, D.J., Holloway, P.J. and Richmond, D.V.**, Fatty acid and hydrocarbon constituents of the surface and wall lipids of some fungal spores, *J. Gen. Microbiol.*, 72, 71, 1972.
130. **Fisher, D.J., Brown, G.A. and Holloway, P.J.**, Influence of growth medium on surface and wall lipid of fungal spores, *Phytochemistry*, 17, 85, 1978.
131. **Carmack, J.L., Weete, J.D.G. and Kelley, W.D.**, Hydrocarbons, fatty acids and sterols of *Cronartium fusiforme* aeciospores, *Physiol. Plant. Pathol.*, 8, 43, 1976.
132. **Jackson, L.I., Dobbs, S.L., Hildebrand, A. and Yokiel, R.**, Surface lipids of wheat stripe rust uredospores, *Puccinia striiformis*, compared with those of the host species, *Phytochemistry*, 12, 2233, 1973.

133. **Laseter, J.L., Hess, W.M., Weete, J.D., Stocks, D.L. and Weber, D.J.,** Cytotaxonomic and ultrastructural studies on three species of *Tilletia* occurring on wheat, *Can. J. Microbiol.*, 14, 1149, 1968.
134. **Weber, D.J. and Trione, E.J.,** Lipid changes occurring during the germination of teliospores of the common bunt fungus *Tilletia caries*, *Can. J. Bot.*, 58, 2263, 1980.
135. **Clark, T. and Watkins, D.A.M.,** Surface lipids of *Sphaerotheca fuliginea* spores, *Phytochemistry*, 17, 943, 1978.
136. **Johnson, D., Weber, D.J. and Hess, W.M.,** Lipid from the conidia of *Erysiphe graminis tritici* (powdery mildew), *Trans. Brit. Mycol. Soc.*, 66, 35, 1976.
137. **Jones, J.G.,** Studies on lipids of soil microorganisms with particular reference to hydrocarbons, *J. Gen. Microbiol.*, 59, 145, 1969.
138. **Merdinger, E. and Devine, E.M., Jr.,** Lipids of *Debaryomyces hansenii*, *J. Bacteriol.*, 89, 1488, 1965.
139. **Weete, J.D., Weber, D.J. and Lester, J.L.,** Lipids of *Rhizopus arrhizus Fischer*, *J. Bacteriol.*, 103, 536, 1970.
140. **Elliott, C.G.,** Sterols in fungi. Their function in fungal growth and reproduction, *Adv. Microb. Physiol.*, 15, 121, 1977.
141. **Rodriguez, R.J. and Parks, L.W.,** Structural and physiological features of sterols necessary to satisfy bulk membrane and sparking requirements in yeast sterol auxotrophs, *Arch. Biochem. Biophys.*, 225, 861, 1983.
142. **O'Day, D.H. and Horgen, P.A.,** *Sexual Interactions*, in *Eukaryotic Microbes*, O'Day, D.H. and Horgen, P.A., Eds., Academic Press, London and San Diego, 1981.
143. **Parks, L.W., Lorenz, R.T., Casey, W.M. and Keesler, C.A.,** Studies on the functions and regulation of sterol biosynthesis in *Saccharomyces cerevisiae*, in *The Metabolism, Structure and Utilization of Plant Lipids*, Cherif, A., Miled-Daoud, D.B., Mazouk, R., Smaoui, A. and Zarrouk, M., Eds., Centre National Pédagogique, Tunisia, 235, 1992.
144. **Parks, L.W., Smith, S.J., Tomeo, M. and Crowley, J.H.,** Regulation and function of sterols in fungi, in *Plant Lipid Metabolism*, Kader, J.C. and Mazliak, P., Eds., Dordecht, 347, 1995.
145. **Van den Ende, H.,** Sexual interactions in the lower filamentous fungi, in *Encyclopedia of Plant Physiology*, Vol. 17, Liskens, H.F. and Heslop-Harrison, J., Eds., Springer, Berlin, 333, 1984.
146. **Kerwin, J.L. and Duddles, N.D.,** Reassessment of the role of phospholipids in sexual reproduction by sterol-auxotrophic fungi, *J. Bacteriol.*, 38, 3831, 1989.
147. **Fox, N.C., Coniglio, J.G. and Wolf, F.T.,** Lipid composition and metabolism in oospores and oospheres of *Achlya americana*, *Exp. Mycol.*, 7, 216, 1983.

148. **Dexter, Y. and Cooke, R.C.,** Fatty acids, sterols and carotenoids of the psychrophile *Mucor strictus* and some mesophilic *Mucor* species, *Trans. Brit. Mycol. Soc.*, 83, 455, 1984.
149. **Grindle, M.,** Sterol content and enzyme defects of nystatin resistant mutants of *Neurospora crassa*, *Mol. Gen. Genet.*, 165, 305, 1978.
150. **Body, D.R. and Bauchop, T.,** Lipid composition of an obligately anaerobic fungus *Neocallimastix frontalis* isolated from a bovine rumen, *Can. J. Microbiol.*, 31, 463, 1985.
151. **Safe, S.,** The effect of the environment on the free and hydrosoluble sterols of *Mucor rouxii*, *Biochim. Biophys. Acta*, 326, 471, 1973.
152. **Sancholle, M., Weete, J.D. and Montant, C.H.,** Effects of triazoles on fungi. I. Growth and cellular permeability, *Pestic. Biochem. Physiol.*, 21, 31, 1984.
153. **Siegel, M.R.,** Sterol inhibiting fungicides: effects on sterol biosynthesis and sites of action, *Plant Dis.*, 65, 986, 1981.
154. **Kim, S.J., Kwon-Chung, K.J., Milne, G.W.A., Hill, W.B. and Patterson, G.,** Relation between polyene resistance and sterol composition in *Cryptococcus neoformans*, *Antimicrob. Agents Chemother.*, 7, 99, 1975.
155. **Pierce, A.M., Pierce, H.D., Jr., Unrau, A.M. and Oelschlager, A.C.,** Lipid composition and polyene antibiotic resistance of *Candida albicans* mutants, *Can. J. Biochem.*, 56, 135, 1978.
156. **Barton, D.H.R., Gunatilaka, A.A.L., Jarman, T.R., Widdowson, D.A., Bard, M. and Woods, R.A.,** Biosynthesis of terpenes and steroids. Part X. Yeast mutants doubly deficient in ergosterol biosynthesis, *J. Chem. Soc. Perkin Trans.*, 1, 88, 1975.
157. **Weete, J.D., Sancholle, M. and Montant, C.H.,** Effects of triazoles on fungi. II. Lipid composition of *Taphrina deformans*, *Biochim. Biophys. Acta*, 752, 19, 1983.
158. **Sancholle, M., Weete, J.D. and Touze-Soulet, J.M.,** Composition of a plasma membrane enriched fraction from *Taphrina deformans*. Effects of propiconazole, in *Structure, Function and Metabolism of Plant Lipids*, Siegenthaler, P.A. and Eichenberger, W., Eds., Elsevier Science B.B, 347, 1984.
159. **Weete, J.D., Sancholle, M., Touze-Soulet, J.M., Bradley, J. and Dargent, R.,** Effects of triazoles on fungi. III. Composition of a plasma membrane enriched fraction of *Taphrina deformans*, *Biochim. Biophys. Acta*, 812, 633, 1985.
160. **Sancholle, M., Dargent, R., Weete, J.D., Rushing, A.E., Miller, K.S. and Montant, D.,** Effects of triazoles on fungi. IV. Ultrastructure of *Taphrina deformans*, *Mycologia*, 80, 162, 1988.
161. **Weete, J.D., Sancholle, M., Patterson, K.A., Miller, K.S., Huang, M.Q., Campbell, F. and Van den Reek, M.,** Fatty acid metabolism in *Taphrina deformans* treated with sterol biosynthesis inhibitors, *Lipids*, 26, 669, 1991.

162. **Beilby, J.P. and Kidby, D.K.**, Biochemistry of ungerminated and germinated spores of the vesicular arbuscular mycorrhizal fungus *Glomus caledonicus*: changes in neutral and polar lipids, *J. Lipid Res.*, 21, 739, 1980.
163. **David, M.H. and Kirsop, B.H.**, Yeast growth in relation to the dissolved oxygen content and sterol content of wort, *J. Inst. Brew*, 79, 20, 1973.
164. **Lin, H., Langenbach, R.J. and Knoche, H.W.**, Sterols of *Uromyces phaseoli* uredospores, *Phytochemistry*, 11, 2319, 1973.
165. **Kramer, R., Knopp, F., Niedermeyer, W. and Fuhrmann, G.F.**, Comparative studies of the lipid composition of the plasmalemma and tonoplast of *Saccharomyces cerevisiae*, *Biochim. Biophys. Acta*, 507, 369, 1978.
166. **Elliott, C.G., Hendrie, M.R., Knights, B.A. and Parker, W.**, A steroid growth factor requirement in a fungus, *Nature*, 203, 427, 1964.
167. **Shapiro, B.E. and Gealt, M.A.**, Ergosterol and lanosterol from *Aspergillus nidulans*, *J. Gen. Microbiol.*, 128, 1053, 1982.
168. **Manocha, M.S.**, Role of lipid in host-parasite relations of a mycoparasite, *Physiol. Plant Pathol.*, 17, 319, 1980.
169. **Valadon, L.R.G.**, Carotenoids as additional taxonomic markers in fungi, *Trans. Brit. Mycol. Soc.*, 67, 1, 1976.
170. **Goodwin, T.W.**, *The Biochemistry of the Carotenoids*, 2nd. ed., Chapman and Hall, London, 1980.
171. **Ramadan-Talib, Z. and Prebble, J.**, Photosensitivity of respiration in *Neurospora* mitochondria. A protective role for carotenoid, *Biochem. J.*, 176, 767, 1978.

Chapter 3

FUNGAL STEROLS

L. W. Parks and W. M. Casey

CONTENTS

0-8493-4794-7/96/$0.00+$.50

I. INTRODUCTION

Fungal sterols consist of the cyclopentanoperhydrophenanthrene tetracyclic nucleus with an equatorial 3β-hydroxy group, α-methyl groups at C-10 and C-13 and an eight to ten carbon side chain at C-17. The principal fungal sterol is ergosterol. It was first isolated in 1889,[1] and has been shown to be produced by 558 fungal cultures covering 60 species in 20 genera.[2] Ergosterol differs from cholesterol, the principal animal sterol, in that the fungal sterol has double bonds at C-7 and C-22 and a methyl substituent at C-24 with the configuration, 24R. Ergosterol is thus named 24-methyl-cholesta-5,7,22-(trans)-trien-3β-ol. The ergosterol structure with numbering and space-filling models is shown in Figure 1. In the limited space that is available for this review, it is impossible to detail the myriad of sterols found in fungi or to discuss the individual fungal species. Excellent treatments of these subjects have been reported and should be consulted for definitive work on sterol synthesis and taxonomic distribution.[3,4]

With the preparation in quantity and high purity of ergosterol from yeast by Smedley-MacLean and Thomas,[5] that sterol gained a lot of attention when it was recognized as an analogue of the animal precursor of vitamin D.[6] Ergocalciferol, obtained by the irradiation of ergosterol, can substitute for cholecalciferol in the diet of some animals. Since 7-dehydrocholesterol was very difficult to obtain in quantity for its conversion to cholecalciferol, ergosterol became an important source of vitamin D_2. However, vitamin D_2 is nutritionally unacceptable for poultry. The poultry industry represents an enormous market for that product. Apparently, the presence of the C-22 unsaturation and/or the C-28 methyl group interferes with the vitamin D metabolism in those animals. An interesting competition exists between several biotechnology firms to develop genetically engineered yeast mutants as an economical source of 7-dehydrocholesterol. Theoretically, it should be possible for the yeast cells to make cholesta-5,7-dienol if the cells had inactive *ERG5* (C-22 desaturase) and *ERG6* (C-24 transmethylase) structural genes. Since yeast lack the Δ^{24}-reductase of animal cells, retention of the C24=25 in the double deficient yeast cells could interfere with the C-25 hydroxylation normally carried out on cholecalciferol.

Ergosterol biosynthesis has generated much more interest recently as a target for antifungal intercession in the growth of fungi. Numerous inhibitors of fungal growth have been identified. Some, e.g. polyenes, act directly on the sterols in the plasma membrane of the organism. Various inhibitors of ergosterol biosynthesis are also antifungal compounds. Three general groups of compounds are seen: inhibitors of squalene synthesis, epoxidation or cyclization; inhibitors of the demethylation of C-14 methyl group; and inhibitors of reduction of the Δ^{14} double bond, resulting in the formation of ignosterol (ergosta-8,14-dienol). An abbreviated pathway is

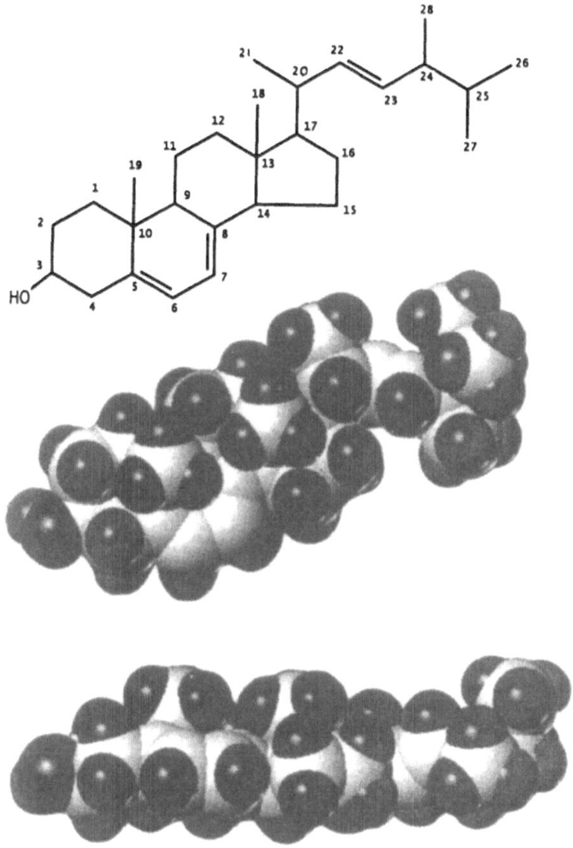

Figure 1. Structure and space filling models of ergosterol.

shown in Figure 2. Selected sterol structures, genetic designations for some steps in the biosynthetic reactions and sites of antifungal action are also shown. The relevance of these compounds in sterol biosynthesis vis-a-vis their effect on fungal growth is the subject of other chapters in this book (Chapters 11 and 12). We shall, therefore, confine our discussion to the role of sterols in various physiological activities of the cell. This may help in understanding as to why cell growth is affected by the sterol biosynthetic inhibitors, and in identifying the biochemical processes in fungi in which sterols are involved.

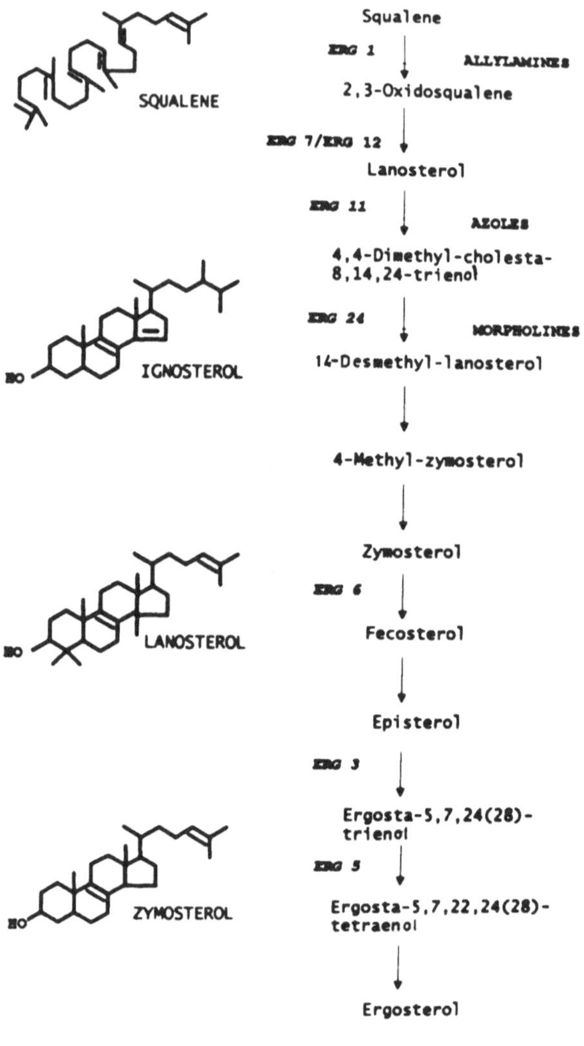

Figure 2. Biosynthetic pathway of ergosterol.

II. ASSESSING STEROL FUNCTIONS

A concern for the functions of sterol in fungi was derived from experiments attempting to define the growth characteristics of mutants that failed to accumulate ergosterol. It became clear that there was no single explanation for the role of sterols in the physiology of the cell. It was concluded that there must be multiple roles, but a mechanism to establish this was needed. In designing experiments to show multiple roles we predicated our work on three assumptions.[7,8] Firstly, we reasoned that if there are multiple functions for sterols in fungi, it is highly unlikely that the amount of sterol required to satisfy the different functions would be the same. Secondly, we proposed that different functions would have different requirements for the unique structural features of ergosterol and that ergosterol must be a major sterol in yeast, since it is the only principal end product of the sterol biosynthetic pathway. The ease of culture, well-defined genetic mechanisms and availability of a library of mutants made *Saccharomyces cerevisiae* ideal for our studies.

Extensive studies with a yeast sterol auxotroph indicated that there are at least four functions of sterols in that organism.[8] These were designated sparking, critical domain, domain and bulk. These assignments were based on the qualitative features of the sterol that permitted growth, and the amounts of the sterol that were required.

Studies have been made of the specific features that a sterol must have to permit growth of yeast. In the sparking function, a Δ^5-sterol is required to satisfy this highly specific requirement, if cholesterol is available to support the other functions. In oxygen-deprived cells, it has been reported that a 24-β-methyl group is needed for sparking.[9] The critical domain role for sterols is characterized by growth on lanosterol, but with ergosterol at ten times the microquantity needed for sparking function. The domain and bulk functions represent the increasing quantitative requirements for sterol.

A study of the bulk function has been made using the sterol auxotroph GL7.[10] Essential features besides the basic sterol nucleus to fulfill that function in GL7 include a 3β-OH group, absence of methylation at C-4 and a methylene segment extending from C-20 not to exceed six carbon atoms. Methylation at C-14 and nuclear unsaturation were not essential for the bulk function. The differences seen in sterol features to satisfy bulk function in this organism and the earlier report[8] emphasize the importance of culture conditions and genetic differences in strains used for the studies.

A definitive understanding of the role(s) of sterols in the physiology of fungi demands that there should be techniques for regulating the amounts of sterols available to the specific activity. Then a quantitative relationship could be established between the availability of a particular sterol or sterolic feature and the consequent physiological activity. Several experiments have been attempted to provide this quantitative relationship.

Artificial mixtures of sterols and cellular components have been made and their biophysical and enzymatic analyses were done. Using monolayers of phospholipids, it was shown almost seventy years ago that cholesterol elicited a "condensing effect" on the monolayer.[11] In comparison to natural membranes, experiments of this type with synthetic mixtures of sterol with other lipids and individual proteins provide a relatively non-complex milieu to isolate the measured response. Temperature variations eliciting phase alterations are more easily studied in such systems. A canonical view of at least one function of sterols has emerged from such studies. Sterols elicit an enhanced ordering effect on membrane phospholipids at temperatures above that of the lipid phase transition. At lower temperatures, the presence of sterols causes disorder in the membrane. Consequently, as poikilothermic organisms experience temperature variations, the sterols "buffer" the membranes against abrupt temperature-dependent physical and biochemical alterations attributable to the lipid phase transitions. It can be seen that this could be of critical importance in the fungi that are naturally subjected to temperature variations.

Mutants defective in sterol biosynthesis provide whole cells and cellular fractions that contain aberrant sterols. Using genetic and molecular biological techniques, it is possible to construct organisms having specific deficiencies in sterol biosynthesis. In yeast, the terminal biosynthetic reactions in ergosterol biosynthesis seem to be relatively non-specific to the substrate used in the reaction. As a consequence, if there is a deficiency at a particular step, in a late reaction of sterol biosynthesis, the organisms will generally complete the remaining steps in the transformation of the molecule. A sterol is then produced that is similar to ergosterol, but would lack one feature mediated in wild-type cells by the missing enzyme. An important constraint must be recognized in this process, however. Only those sterols that are "acceptable" to the organism will be present in such mutants. The acceptability of sterols implies that the organisms are able to use the altered sterol in all growth-sensitive reactions. The logic is simple, for if an "unacceptable" sterol were to be produced, the organism would not grow and never be seen in a selection process. It must be clear that an "acceptable" sterol is one that merely lets the organism survive under the conditions of the experiment.

Failure to obtain certain mutant classes in yeast may be indicative of deficiencies which result in the generation of "unacceptable" sterols. Mutants have not been seen which are defective only in the removal of 4,4'-dimethyl groups of the ergosterol precursors. Of course, the possibility remains that such mutations have never been induced. This seems unlikely, since there is no reason to presume that the genes involved in that transformation are anymore refractory to alterations than any of the other sterol biosynthetic genes. The failure to obtain certain mutant classes may arise either in the mutant selection process or in the acceptability of the altered sterols for the growth of the organism. The latter seems most reasonable, but until such mutants are available for study, their coquettishness remains enigmatic. During this discussion, we shall use the term ergosterol mutant to represent yeast clones that are defective in sterol biosynthesis, but produce a physiologically acceptable sterol and can grow. Endogenously generated sterols are sufficient for the growth of ergosterol mutants. We shall use the term ergosterol auxotroph to describe those strains that must be provided with preformed sterols in the medium for growth. Generally, ergosterol auxotrophs have genetic defects resulting in biochemical losses early in the sterol biosynthetic pathway. In contrast, ergosterol mutants have deficiencies in the later enzymes involved in the synthesis of ergosterol.

If certain sterols are able to sustain the growth of an organism in a laboratory environment it does not imply that those sterols are also equivalent in the various functions for sterols. Although yeast cells can grow well vegetatively on glucose as a sole carbon source with a variety of sterols, various specialized functions are not uniformly responsive; an example of this is seen in the haploid conjugation during yeast mating. Sterol auxotrophic strains were grown and allowed to conjugate on media supplemented with a variety of sterols. All were considered acceptable by the criterion of growth on glucose. However, the mating efficiency of the auxotrophs was severely perturbed by sterols other than ergosterol. Although the mated pairs became adherent, they frequently failed to undergo cytoplasmic fusion.[12] Ergosterol could rescue impaired mating when the pairs had been grown on other sterols.

The nature of the genetic construction of the parental organism can have a profound effect on the sterol mutants that are isolated. We attempted to obtain *erg*6 mutants in a parental strain that had a defect in tryptophan biosynthesis. While this mutant, which has a defect in the C-24 transmethylase, is one of the most abundant types found in the selection of sterol mutants, it has never been observed in our organism. We did recover many *erg*3 (lacking C-5 desaturase) mutants in that background and they became important in our studies. The failure to obtain *erg*6 mutants was frustrating. It became clear when it was reported that *erg*6 mutants have an inactive tryptophan permease.[13] Thus, in a Trp^- background, sterols lacking the C-28 methyl group are "unacceptable." Recently, mutants with enhanced sensitivity to monovalent cations were selected.[14] The structural gene for the wild-type allele of the recessive mutant was cloned and sequenced. The gene was shown to have 99% sequence identity to *ERG6*.

Yeast strains incapable of removing the 14-methyl group of lanosterol are quite easily isolated by selecting for resistance to azole antifungal agents. The sterol that accumulates, 14α-methyl fecosterol (Figure 3), is an "acceptable" sterol and is produced only if the mutant has a second deficiency in ergosterol biosynthesis.[15] This deficiency is in the *ERG3* structural gene, which has the information for 5,6-desaturase. Without *erg*3, *erg*11 mutants produce 14α-methyl-ergosta-8-ene-3,6-diol (Figure 4). This is apparently an unacceptable sterol in *Saccharomyces*, although a similar compound is produced in some mutants of *Candida*[16-18] without any lethal effect. It seems reasonable to assume that C-5-desaturase is responsible for the formation of the diol. Whether the diol is a normal intermediate in the desaturation at C-5 has not been established. Mutants lacking 5,6-desaturase can grow on glucose with 14α-methyl fecosterol. Thus, the *erg3* compensating mutant allows *erg11* to grow aerobically. Under anaerobic conditions, where yeast can not synthesize sterols, the effect of the *erg11*

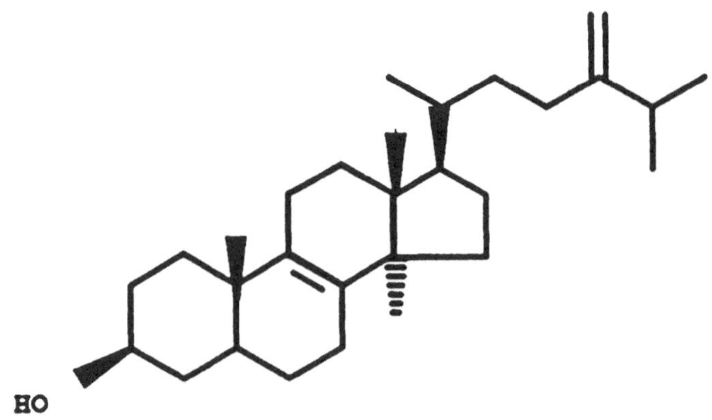

Figure 3. 14α-methyl fecosterol.

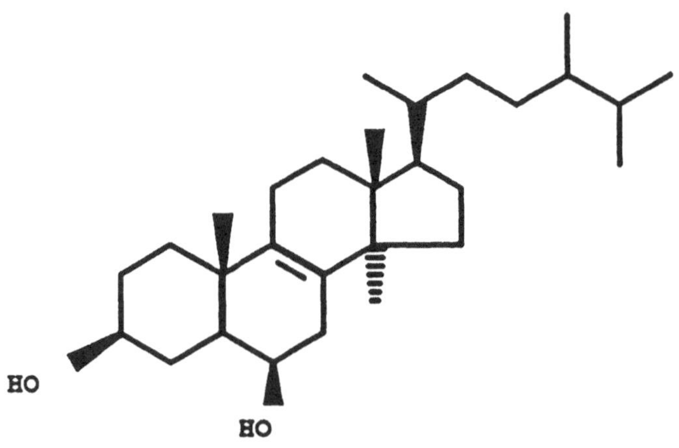

Figure 4. 14α-methyl ergosta-8-ene-3,6-diol.

defect is not demonstrable. Those cells, like all yeast strains, must be provided with a suitable source of sterols to permit growth in the absence of oxygen. It is somewhat surprising that *Candida* can survive with the *erg*11 defect without compensating mutations while *Saccharomyces* can not. The variety of sterol mutants that have been reported in *Saccharomyces* is considerably greater than that in *Candida*. This suggests that *Saccharomyces* might be more tolerant of sterol alterations as compared to *Candida*. Alternatively, the *Candida* strains with the *erg11* defect may have a silent suppressor (see below) that compensates for the aberrant sterols without affecting the production of that sterol.

In its several functions, ergosterol may play various roles such as a structural component, cofactor, regulatory element, etc. As studies are initiated to define specific functions, it is difficult to anticipate in which biochemical capacities ergosterol may be acting. To try randomly to define the strategic target reactions for sterol functions is not practical. However, a convenient genetic approach can markedly reduce the number of possibilities. Assuming that the sterol-dependent function is essential for growth, it is possible to plate a sterol deficient clone in high numbers under conditions where the target function is growth limiting. Using mutagens or spontaneous mutations, there often appear isolates that can grow under otherwise restrictive conditions. The aim of the experiment is to obtain strains with a second mutation, occurring in the sterol targeted activity, which either eliminates the requirement for the sterol or has a decreased specificity for the features of ergosterol, allowing the organism to grow with the altered sterol. These extragenic suppressors (occurring outside the original structural gene for the ergosterol mutant) can help us to define the specific reactions in which ergosterol functions. Several groups of mutants are possible that return the culture to the wild phenotype, but are of little interest in the studies of sterol functions. These include the up-regulation of a minor pathway that also provides the defective sterol; informational suppression where the translational fidelity of the mutant is affected; and the activation of some previously cryptic pathway that also provides ergosterol. Again, these are easily recognized, since they all would result in ergosterol formation, at least in part. Two general modes of extragenic suppression are particularly valuable in studies of ergosterol function: alteration of the metabolic constituent that interacts with sterol or is regulated by it so that the sensitive function now accepts the defective sterol; or a by-pass mechanism whereby the organism develops an alternative pathway, simply removing the requirement for the sterol-dependent function. Thus, a detailed analysis of the suppressed strain can provide valuable insight into the essential physiological functions in which ergosterol participates. Rarely, a true reversion to the wild-type genome is obtained. Appropriate genetic and biochemical analyses must be performed to authenticate the nature of the suppression. In this discussion, we shall reserve the term suppressor to that of the second-site compensating mutation that returns the organism to a wild phenotype.

Certain enabling mutants are in fact suppressors that were obtained before the actual mutation they were supposed to suppress. We were unable to obtain mutants defective in *ERG24*, the structural gene for the C14=15 reductase. We reasoned that such a mutant should accumulate ignosterol, and our previous work with azasterols that inhibit C14=15 reductase indicated that ignosterol was an unacceptable sterol.[19,20] The morpholine, fenpropimorph, also inhibits C14=15 reductase and variants (*fen1* and *fen2*) of yeast resistant to it were obtained.[21] These were then used in a separate screening regimen to obtain the desired *erg24* mutant, which we subsequently cloned and sequenced.[22]

III. ESTERIFICATION

Ergosterol exists in two forms in yeast, as the 3β-OH free alcohol or esterified to long-chain fatty acids. We have shown that the esterification of ergosterol increases dramatically upon entry of the culture into the stationary phase of growth.[23] However, when the culture is returned to fresh medium, there is a rapid hydrolysis of the ester to produce free ergosterol.[24] Free ergosterol is principally localized in membranes. There it serves to modulate the physical properties of the membranes, regulate enzymatic activity, and mediate the anchoring of other membrane components. Although ergosterol has multiple functions in the physiology of the cell (see above),[8] sterols have received the maximum attention as a structural unit of membranes. We have been intrigued by the cycling of ergosterol into and out of the ester fraction as a function of the culture cycle of the organism. Since ergosterol is a metabolically expensive molecule for the organism to synthesize, we assume that the esterification of ergosterol during growth retardation conserves the sterol for reuse during active growth. However, there are other efficient feed-back mechanisms for the cell to use for the regulation of biosynthetic activity, normally avoiding the overproduction of almost all cellular constituents. We have demonstrated such regulation of sterol biosynthesis in yeast.[25] It is antithetical to suppose that the cell would use precious cellular components and energy to synthesize ergosterol unless and until it is needed for essential cellular functions. To support this notion further, there is a limit in yeast on the amount of free sterol that membranes can contain,[26] and the cell must expend additional energy for the synthesis of fatty acids to be used in the esterification and for synthesis of the ester as well. Therefore, we are compelled to propose that the "excess" of ergosterol within the cells as growth ceases is a consequence of a mandatory but transient need for an abundance of free sterol in the synthesis or function of a cell membrane during proliferation. We may assume that there is an ergosterol-rich membrane component that is critical and transient in the actively growing cell. Evidence for such a sterol-rich component has been obtained in cultured human fibroblasts. A cholesterol-rich intracellular precursor of the plasma membrane has been described.[27]

Sterol interconversion in yeast, between the free and esterified forms, is directed towards maintaining the membrane requirements for free sterol. Changes in the turnover rate of sterols and sterol esters appears to affect the sensitivity of *Aspergillus fumigatus* to polyene antibiotics.[28] Our challenge is to explain the free sterol:sterol ester interconversion and define its role in the physiology of the cell.

The enzyme for sterol esterification is acyl-coenzyme A:ergosterol acyltransferase; we have given this enzyme a trivial name of sterol ester synthase (SES). Our *in vitro* studies on the substrate requirements for that assay are consistent with such an enzyme. However, both *in vitro* and *in vivo* experiments in our laboratory have failed to demonstrate any sensitivity of SES to the substances that inhibit the homologous animal enzyme, acyl-coenzyme A:cholesterol acyltransferase (ACAT). This is surprising because of the anticipated similarity in mechanisms for the two enzymes.

Conditions for growing cultures can affect SES activity in yeast. Four-fold increase in the stimulation of esterification was observed in heme competent cells in comparison to heme deficient ones,[29] and was seen on both respiratory and fermentative carbon sources. SES activity increases as the cells enter stationary phase of growth. Although esterification continues throughout the culture cycle, sterols lacking the Δ^7, Δ^{22} and C-24-methyl groups are preferentially esterified during active growth.[30] All cellular forms of sterols are esterified in a stationary phase. SES may have two physiological roles. The first one is to preserve ergosterol for subsequent re-use when its demand is high. Second, SES may have a "proofing" function. Non-ergosterol sterols would be sequestered into the microdroplets of lipids. This physiological incarceration would assure non-participation of undesirable sterols in the critical cellular functions of ergosterol. This also could explain the greater abundance of lanosterol than ergosterol in the sterol esters of *Aspergillus nidulans*.[31]

We have selected a mutant (*upc2*) in which the culture cycle dependent interconversion of sterols and sterol esters is defective under certain growth conditions.[32] This defect is found concurrently in cells that have a modified protein that cross-reacts immunologically with antibodies raised against rat apolipoproteins.[33] Independent extragenic suppressors of this mutation return the organism's phenotype to that of the parental wild type. The aberrant immunoblot banding pattern of the mutant is also returned to that of the wild type. Unfortunately, the *UPC2* gene has neither been cloned nor its insertionally inactivated (*upc*2) construction been made. Because of the high reversion rate and complex patterns of phenotypic reversion, experiments with this organism must be carefully scrutinized to assure retention of the original *upc2* mutation.

IV. TRANSALKYLATION

Transmethylation at C-24 is characteristic of fungal and plant sterol biosynthesis. In fungi, of course, a C_1 group is transferred from S-adenosylmethionine,[34] while a subsequent transmethylation generates an ethyl substituent in some plant tissues.[35] Within the fungi, the substrate for methylation is either lanosterol or zymosterol. As a general rule, filamentous fungi and yeast forms differ at the step of alkylation relative to nuclear demethylation. In the filamentous forms alkylation at C-24 occurs prior to demethylation, while it is subsequent to demethylation in yeast. In this regard, the filamentous forms are more like plants.

Gaylor's group has isolated, purified, and characterized the transmethylase from yeast,[36] where it was convincingly demonstrated that zymosterol is the preferred substrate. We have found a low level of transmethylation of 4-α-methylzymosterol in aerobically adapting cells.[37] The availability of the other substrate, S-adenosylmethionine, may be important in the accumulation of ergosterol in some yeasts.[38]

Transmethylation at C-24 occurs after nuclear demethylation in *C. albicans*, *C. parapsilosis*, and *Trichophyton mentagrophytes*.[39] The sterol methyltransferase from *C. albicans* has been solubilized and shown *in vitro* to prefer zymosterol as its substrate.[40] However, an azole-resistant mutant accumulates both lanosterol and 24-methylene-24,25-dihydrolanosterol,[41] while a polyene-resistant mutant was found to have obtusifoliol, which retains a C4-methyl following side-chain alkylation.[42] Feeding studies using a *Candida* mutant showed that zymosterol was the preferred substrate but 4-α-methylzymosterol and 14-α-methyl zymosterol were also methylated at C-24 *in vivo*;[43] however, 4,4′-dimethylzymosterol was not a substrate.

While the transmethylase has not been purified from most fungi, evidence for the timing of transmethylation in the biosynthetic process has come from the accumulation of alkylated precursors of sterols following chemical inhibition. *Cryptococcus neoformans* accumulates an obtusifoliol derivative and eburicol.[44] *T. mentagraphytes* accumulates 24-methylene lanosterol following exposure to miconazole.[45] Following azole treatment *C. albicans*, *Microsporum canis*, *T. mentagrophytes*, and *Epidermophyton floccosum* accumulated 24-methylenedihydrolanosterol, while *Torulopsis glabrata* formed lanosterol.[46] Thus, the results of these studies are often inconsistent with *in vitro* studies on the same organism. This is probably due to the imprecise substrate specificity of the transmethylase and the accumulation of abnormal levels of certain sterol intermediates as a consequence of the inhibition.

The transmethylation of sterols using S-adenosylmethionine is one of the most "expensive" biochemical reactions in the cell. It can be estimated that each methyl transfer costs 12 - 14 ATP equivalents.[47] We must assume that this is an important sterolic feature; otherwise this activity would have been selected out through evolution. Our understanding of the functions of transmethylation is extremely limited.

Methylation at C-24 is necessary for oxygenated steroid hormone synthesis in *Achlya*.[48] It has been proposed that the regulation of sterol transalkylation may in turn regulate the life cycle of *Gibberella*.[49] Inhibition of the transmethylase activity in *C. albicans*, *C. tropicalis* and *T. glabrata* showed mild antifungal behavior,[50] suggesting an important metabolic function of the C-28-methyl group. In *Saccharomyces*, *erg*6 mutants lack sterol transmethylase. In such mutants, impaired ion and tryptophan transport is seen.[13,14] The mutants are also resistant to polyenes, but are more sensitive to cycloheximide. We have shown that a functional *erg6* gene is not required for respiratory development.

V. COORDINATION OF STEROLS, PHOSPHOLIPIDS AND MEMBRANE FUNCTIONS

Because free sterols are principally membrane components, it is interesting to look at the interrelations between sterols and phospholipids. The physical effects of sterols in artificial membranes have been discussed earlier. We have compared the properties of yeast membranes that had altered sterol composition either through mutant sterol accumulation or by feeding sterols to sterol auxotrophs. As a probe, the fluorescence anisotropic behavior of 1,6-diphenyl-1,3,5-hexatriene was monitored in different membrane preparations. We have observed that when auxotrophs were fed with different sterols, there was a substantial accommodation by the cells[51] of different sterols and plasma membranes were isolated from these cells. Each of these preparations showed no discontinuity in the plots of the steady-state fluorescence anisotropy of the probe. Liposomes were prepared from the phospholipids that were produced by the auxotroph. The specific sterol that had been fed to the yeast showed patterns of anisotropy that were similar to that of membrane preparations, having no discontinuities. A contrasting result was obtained when liposomes were prepared from phospholipids extracted from a culture but were mixed with a sterol which was different from the one that was used to grow the culture. Discontinuities were found. This was true of each of the sterols tested, when they were mixed with phospholipids prepared from cells grown on a different sterol.

Thus, accommodating changes were made in the phospholipids as a function of the sterol in the growth medium, which were specific for the sterol provided to grow the auxotroph. Only the sterol on which the culture was grown eliminated the discontinuity in the anisotropic measurements of the probe in the phospholipids prepared from the different cultures.

Quantitation of the lipids prepared from the auxotroph grown on different sterols showed no major differences in the total amount of sterol present. There were changes in the percentages of individual phospholipids and in their esterified fatty acids. These were dependent on the nature of sterol provided. These results showed that the membrane phospholipid composition could be altered in response to the sterol available to the cell. An interesting physiological dilemma exists: how does the cell discriminate and respond to the particular features of the sterol molecule and produce an appropriate head-group and fatty acid composition of the phospholipids that is distinctive to the sterol structure? This is a continuing project, but we found that ergosterol increases the transmethylation rate of intermediates in the synthesis of phosphatidylcholine.[52] We have also shown that there is a further coordination of sterol[53] and phospholipid[54] synthesis by the available fatty acids in the yeast cell.

Sterol mutants of *C. albicans*[55] showed an increase in phosphatidylinositol and phosphatidic acid, and a decrease in phosphatidylcholine and phosphatidylserine. Slight changes in fatty acids were also found in this organism,[56] although the effect may not be related directly to ergosterol depletion. The phospholipid composition of various azole resistant mutants was analyzed in a separate study, and many differences were found to be strain specific.[57] Mutants of *Torulopsis glabrata* with reduced levels of ergosterol but enhanced sterol precursors contained fatty acids with shorter and more saturated chains.[58] Like the situation in *Saccharomyces*, it has not been possible to demonstrate which sterol feature elicited these individual phospholipid and fatty acyl alterations. A temperature-dependent coordinated depletion of ergosterol and cardiolipin in *Aspergillus niger* has been observed.[59] Since cardiolipin is found exclusively in the mitochondria, a possible role for those lipids in mitochondriogenesis was proposed.

The interactive regulation by sterols, phospholipids and fatty acids combine to make a highly sophisticated control network. Obviously, this has evolved to make the cell optimally responsive to alterations in its chemical and physical environment.

There are numerous observed biochemical consequences of altering the membrane sterol composition of the fungi. For most of these, it has not been established if the effect is a direct consequence of the sterol *per se* or if the biochemical alterations are secondary to a sterol-elicited change. Because the membrane is such a complex structure, direct causal relationships are difficult to establish. In particular, various effects of sterol alterations on respiration and the utilization of respiratory substrates have been described.[59-64] An interesting inverse relationship between chitin synthesis and ergosterol content was observed in four *Candida* species.[65] Whether this was a direct effect of diminished ergosterol, increased lanosterol or the presence of the azole inhibitor still needs to be resolved.

Amino acid uptake has been shown to be affected by conditions that elicit sterol modifications. In addition to the work mentioned earlier involving *erg*6, changes in the accumulation rates of specific amino acids[66-68] and amino acid uptake in general[69,70] have been described.

VI. CONCLUSION

Ergosterol biosynthesis is a physiologically expensive process in fungi, involving at least 25 reactions, 10 ATP and 16 NADPH molecules per molecule of ergosterol formed.[71] Study of ergosterol synthesis and function provides an interesting vehicle to probe fundamental problems in the physiology of the cell. It is clear that ergosterol is critical to normal cell function. Understanding the various roles of sterol is not only an intellectually challenging and exciting exercise, but it has important practical ramifications also. It is already known that chemical intercession of sterol synthesis is important in antifungal therapy. By understanding the various roles of ergosterol, it may be possible through biorational design to develop fungal controlling agents that are more specific and directly targeted. The potential import of such a development in agriculture or medicine can not be overestimated.

VII. REFERENCES

1. **Tanret, C.**, Sur un noveau principe immediat de l'ergot de seigle Pergosterine, *C.R. Seances Acad. Sci.*, 108, 98, 1889.
2. **Delaney, E.L., Staply, E.O. and Simpf, K.**, Studies on ergosterol production by yeast, *Appl. Microbiol.*, 2, 371, 1954.
3. **Weete, J.D.**, Structure and function of sterols in fungi, *Adv. Lipid Res.*, 23, 115, 1989.
4. **Weete, J.D.**, *Lipid Biochemistry of Fungi and Other Organisms*, Plenum Press, New York, 1980.
5. **Smedley-MacLean, I. and Thomas, E.M.**, The nature of yeast fat, *Biochem J.*, 14, 483, 1920.
6. **Rosenheim, O. and Webster, T.A.**, The parent substance of vitamin D, *Biochem J.*, 21, 389, 1927.
7. **Rodriguez, R.J., Turner, F.R. and Parks, L.W.**, A requirement for ergosterol to permit growth of yeast sterol auxotrophs on cholestanol, *Biochem. Biophys. Res. Comm.*, 106, 435, 1982.
8. **Rodriguez, R.J., Low, C., Bottema, C.D.K. and Parks, L.W.**, Multiple functions for sterols in *Saccharomyces cerevisiae*, *Biochim. Biophys. Acta.*, 837, 336, 1985.
9. **Pinto, W.J. and Nes, W.R.**, Stereochemical specificity for sterols in *Saccharomyces cerevisiae*, *J. Biol. Chem.*, 258, 4472, 1983.
10. **Nes, W.R., Janssen, G.G., Crumley, F.G., Kalinowska, M. and Akihisa, T.**, The structural requirements of sterols for membrane function in *Saccharomyces cerevisiae*, *Arch. Biochem. Biophys.*, 300, 724, 1993.
11. **Leathes, J.B.**, The role of fats in vital phenomena, *Lancet*, 1, 853, 1925.
12. **Tomeo, M.E., Fenner, G., Tove, S. and Parks, L.W.**, Effect of sterol alterations on conjugation in *Saccharomyces cerevisiae*, *Yeast*, 8, 1015, 1992.

13. **Gaber, R.F., Copple, D.M., Kennedy, B.K., Vidal, M. and Bard, M.,** The yeast gene *ERG6* is required for normal membrane function but is not essential for biosynthesis of the cell-cycle-sparking sterol, *Mol. Cell. Biol.*, 9, 3447, 1989.
14. **Welihinda, A.A., Beavis, A.D. and Trumbly, R.J.,** Mutations in *LIS1* (*ERG6*) gene confer increased sodium and lithium uptake in *Saccharomyces cerevisiae*, *Appl. Environ. Microbiol.*, 56, 2853, 1994.
15. **Taylor, F.R., Rodriguez, R.J. and Parks, L.W.,** A requirement for a second sterol biosynthetic mutation to allow for viability of a sterol C-14 demethylation defect in *Saccharomyces cerevisiae*, *J. Bacteriol.*, 155, 64, 1983.
16. **Shimokawa, O., Kato Y., Kawano, K. and Nakayama, H.,** Accumulation of 14-α-methylergosterol-8,24(28)-dien-3-β, 6-α-diol in 14α-demethylation mutants of *Candida albicans*: genetic evidence for involvement of 5-desaturase, *Biochim. Biophys. Acta.*, 1003, 15, 1989.
17. **Lees, N.D., Broughton, M.C., Sanglard, D. and Bard, M.,** Azole susceptibility and hyphal formation in a cytochrome P450-deficient mutant of *Candida albicans*, *Antimicrob. Agents Chemother.*, 34, 831, 1990.
18. **Bard, M., Lees, N.D., Turi, T., Craft, D., Cofrin, L., Barbuch, R., Koegel, C. and Loper, J.C.,** Sterol synthesis and viability of *erg11* (cytochrome-P450 lanosterol demethylase) mutations in *Saccharomyces cerevisiae* and *Candida albicans*, *Lipids*, 28, 963, 1993.
19. **Hays, P.R., Neal, W.D. and Parks, L.W.,** Physiological effects of an antimycotic azasterol on cultures of *Saccharomyces cerevisiae*, *Antimicrob. Agents Chemother.*, 12, 185, 1977.
20. **Hays, P.R., Parks, L.W., Pierce, H.D. and Oehlschlager, A.C.,** Accumulation of ergosta-8,14-3ß-ol by *Saccharomyces cerevisiae* cultured with an azasterol antimycotic agent, *Lipids*, 12, 666, 1977.
21. **Lorenz, R.T. and Parks, L.W.,** Physiological effects of fenpropimorph on wild-type *Saccharomyces cerevisiae* and fenpropimorph-resistant mutants, *Antimicrob. Agents Chemother.*, 35, 1532, 1991.
22. **Lorenz, R.T. and Parks, L.W.,** Cloning, sequencing, and disruption of the gene encoding sterol C-14 reductase in *Saccharomyces cerevisiae*, *DNA Cell Biol.*, 11, 685, 1992.
23. **Bailey, R.B. and Parks, L.W.,** Yeast sterol esters and their relationship to the growth of yeast, *J. Bacteriol.*, 124, 606, 1975.
24. **Taylor, F.R. and Parks, L.W.,** Metabolic conversion of free sterols and steryl esters in *Saccharomyces cerevisiae*, *J. Bacteriol.*, 136, 531, 1978.
25. **Casey, W.M., Burgess, J.P. and Parks, L.W.,** Effect of sterol side-chain structure on the feed-back control of sterol biosynthesis in yeast, *Biochim. Biophys. Acta.*, 1081, 279, 1990.

26. **Lewis, T.A., Rodriguez, R.J. and Parks, L.W.,** Relationship between intracellular sterol content and sterol esterification and hydrolysis in *Saccharomyces cerevisiae*, *Biochim. Biophys. Acta.*, 921, 205, 1987.
27. **Lange, Y. and Steck, T.L.,** Cholesterol-rich intracellular membranes: a precursor of the plasma membrane, *J. Biol. Chem.*, 260, 15592, 1985.
28. **Russell, N.J., Kerridge, D. and Bokar, J.T.,** Sterol metabolism during germination of the conidia of *Aspergillus fumigatus*, *J. Gen. Microbiol.*, 101, 197, 1977.
29. **Keesler, G.A., Casey, W.M. and Parks, L.W.,** Stimulation by heme of steryl ester synthase and aerobic sterol exclusion in the yeast, *Saccharomyces cerevisiae*, *Arch. Biochem. Biophys.*, 296, 474, 1992.
30. **Taylor, F.R. and Parks, L.W.,** An assessment of the specificity of sterol uptake and esterification in *Saccharomyces cerevisiae*, *J. Biol. Chem.*, 256, 13048, 1981.
31. **Shapiro, B.E. and Gealt, M.A.,** Ergosterol and lanosterol from *Aspergillus nidulans*, *J. Gen. Microbiol.*, 128, 1053, 1982.
32. **Keesler, G.A., Laster, S.M. and Parks, L.W.,** A defect in the sterol: steryl ester interconversion in a mutant of the yeast, *Saccharomyces cerevisiae*, *Biochim. Biophys. Acta.*, 1123, 127, 1992.
33. **Keesler, G.A., Moore, S., Usher, D.C. and Parks, L.W.,** Yeast proteins with reactivity to antibodies elicited against mammalian apolipoproteins, *Biochem. Biophys. Res. Comm.*, 174, 631, 1991.
34. **Parks, L.W.,** S-adenosylmethionine and ergosterol synthesis, *J. Am. Chem. Soc.*, 80, 2023, 1958.
35. **Van Aller, R.T., Chikamatsu, H., de Sousa, N.J., John, J.P. and Nes, W.R.,** The mechanism of introduction of alkyl groups at carbon 24 of sterols. III. The second one-carbon transfer and reduction, *J. Biol. Chem.*, 244, 6645, 1969.
36. **Moore, J.T., Jr. and Gaylor, J.L.,** Isolation and purification of an S-adenosylmethionine:Δ^{24}-sterol methyltransferase from yeast, *J. Biol. Chem.*, 244, 6334, 1969.
37. **Parks, L.W., Anding, C. and Ourisson, G.,** Sterol transmethylation during aerobic adaptation of yeast, *Eur. J. Biochem.*, 43, 451, 1974.
38. **Avram, D. and Stan, R.,** Ergosterol levels in two L-methionine-enriched mutants of the methylotrophic yeast, *Candida boidinii* ICCF26, *FEMS Microbiol. Lett.*, 77, 133, 1992.
39. **Ryder, N.S.,** Effect of allylamine antimycotic agents on fungal sterol biosynthesis measured by sterol side-chain methylation, *J. Gen. Microbiol.*, 131, 1595, 1985.
40. **Ator, M.A., Schmidt, S.J., Adams, J.L. and Dolle, R.E.,** Mechanism and inhibition of Δ^{24}-sterol methyltransferase from *Candida albicans* and *Candida tropicalis*, *Biochem.*, 28, 9633, 1989.

41. **Hitchcock, C.A., Barnett-Bee, K.J. and Russell, N.J.,** The lipid composition and permeability to azole of an azole- and polyene-resistant mutant of *Candida albicans*, *J. Med. Vet. Mycol.*, 25, 29, 1987.
42. **Subden, R.E., Safe, L., Morris, D.C., Brown, R.G. and Safe, S.,** Eburicol, lichesterol, ergosterol and obtusifoliol from polyene-resistant mutants of *Candida albicans*, *Can. J. Microbiol.*, 23, 751, 1977.
43. **Pierce, A.M., Mueller, R.B., Unrau, A.M. and Oehlschlager, A.C.,** Metabolism of Δ^{24}-sterols by yeast mutants blocked in removal of the C-14 methyl group, *Can. J. Biochem.*, 56, 794, 1978.
44. **Vanden Bossche, H., Marichal, P., Le Jeune, L., Coene, M.C., Gorrens, J. and Cools, W.,** Effects of itraconazole on cytochrome P450-dependent sterol 14α-demethylation and reduction of 3-ketosteroids in *Cryptococcus neoformans*, *Antimicrob. Agents Chemother.*, 37, 2101, 1993.
45. **Morita, T. and Nozawa, Y.,** Effects of antifungal agents on ergosterol biosynthesis in *Candida albicans* and *Trichophyton mentagrophytes*, *J. Invest. Dermatol.*, 85, 434, 1985.
46. **Berg, D., Regel, E., Harenberg, H.E. and Plempel, M.,** Bifonazole and clotrimazole. Their mode of action and the possible reason for the fungicidal behavior of bifonazole, *Arzneimittelforschung*, 34, 139, 1984.
47. **Atkinson, D.E.,** *Cellular Energy Metabolism and its Regulation*, Academic Press, New York, 1977.
48. **McMorris, T.C.,** Sex hormones of the aquatic fungus, *Achlya*, *Lipids*, 13, 716, 1978.
49. **Nes, W.D., Hanners, P.K. and Parish, E.J.,** Control of fungal sterol C-24 transalkylation: importance to developmental regulation, *Biochem. Biophys. Res. Comm.*, 139, 410, 1986.
50. **Ator, M.A., Schmidt, S.J., Adams, J.L., Dolle, R.E., Kruse, L.I., Frey, C.L. and Barone, J.M.,** Synthesis, specificity and antifungal activity of the inhibitors of the *Candida albicans* Δ^{24}-sterol methyltransferase, *J. Med. Vet. Mycol.*, 35, 100, 1992.
51. **Low, C., Rodriguez, R.J. and Parks, L.W.,** Modulation of yeast plasma membrane composition of a yeast sterol auxotroph as a function of exogenous sterol, *Arch. Biochem. Biophys.*, 240, 530, 1985.
52. **Lewis, T.A., Rodriguez, R.J. and Parks, L.W.,** Relationship between intracellular sterol content and sterol esterification and hydrolysis in *Saccharomyces cerevisiae*, *Biochim. Biophys. Acta.*, 921, 205, 1987.
53. **Casey, W.M., Keesler, G.A. and Parks, L.W.,** Regulation of partitioned sterol biosynthesis in the yeast, *Saccharomyces cerevisiae*, *J. Bacteriol.*, 174, 7283, 1992.

54. **Casey, W.M., Rolph, C.E., Tomeo, M.E. and Parks, L.W.,** Effects of unsaturated fatty acid supplementation, on phospholipid and triacylglycerol biosynthesis in *Saccharomyces cerevisiae*, *Biochem. Biophys. Res. Comm.*, 193, 1297, 1993.
55. **Pesti, M., Horvath, L., Vigh, L. and Farakas, T.,** Lipid content and ESR determination of plasma membrane order parameter in *Candida albicans* sterol mutants, *Acta Microbiol. Hungary*, 32, 305, 1985.
56. **Georgopapadakou, N.H., Dix, B.A., Smith, S.A., Freudenberger, J. and Funke, P.T.,** Effect of antifungal agents on lipid biosynthesis and membrane integrity in *Candida albicans*, *Antimicrob. Agents Chemother.*, 31, 46, 1987.
57. **Hitchcock, C.A., Barnett-Bee, K.J. and Russell, N.J.,** The lipid composition of azole-sensitive and azole-resistant strains of *Candida albicans*, *J. Gen. Microbiol.*, 132, 2421, 1986.
58. **Lomb, M., Fryberg, M., Oehlschlager, A.C. and Unrau, A.M.,** Sterol and fatty acid composition of polyene macrolide antibiotic resistant *Torulopsis glabrata*, *Can. J. Biochem.*, 53, 1309, 1975.
59. **Mandal, S.B., Sen, P.C., Chakrabarti, P. and Sen, K.,** Effect of respiratory deficiency and temperature on the mitochondrial lipid metabolism of *Aspergillus niger*, *Can. J. Microbiol.*, 24, 586, 1978.
60. **Smith, S.J. and Parks, L.W.,** The *ERG3* gene in *Saccharomyces cerevisiae* is required for the utilization of respiratory substrates and in heme-deficient cells, *Yeast*, 9, 1177, 1993.
61. **Shimokawa, O. and Nakayama, H.,** Deficient utilization of succinate in a sterol 14-α-demethylation mutant of *Candida albicans*, *J. Med. Vet. Mycol.*, 29, 117, 1991.
62. **Shimokawa, O. and Nakayama, H.,** Phenotypes of *Candida albicans* sterol mutants deficient in $\Delta^{8\text{-}7}$-isomerization or Δ^5-desaturation, *J. Med. Vet. Mycol.*, 29, 53, 1991.
63. **Shimokawa, O. and Nakayama, H.,** A *Candida albicans* mutant conditionally defective in sterol 14-α-demethylation, *J. Med. Vet. Mycol.*, 27, 121, 1989.
64. **Pesti, M. and Novak, R.,** Decreased permeability of glycerol in an ergosterol-less mutant of *Candida albicans*, *Acta Microbiol. Hungary*, 31, 81, 1984.
65. **Pfaller, M. and Riley, J.,** Effects of fluconazole on the sterol and carbohydrate composition of four species of *Candida*, *Eur. J. Microbiol. Infect. Dis.*, 11, 152, 1992.
66. **Madhosingh, C. and Orr, W.,** Antimicrobial effects of clofibrate on the wheat pathogen, *Fusarium*, *J. Environ. Sci. Health*, 16, 587, 1981.
67. **Singh, M., Jayakumar, A. and Prasad, R.,** The effect of altered ergosterol content on the transport of various amino acids in *Candida albicans*, *Biochim. Biophys. Acta.*, 555, 42, 1979.

68. **Alonso, A., Martinez, E., Struzinsky, R., Michaljanicova, D. and Kotyk, A.,** Interaction of nystatin with nystatin-resistant *Candida tropicalis*, *Folia Microbiol.*, 28, 57, 1983.
69. **Mazumder, C., Kundu, M., Basu, J. and Chakrabarti, P.,** Lipid composition and amino acid transport in a nystatin-resistant mutant of *Aspergillus niger*, *Lipids*, 22, 609, 1987.
70. **Barrett-Bee, K., Newboult, L. and Pinder, P.,** Biochemical changes associated with the antifungal action of the triazole ICI 153066 on *Candida albicans* and *Trichophyton quinckeanum*, *FEMS Microbiol. Lett.*, 63, 127, 1991.
71. **Parks, L.W.,** Metabolism of sterols in yeast, *CRC Crit. Rev. Microbiol.*, 6, 301, 1978.

Chapter 4

LIPIDS OF *HISTOPLASMA CAPSULATUM*

K. Barr

CONTENTS

0-8493-4794-7/96/$0.00+$.50

I. INTRODUCTION

Histoplasma capsulatum, a pathogenic dimorphic fungus, is the causative agent of histoplasmosis. In the soil, the fungus exists as a mycelium which gets converted to a yeast-like form in host tissues. The phase transition can be mimicked in the laboratory by incubation at 25°C of the mycelial phase or by incubation at 37°C for cultivation of the yeast phase.[1] This feature of dimorphism is shared with many pathogenic fungi.

The mycelial phase of *H. capsulatum* grows in soil with high nitrogen content, associated with the guano of birds and bats.[2] In the laboratory, sporulation starts 7-10 days after incubation of mycelia. Regardless of the environment, two kinds of spores are produced: macroconidia and microconidia. Under laboratory conditions, sporulation is dependent upon the strain, the medium and the environmental conditions.[2]

The yeast, or parasitic phase, is a budding yeast consisting of oval bodies, which are approximately 1.5 to 2.0 μm by 3.0 to 3.5 μm in diameter. Yeast cells reproduce by budding.[3] Both phases are strictly aerobic. While the temperature range for cultivation of the mycelial phase is 25-30°C, the yeast form grows at temperatures between 34-37°C. The optimum pH for the mycelial phase is 6.5, although the yeast phase prefers a pH range between 6.5-7.5.[3] Although several phenomena have been observed in association with this temperature-induced phase shift, the basis of morphogenesis of *H. capsulatum* is still unknown. Temperature alone is not important for the phase transition;[3,4] rather, the concentration of sulfhydryl groups or redox potential is temperature-dependent, which in turn influences the shift. Concentrations of cAMP vary five folds between the mycelial and yeast phase.[5] Changes in cell wall glycans have been observed as the organism undergoes morphogenesis.[6,7] These phenomena would implicate surface components in the transition.

Histoplasmosis is initiated by the inhalation of a sufficient quantity of spores or conidia. The primary site of infection is lungs.[2] In its pathology, histoplasmosis resembles tuberculosis, which is caused by *Mycobacterium tuberculosis*. Studies in the 1950's on the pathology of tuberculosis focused on a lipid-like component which was isolated from the organism and was toxic to certain strains of mice.[8] This factor was called "cord factor" and was found to be trehalose-6,6′-dimycolate.[9] This suggested that lipid associated with the bacilli could be involved in the pathogenicity of the organism. Due to its similarity to tuberculosis, an analysis of the lipids of both phases of *H. capsulatum* was undertaken in an effort to determine what role such compounds could play in the pathogenicity or in the dimorphism of the organism. This chapter will present our current knowledge of the lipids of *H. capsulatum*, with emphasis on novel phosphoinositol SPH isolated from the yeast phase which reacted with sera from patients with histoplasmosis.

II. FREE LIPIDS AND PHOSPHOLIPIDS

Studies were initiated by Al-Doory[10] to examine the free lipids and phospholipid phosphorus of both phases of a single strain of *H. capsulatum*. The free lipids were extracted with chloroform:methanol (2:1, v/v) and the percentage of free lipids averaged 19% for the yeast phase and either 2% or 10% for the mycelial phase, depending on whether the cultures were shake or static, respectively. Likewise, the phospholipid phosphorous was also higher in the yeast phase, 0.2% versus 0.035% in the mycelial phase. Mycelial "still cultures" (containing mycelia and spores) had significantly higher chloroform-methanol extractable lipids and phospholipid phosphorous when compared to mycelial "shake cultures" which were reported to contain almost exclusively mycelium.[10] The author attributed the increase in lipids in the still culture to the presence of spores. This was the first demonstration that there were quantitative lipid differences between the phases of *H. capsulatum*, indicating that the yeast phase, or pathogenic phase, had more lipid than the mycelial phase.

Further studies examined total lipid, acetone soluble lipid, and the phospholipid fractions from *H. capsulatum* yeast phase cells grown on solid medium, mycelial cells grown in still cultures and mycelial cells grown in shake cultures.[11] On a dry weight basis, the yeast phase contained the highest percentage of total lipids, approximately 35%. This was almost two fold higher than that observed in the previous study[10] but it is probably due to longer extraction with chloroform:methanol (2:1, v/v) and differences in growth conditions.

Equivalent amounts of phospholipids were extracted from the yeast and mycelial phases while the percentage of acetone-soluble lipid was found to be higher in the yeast phase of *H. capsulatum.*[11]

III. YEAST PHASE LIPIDS

Nielsen[12] examined the yeast phase of six different isolates of *H. capsulatum*. The isolates were analyzed for total lipids, acetone-soluble lipids, phospholipids and for their virulence properties in mice to ascertain if differences in lipids among the isolates could be correlated with an increase in the virulence of the isolate for mice. Virulence was assessed in a mouse model by using two criteria: recovery of *H. capsulatum* from cultured organs and by animal deaths in 20 days. Of the isolates tested, strains H-100 and G-8, both from human isolates, were the most virulent in mice. For both strains, 80 to 100 % of the mice died within 20 days and the organism was recovered from 50 to 100 % of the cultured organs. The only isolate from dogs, G-12, and strains Strauser, 6651 and G-17 and all human isolates

Table 1
Lipid Content of Yeast Cells of Six Isolates of *H. capsulatum*

Isolate	Total Lipid %	Acetone-soluble lipid %	Phospholipid %
H-100	10.25	9.00	1.25
	10.30	9.20	1.10
G-8	6.31	5.40	0.91
	6.00	5.20	0.80
Strauser	9.56	8.60	0.96
	9.40	8.50	0.90
G-12	5.41	4.90	0.51
	5.80	5.20	0.60
6651	7.26	7.00	0.26
	7.40	7.10	0.30
G-17	6.50	5.60	0.90
	6.70	5.80	0.90

From Nielsen, H.S., *J. Bacteriol.*, 91, 273, 1966. With permission.

showed extreme variability in virulence properties, both in terms of animal death and in recovery of the fungus from infected organs. None of the strains was as virulent as isolate H-100 and G-8.

Lipids were extracted with ethanol:diethyl ether (3:1, v/v), and phospholipids were separated on a low performance column of silicic acid, according to the procedure of Hanahan et al.[13] There was little variability in the amount of total lipids isolated from each isolate in replicate samples. As shown in Table 1, lipids accounted for 10% of the total weight of isolate H-100, and the values for the other isolates ranged from 5.6% to 10%. Isolate G-8, which was highly virulent in mouse, had almost half the amount of lipid as in H-100. These percentages are much lower than those reported by Al-Doory[10,11] for the yeast phase. The amount of acetone-soluble lipid paralleled the amount of total lipid extracted. Accordingly, H-100 had the highest percentage of acetone-soluble lipid (9-10%) and G-8 contained significantly lower amounts of acetone-extractable lipids. Nielsen[12] found that the percentage of phospholipids varied from 0.26% in isolate 6651 to 1.25% in H-100. When the ratio of phospholipid to acetone-soluble lipid was calculated, 6651 had only 4% of the total lipid fraction as phospholipid. Although isolate H-100 had the highest amount of total lipid, phospholipids accounted for 13% of the total lipid, which was less than that in G-8, in which the phospholipid fraction represented 16% of the total lipid extracted.

The phospholipid fractions of H-100 and G-12 were examined in more detail by chromatography on a silicic acid column to resolve the individual phospholipids. The phosphoinositide fraction of both isolates represented almost 50% of the total phospholipid phosphorous. Both strains had similar percentages of PS (phosphatidylserine) and SPH (sphingolipids). PS constituted almost 20% of the phospholipid fraction. SPH and PE (phosphatidylethanolamine) each represented about 10% of the mixture. The striking difference was observed in the amount of PC (phosphatidylcholine) recovered from each isolate. PC was absent from G-12, yet it represented about 8% of the phospholipid phosphorous of H-100, a highly virulent strain.[12]

There was no direct correlation between virulence and total extractable lipid in *H. capsulatum*, as had been suggested for other human pathogenic fungi such as *Blastomyces dermatitidis.*[12,14] Isolate H-100, which produced high mortality rates in mice, did contain the largest percentage of total lipid extracted. This relationship was not observed for G-8 which had the same mortality rate in mice as H-100, yet the lowest percentage of total extractable lipid of the six isolates tested. G-17, which killed only 50% of the inoculated mice, had a similar amount of phospholipid as G-8. Nevertheless, this study demonstrated that the yeast phase of *Histoplasma* contains a full complement of phospholipids which include PS, PC, PE, PI's and SPH.[12] Nielsen observed a difference in phospholipid profiles between a highly virulent isolate, H-100 and G-12, a less virulent isolate. However, the significance of this observation was not further investigated by the author.

A. LIPIDS OF CELL WALL AND CELL SAP OF MYCELIAL AND YEAST PHASE

Subsequent studies were performed by Domer and Hamilton[15] on readily extractable lipids of the cell wall and cell sap (defined by the authors as a lyophilized preparation containing all the cell constituents except the cell wall) of the yeast and mycelial phase of *H. capsulatum*. This was an extension of their earlier studies on differences in cell wall glycans between the two phases.[6] An analysis of the lipids of the cell sap was included to determine whether readily extractable lipids are true constituents of the cell wall or simply represent contamination with cell sap material.

After breaking the washed cells in a Braun homogenizer, cell walls were separated from cell sap by centrifugation and both fractions were lyophilized. The lipids were subjected to chloroform:methanol (2:1, v/v) extraction. On a dry weight basis, the lipids constituted 12-18% of the yeast and mycelial cell sap preparations, with no significant differences between the two phases. Cell wall preparations of the yeast phase contained 1.5-2% lipids, whereas mycelial phase cell wall material contained proportionately less lipids. By thin layer chromatography, the major components were identified as PE, PC, PS, triacylglycerols, diacylglycerols, sterols, sterol esters and free fatty acids. Neither phase-specific nor cell preparation differences were observed.[15] There were qualitative differences in triacylglycerol levels in yeast cells, depending on the medium used. Growth in Soy medium (soy dialysate medium) produced a greater amount of triacylglycerols than GYE medium (glucose-yeast extract medium) although the amount of phospholipid remained constant. Triacylglycerol levels were higher in both media as incubation time was increased from 2 to 6 days.

Domer and Hamilton analyzed the fatty acid distribution patterns of both phases and preparations.[15] *H. capsulatum* did exhibit a phase distinction in fatty acid content. Oleic acid was the most prevalent fatty acid of yeast phase extracts, accounting for 50-70% of the total fatty acids. The second most abundant fatty acid was linoleic, which represented 13-30% of the total fatty acids. In mycelial preparations, linoleic and oleic acids were present approximately in the same proportion. Consistently, more oleic acid and less linoleic acid were found in cell wall preparations when compared to the corresponding cell sap preparations. Fatty acid profile of yeast phase cultures changed as a function of time of incubation. Levels of unsaturated fatty acids increased with age with a corresponding decrease in the level of saturated fatty acids.[15]

The distribution of fatty acids within the different classes of lipids was extremely variable. Among the phospholipids, which included PS, PE and PC, there were differences in the percentage of palmitic acid between yeast phase cell walls and cell sap. For cell sap preparations, palmitic acid was lowest in PC and highest in PS with an intermediate level in PE. This pattern was not observed in cell wall preparations. The level of palmitic acid in both cell walls and cell sap of both phases of *H. capsulatum* was higher in the diacylglycerol fraction than in the triacylglycerol fraction.[15]

Oleic acid distribution patterns were opposite to those of palmitic acid. It was highest in PC of yeast phase cell wall and cell sap and was lowest in the PS fraction of the same preparations. In the mycelial phase cell wall and cell sap, oleic acid levels were highest in PS. Linoleic acid distribution patterns were distinct from those of oleic acid. This fatty acid was highest in PS of the yeast phase cell sap although it was lowest in that of the yeast cell wall. Linoleic acid was highest in PC of both preparations from the mycelial phase.

There were marked differences in fatty acid distribution among sterol ester extracts. The levels of oleic acid found in sterol esters of mycelial and yeast phase cell wall extracts were higher than that of linoleic acid in cell wall extracts from either phase. The ratio was reversed when mycelial phase cell sap was examined. Palmitoleic and stearic acids were found in all the lipid classes but together they accounted for less than 10% of the total fatty acids.

The complete absence of PI and SPH in the preparations[15] is in direct contrast to the earlier findings[12] where at least 50% of the phospholipids of the yeast phase were PIs. With *Saccharomyces cerevisiae* and *Neurospora crassa*, Hanson and Lester[16] demonstrated that chloroform:methanol (2:1, v/v) does not quantitatively extract inositol and choline-containing lipids, which may explain the absence of PI and SPH in the preparations of Domer and Hamilton.[15]

Although Domer and Hamilton did not find phase differences in total lipids extracted from cell sap and cell wall preparations of *H. capsulatum*, they did observe differences in fatty acid profiles between the two phases. Oleic acid was most predominant in the yeast phase extracts whereas linoleic acid was slightly greater in the mycelial phase preparations.[15]

SPH fraction from the yeast phase of six isolates of *H. capsulatum* was isolated. SPHs are a class of complex lipids characterized by the presence of a long-chain alcohol, or long-chain base.[17,18] Sphingosine was the first of this class to be purified and analyzed. Long-chain bases exist as a homologous series which vary in length of the carbon chain, the organization of the chain (straight vs. branched) and in the degree of saturation or hydroxylation of the chain.[19] Long-chain bases generally do not occur free in nature but are found covalently bound to fatty acids, phosphate, simple sugars or oligosaccharides.[18] The N-fatty acyl derivatives are referred to as ceramides. Since the fatty acid is linked to the base by an amide linkage, SPH are stable to mild alkali. The alkaline stability of these compounds is often exploited to separate them from the alkali labile ester lipids in lipid extracts. The fatty acids released upon hydrolysis of SPH range in length from 12 to 26 carbons, and they are often hydroxylated at the 2nd position.

Higher plants[20] and fungi contain a unique group of SPH, not found in animals, with a ceramide consisting of a long-chain base, usually hydroxysphinganine (phytosphingosine), N-acylated with hydroxy and non-hydroxy fatty acids and with polar head groups consisting of inositol, phosphate and carbohydrate. Smith and Lester[21] isolated three classes of SPH from *S. cerevisiae* which varied according to the polar head group: inositolphosphoceramide, mannosylinositolphosphoceramide and mannosyl-di-(inositolphospho) ceramide. Preliminary analysis of the SPH of the yeast phase of *H. capsulatum* indicated that they were different from *S. cerevisiae*, probably due to differences in glycosyl moieties. When tested by immunodiffusion in agarose gels, a crude SPH mixture reacted with the serum from a patient with histoplasmosis.[22]

B. ALKALI-STABLE LIPIDS OF THE YEAST PHASE

Due to their novel nature and their apparent immunoreactivity, the major alkali-stable phospholipids from the yeast phase of *H. capsulatum* were purified and the composition and major structural features of each were determined.[22] Initial studies examined the alkali-stable lipids when cells were cultured with [^{32}P]-Pi and [^{3}H]-inositol for 2 and 4 days. Four major compounds, evident by thin layer chromatography (Figure 1, lanes 1 and 6) and by liquid chromatography (Figure 2), were designated compounds III, V, VI and VIII. The 2-day profile was identical to the one obtained from lipids cultured for 4 days with [^{32}P]-Pi. Liquid chromatography was employed to isolate pure lipids from a mixture of alkali-stable lipids labeled for 4 days separately with [^{3}H]-inositol and [^{32}P]-Pi. Each peak was pooled

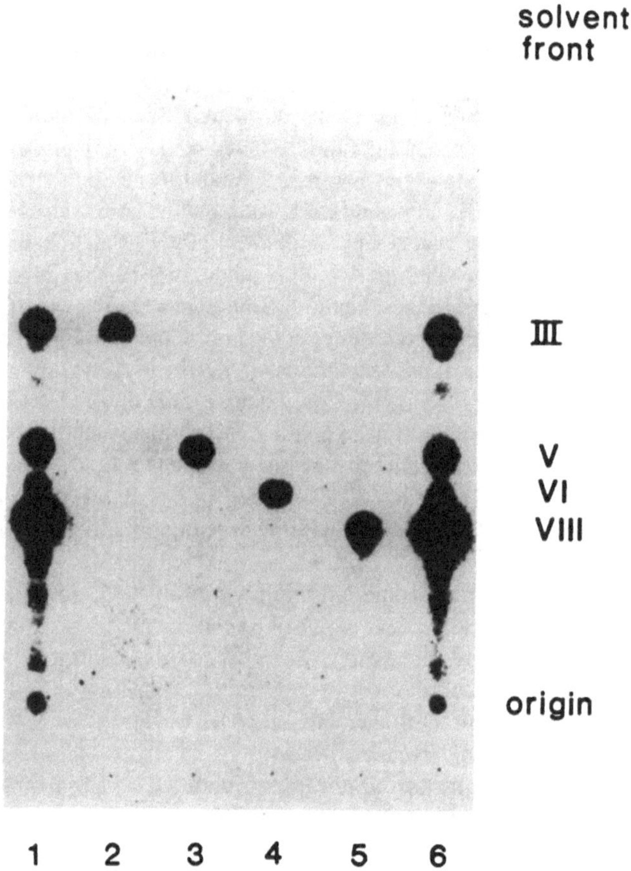

Figure 1. Thin layer chromatography of [^{3}H]-inositol, [^{32}P]-labeled alkali-stable lipids from *H. capsulatum*. [^{3}H]-inositol and [^{32}P]-labeled alkali-stable lipids were obtained from a 4 day culture of *H. capsulatum*. Silica gel thin layer chromatography was carried out with chloroform:methanol:conc.ammonium hydroxide:water (50:36.8:6:7.2, v/v) plus 0.6 g/l NH_4Cl, followed by autoradiography. Total alkali-stable lipids were applied to lanes 1 and 6; compounds III, V, VI and VIII, obtained by HPLC, were applied to lanes 2, 3, 4, and 5, respectively.

which then exhibited a single spot by thin layer chromatography (Figure 1, lanes 2-5).

To find out if the major *Histoplasma* components were the same as one or more of the already characterized SPH from *S. cerevisiae,*[21] alkali-stable [^{32}P]-labeled lipids from a semianaerobic culture of *S. cerevisiae* strain MC6A were chromatographed with the [^{3}H]-inositol-labeled alkali-stable

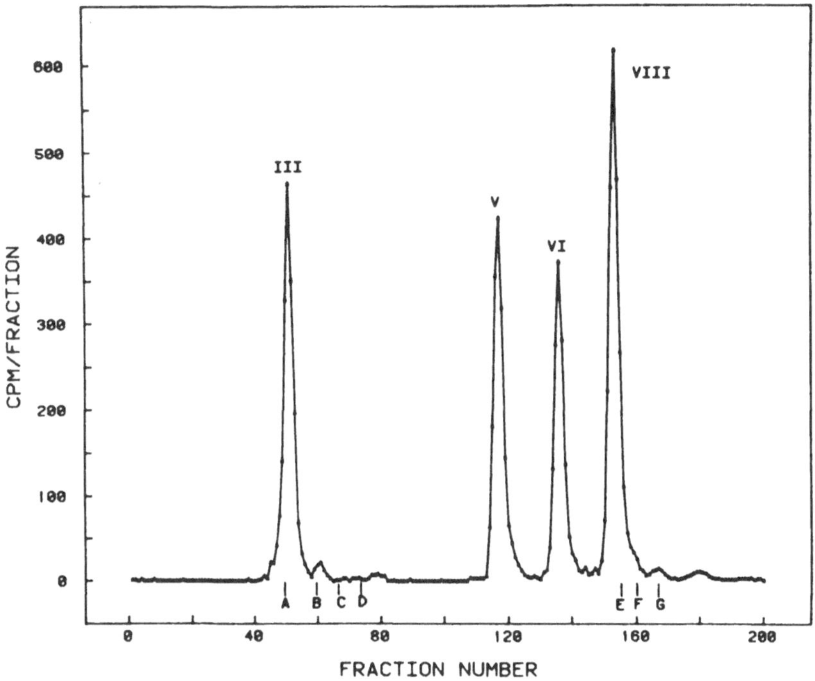

Figure 2. Liquid chromatography of [^{3}H]-inositol labeled alkali-stable lipids from *H. capsulatum* and [^{32}P]-labeled lipids from *S. cerevisiae*. The alkali-stable [^{32}P]-labeled lipids from a semi-anaerobic culture of *S. cerevisiae* were chromatographed with alkali-stable lipids from *H. capsulatum*, cultured with [^{3}H]-inositol, on Lichrosorb Si60 (5μ) column. The tritium profile is denoted with solid triangles and letters represent the location of the [^{32}P] peaks. The latter were A, inositol-P-ceramide (hydroxy fatty acid, phytosphingosine); B, mannose-P-inositol-ceramide (non-hydroxy fatty acid, dihydrosphingosine); C, mannose-inositol-P-ceramide (hydroxy fatty acid, dihydrosphingosine); D, mannose-inositol-P-ceramide (hydroxy fatty acid, phytosphingosine); E, mannose-(inositol-P)$_2$-ceramide (non-hydroxy fatty acid, dihydrosphingosine); F, mannose-(inositol-P)$_2$-ceramide (hydroxy fatty acid, dihydro-sphingosine); G, mannose-(inositol-P)$_2$-ceramide (hydroxy fatty acid, phytosphingosine).

lipids from *H. capsulatum*. These results are shown in Figure 2 and the letters indicate the retention time of the *Saccharomyces* lipids. Only *H. capsulatum* compound III appeared to be one of the *S. cerevisiae* SPHs, although it retained slightly more on the column than inositol phosphoceramide from *S. cerevisiae*. Compounds V and VI were more polar than mannosylinositol phosphoceramides (B-D) of *S. cerevisiae*. Compound VIII from *Histoplasma* was less polar than the least polar variant of mannosyl-*bis*-(inositolphospho) ceramide (E, Figure 2), which suggests that *Histoplasma* contains a unique complement of SPH.

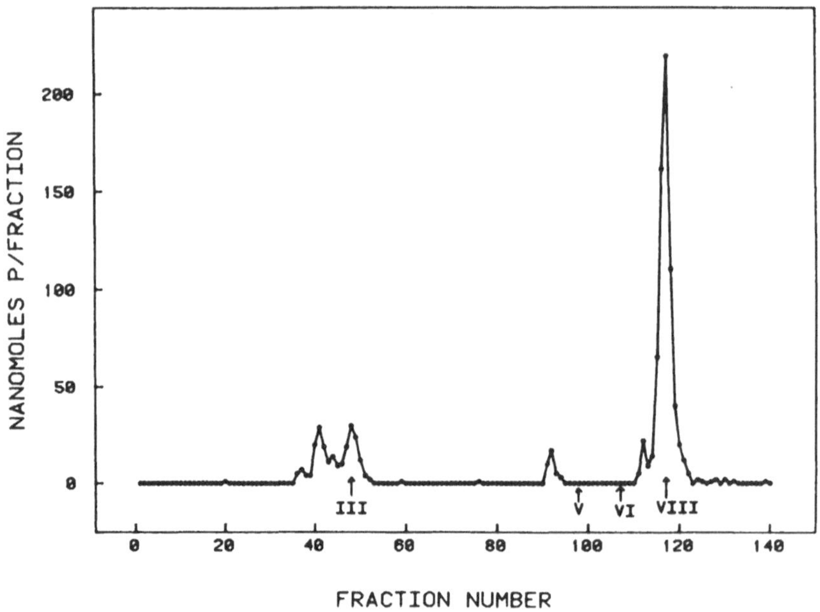

Figure 3. Comparison of the alkali-stable lipids from the yeast and mycelial phase. Alkali-stable lipids from the mycelial phase were chromatographed with yeast [^{3}H]-inositol alkali-stable lipids on Lichrosorb Si60 (5μ) column. Total phosphorous of the mycelial lipids is denoted with solid circles and Roman numerals indicate the peak retention times of the four major yeast phase lipids (Figure 2).

IV. COMPARISON OF THE YEAST AND MYCELIAL PHASE

It was of interest to compare the alkali-stable lipids of the yeast phase with those of the mycelial phase. Unlabeled alkali-stable lipids from the mycelial phase were chromatographed on a larger column with marker [^{3}H]-inositol-labeled lipids of the yeast phase (Figure 3).

The chromatographic pattern obtained from the mycelial lipids, detected as chemical phosphorous, showed a striking absence of compounds V and VI, and the presence of compounds III and VIII in both phases. Compound lipid VIII, or a compound resembling it, appeared to be the most abundant alkali-stable lipid of the mycelial phase.

Since the yeast phase is the pathogenic phase, the alkali-stable lipids were chemically characterized. This necessitated large-scale isolation of the lipids. Therefore, yeast phase *H. capsulatum* was cultured in 100 l of medium, and the lipids were extracted and transesterified. Preparative liquid

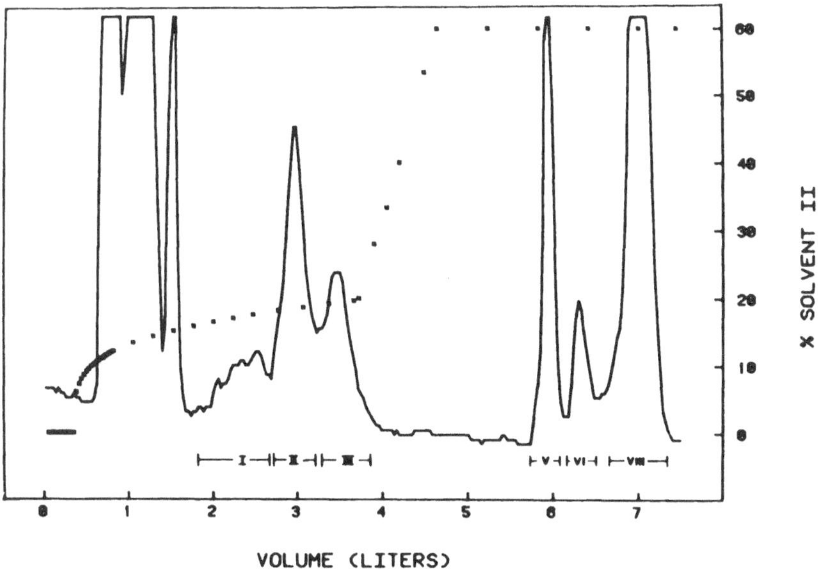

Figure 4. Preparative liquid chromatography of the alkali-stable lipids. Alkali-stable lipid was chromatographed on a 2.5 × 200 cm silica gel column with a non-linear gradient for 6 h followed by a linear gradient for 1.5 h. Solvent I was chloroform:methanol:conc. ammonium hydroxide (65:29:6, v/v). Solvent II was chloroform:methanol:conc. ammonium hydroxide:water (40:42:6:12, v/v) and both solvents contained 0.6 g/l NH_4Cl. The solid lines denote the relative carbon detector response and the dotted line depicts the gradient composition. Peaks were pooled as indicated by brackets. Lipids were extracted and transesterified. Preparative liquid chromatography of the mild alkali-stable lipid mixture shows six major peaks.

chromatography of the mild alkali-stable lipid mixture did show six major peaks in the non-volatile carbon moving-wire detector profile (Figure 4).

Pooled fractions from all six peaks were positive to rhodamine, but only peaks V, VI and VIII were glycolipids since they also gave positive reactions to orcinol-H_2SO_4.[23] Each peak contained phosphorous. Due to contamination with acyl ester lipid, compounds II and III were subjected to an additional transesterification. It is interesting that compound II was never observed in small-scale radiolabeling experiments (Figures 1 and 2) but was present in duplicate large scale culture experiments (Figure 4).

Compound VI was chromatographed isocratically with chloroform: methanol:concentrated ammonium hydroxide:water (52.5:35.5:6:6, v/v) plus 0.6 g/l NH_4Cl. Liquid chromatography gave two glycolipid peaks which were designated as compounds VIA and VIB. These two compounds did not get chromatographed with compound V; still they couldn't be separated from each other.

Table 2
Summary of Quantitative Analyses

Compound	II	III	V	VIA	VIII
			moles/mole P		
Long-chain base	0.84	1.07	1.06	0.83	1.13
Total fatty acid	1.12	1.13	1.04	1.01	1.09
Inositol	1.01	1.09	1.19	0.98	1.04
Total carbohydrate[a]	0.09	0.07	2.14	3.24	3.35

From Barr, K. and Lester, R. L., *Biochem.*, 23, 5581, 1984. With permission.
[a] Mannose was used as a standard in the phenol sulfuric acid analysis on Compounds II, III and V. Mannose and galactose in a ratio of 2 to 1 were the standards for carbohydrate analysis of compounds VIA and VIII.

Liquid chromatography of compounds II, V and VIII confirmed that each represented a pure alkali-stable lipid. Compound III did contain a shoulder with a retention time of compound II. Since compound III did not represent a novel compound but appeared to be similar to the inositol-SPH from *S. cerevisiae*,[21] there was no attempt to purify compound III. Carbon detection of compound VI revealed a peak which was detected neither by rhodamine nor by the orcinol-H_2SO_4 reagents by thin layer chromatography. Since this non-lipid, non-carbohydrate contaminant represented less than 10% of the carbon detector response of VIA, this compound was not further purified.

Each of the compounds was analyzed for fatty acid, long-chain base and inositol. The results are summarized in Table 2. All of the compounds yielded one mol of long-chain base which was identified as phytosphingosine. Additionally, compounds II and III also possessed a small amount of DL-erythrodihydrosphingosine. The thin layer chromatography solvent system of benzene:chloroform:acetic acid (90:10:1, v/v) separates the methyl esters of nonhydroxylated fatty acids from mono- and dihydroxy fatty acids. When this was employed with the methyl esters of the *Histoplasma* compounds, compound II exhibited only non-hydroxy fatty acids whereas only monohydroxy fatty acid methyl esters were evident in the methanolysates of the other four compounds. Each compound exhibited one mol of fatty acid/mol of phosphorous and one mol of inositol/mol of phosphorous (Table 2). The fatty acid moiety of all of the compounds was lignoceric acid (24:0). Compound II had a nonhydroxlyated 24:0 fatty acid in the ceramide moiety whereas the others contained mostly hydroxylated 24:0.

The presence of 1 mol of fatty acid, long-chain base and inositol per phosphorous indicated that these alkali-stable lipids were phosphoinositol-SPH. Compounds V, VI, and VIII were orcinol-H_2SO_4 positive, suggesting that differences in carbohydrate composition were responsible for the chromatographic separation of these compounds. When assayed by the phenol-sulfuric acid method,[24] compound V gave two equivalents of hexose/mol of phosphorous (Table 2). Although separable by thin-layer and

Table 3
Carbohydrate Analysis of SPH and Products Obtained from Ammonolysates

SPH	Inositol	Mannose	Galactose
		moles/mole P	
V		2.20	0.00
VIA		2.15	1.05
VIII		1.95	0.98
Ammonolysates		**moles/mole Inositol**	
V	1.00	1.80	0.00
VIA	1.00	2.10	1.00
VIII	1.00	2.10	1.01

The carbohydrates were converted to alditol acetates and analyzed by gas chromatography. Glucose was added at the beginning of the procedure as an internal standard.
From Barr, K. and Lester, R. L., *Biochem.*, 23, 5581, 1984. With permission.

liquid chromatography, both compounds VI and VIII had three hexose equivalents. No hexose was found associated with either compound II or compound III (Table 2). Analysis by gas-liquid chromatography of the carbohydrate moieties after conversion to alditol acetates demonstrated that compounds VIA and VIII had the identical hexose composition: two mannose and one galactose residues (Table 3). Compound V contained two equivalents of mannose residues (Table 3).

Treatment of SPH with 10 N ammonium hydroxide overnight at 150°C cleaves phosphomonoester and -diester bonds but does not affect glycosidic linkages,[25] a procedure referred to as ammonolysis. Ammonolysis of compounds V, VIA and VIII produced single, unique oligosaccharides detected with the orcinol-H_2SO_4 reagent that could be separated on silica gel thin-layer plates with double development with acetonitrile:water (2:1, v/v). The ammonolysates of compounds V and VIII were acetylated with acetic anhydride:pyridine (1:1, v/v) for 2 h at 100°C. No free inositol hexa-acetate was found to be associated with the ammonolysates of compounds V and VIII when the acetylated preparations were analyzed by thin-layer and gas-liquid chromatography. This demonstrated that the inositol and all of the hexoses of each compound were linked. Analysis of the alditol acetates of the ammonolysates or oligosaccharide moieties confirmed the results obtained with SPH. Compound V consisted of two mannose molecules and one inositol molecule (Table 3). No differences in carbohydrates were found in the ammonolysates of compounds VIA and VIII. Both contained two mols of mannose, 1 mol of galactose and 1 mol of inositol (Table 3).

Compositionally, compound III was very similar to the *S. cerevisiae* inositolphosphoceramides, which were shown to be alkali-labile.[21] On the other hand, the mannosylinositolphosphoceramides from *S. cerevisiae* were stable to such alkali treatment,[21] probably due to mannosylation vicinal to the phosphate group, preventing facile formation of a cyclic inositol

Table 4
Detection of Antibodies to *Histoplasma* SPH in Humans with Histoplasmosis

SPH	V	V	VIA		VIII	
Sera	Ag	AntiAb	Ag	AntiAb	Ag	AntiAb
			Absorbance ($\times 10^3$)			
132	69	144	221	296	189	156
449	238	307	270	243	594	660
399	193	275	204	295	305	392
645	115	196	1,427	1,509	444	507
552	10	0	201	206	32	5
A	15	23	16	30	24	35
B	15	23	49	62	-5	6
C	33	84	-16	38	99	159
D	8	17	17	30	1	14

An enzyme-linked immunosorbent assay (ELISA) was performed with compounds V, VIA and VIII. Control sera were designated by letters and numbers to represent patient sera.
From Barr, K. and Lester, R. L., *Biochem.*, 23, 5581, 1984. With permission.

phosphate. Therefore, the stability of the labeled *H. capsulatum* SPH to alkali was examined. Aliquots of [^{3}H]-labeled and [^{32}P]-labeled lipids were treated with 1 N KOH at 37°C overnight. The reaction mixture was neutralized with glacial acetic acid and examined by chromatography on 589 orange ribbon paper with chloroform:methanol:4.2N ammonium hydroxide (9:7:2, v/v), followed by autoradiography. Only compound III was alkali labile since all of the [^{32}P]- and [^{3}H]-inositol remained at the origin as expected for an inositol monophosphate. However, compounds V, VI and VIII migrated as intact lipids after alkali treatment. The product of alkaline hydrolysis of compound III was completely susceptible to alkaline phosphatase, yielding free inositol as monitored by paper chromatography. Therefore, alkaline susceptibility of compound III indicated that it was an inositolphosphoceramide with a structure similar to the *S. cerevisiae* compound[21] and the alkaline stability of compounds V, VI and VIII indicated that they are substituted vicinal to the phosphate group.

Since complex glyco-SPHs are known to be antigenic in animals, the novel lipids purified from *H. capsulatum* were tested for their reactivity with the sera of individuals who were suspected of histoplasmosis. These individuals showed a positive titer against histoplasmin, the reagent used in complement fixation and immunodiffusion tests. Initially, the interaction of the sera with the *Histoplasma* lipids was examined by double immunodiffusion in agarose gels. Compound VI formed a precipitin band with 1 of 11 different patients' sera tested in this manner; compounds V and VIII did not react with any of the sera. The inclusion of exogenous lipids,[26] such as PC, did not improve the sensitivity. In contrast, an enzyme-linked immunosorbent assay (ELISA) showed that each patient's serum reacted better with one or more of the purified SPH than did the four randomly selected human control sera (Table 4). Compound VI was reactive with sera of all the patients, particularly serum 645. The reactivities of compounds V and VIII were more variable (Table 4). Neither of the lipids reacted with serum 552, but compound VIII was more reactive than VI with serum 449. The absorbance values of the control sera were negligible when compared with those obtained from patients' sera. SPH from *S. cerevisiae*, inositol-phosphoceramide, mannosylinositolphosphoceramide, and mannosyl-*bis*-(inositol-phospho) ceramide[21] did not react in a comparable ELISA experiment, indicating that there may be a specificity of the reaction between the *Histoplasma* lipids and patients' sera.

To summarize at this point, the yeast phase of *H. capsulatum* contains five major phosphoinositol SPH. Compounds II and III are inositol-phosphoceramides, which were similar to those isolated from *S. cerevisiae*[21] and their composition was established. Compounds V, VI and VIII are of novel composition, possessing an identical inositol-P-ceramide core but differing in the glycosyl substitution on the polar head groups:

compound V	ceramide-P-inositol-[mannose$_2$]
compound VI	ceramide-P-inositol-[mannose$_2$, galactose]
compound VIII	ceramide-P-inositol-[mannose$_2$, galactose]

Compound VIII, isomeric to compound VI in an unknown manner, was found in both the mycelial and yeast phases, whereas compounds V and VI were unique to the yeast phase. Due to the novel nature of these SPH, studies of sequence, linkage and anomeric configuration were performed on the carbohydrate moieties of these lipids.

V. ANALYSIS OF OLIGOSACCHARIDES

Strong ammonolysis of lipids V, VI and VIII yielded in each case a single oligosaccharide containing the inositol and the hexoses of the intact lipids. As a verification of the mass and composition of the *Histoplasma* lipids, the permethylated oligosaccharides of compounds V, VI and VIII were examined by chemical ionization mass spectrometry. Permethylation causes any free hydroxyl group to become methylated but those hydroxyl groups participating in ring formation or in the glycosidic linkage will not be methylated.

Permethylated oligosaccharide-V was analyzed by ammonia chemical ionization mass spectrometry. The ammonium addition product molecular ion, [M + 18], had a m/z at 690, which would be expected for a trisaccharide including 2 hexoses and inositol. The mass spectrum of the major peak of permethylated oligosaccharide VIII exhibited an [M + 18] ion at m/z 894 and retention time 6.2 min, consistent with an ammonium addition product composed of three hexoses and inositol. The [M + 18] ion of permethylated oligosaccharide-VI appeared at m/z 894 and retention time of 7 min, identical to the results obtained with oligosaccharide-VIII, demonstrating that VI-oligosaccharide was also a tetrasaccharide composed of three hexose molecules and one inositol; yet it had a distinct retention time. Mass spectrometry confirmed that VI and VIII were isomers, as predicted by the compositional data.[27]

1. Analysis of Oligosaccharide-V

The anomeric configuration of the mannoses of oligosaccharide-V was determined by treatment with α-mannosidase for 24 h. The products of the digestion were examined by thin layer chromatography with acetonitrile:water (2:1, v/v) followed by detection with orcinol-H_2SO_4. Approximately 90% of the mannose of oligosaccharide-V was released by α-mannosidase treatment as judged by the appearance of free mannose and the disappearance of orcinol-positive material which migrated with oligosaccharide-V. The products formed by α-mannosidase incubation with oligosaccharide-V were quantified by gas liquid chromatography. Mannose and inositol were liberated in a ratio of 2.06:1, respectively, indicating that both of the mannoses of oligosaccharide-V were α-linked.

Permethylation of an oligosaccharide and subsequent conversion of the permethylated oligosaccharide to partially methylated alditol and inositol acetates provides information about the linkages of the hexoses within the original oligosaccharide. As explained above, only free hydroxyl groups will be methylated. Hydrolysis of the permethylated oligosaccharide followed by reduction with $NaBD_4$ sets free the hydroxyls which were involved in glycosidic linkage or in ring formation and renders them susceptible to acetylation. The resulting partially methylated alditol and inositol acetates can then be analyzed by gas chromatography/mass spectrometry. The ionized derivatives produce characteristic fragments which delineate the linkage of the hexose. The pattern also distinguishes whether the hexose was in the pyranose or furanose form.

The linkages of sugars in oligosaccharide-V were determined by analysis of the partially methylated alditol acetates of the oligosaccharide by gas chromatography/mass spectrometry. Three peaks were evident in the mass spectrum. Peak I was identified as penta-O-methyl-mono-O-acetylinositol. The ion profile of peak II indicated that it was 1,5-di-O-acetyl-2,3,4,6-tetra-O-methylmannitol, or a terminally-linked mannose. The mass spectrum of peak III supports the assignment of 2,4,6-tri-O-methyl-1,3,5-tri-O-acetyl-mannitol, indicating that the hexose is substituted at the third position. Compound V, VI and VIII were stable to alkaline hydrolysis, and each gave inositol-1-P as the major product after acid hydrolysis.[22] These data would suggest that all three oligosaccharides are linked to inositol at either the second or sixth position. Based on the results of the analysis, the structure of oligosaccharide-V is proposed as follows:

man (α1→3) man (α1→2 or 6) *myo*-inositol.

2. Analysis of Oligosaccharide-VI

Oligosaccharide-VI was incubated with α-mannosidase for 70 h at 37°C. The reaction mixture was chromatographed with acetonitrile:water (2:1, v/v) and the sugars were detected with orcinol-H_2SO_4. None of the starting oligosaccharide-VI remained; the products were free mannose and an orcinol-positive spot which migrated to a position between that of galactinol (galactosyl-inositol) and oligosaccharide-V (dimannosylinositol). Thus, at least one of the mannoses of oligosaccharide-VI was α-linked. Galactose was not released upon incubation of oligosaccharide-VI with α-mannosidase in combination with an α- or β-galactosidase.

Alternatively, anomeric configuration can be established from treatment of acetylated oligosaccharides with CrO_3 in glacial acetic acid. Acetylated hexopyranosides in an α-linkage are resistant to CrO_3 oxidation whereas those in a β-linkage are destroyed by the same treatment.[28,29] Oligosaccharide-VI was acetylated and subjected to 45 min CrO_3 oxidation. When compared with the zero time point, 74% of the mannose and 11% of the galactose survived the CrO_3 treatment. Thus, galactose was destroyed by CrO_3 treatment; at least half of the mannose was in the α-configuration and the remaining mannose was susceptible to CrO_3 oxidation.

Oligosaccharide-VI was converted to partially methylated alditol acetates and the products were analyzed by gas chromatography/mass spectrometry. Four peaks were evident in the mass spectrum of oligosaccharide-VI. Peak I was identified as penta-O-methyl-mono-O-acetylinositol. The ion profile of peak II indicated that it was 1,5-di-O-acetyl-2,3,4,6-tetra-O-methyl-mannitol, or an unsubstituted mannopyranoside. Peak IV had the same retention time and mass spectrum as authentic 2,4-di-O-methyl-1,3,5,6-tetra-O-acetylmannitol.

Peak III, which eluted very close to peak II, exhibited a very different spectrum. The ion profile indicated that it was a derivative of a terminal hexose but one in which the hexose was unsubstituted at positions 2 and 3 and acetylated at carbon 4. The retention time of peak III matched the retention time of 1,4-di-O-acetyl-2,3,5,6-tetra-O-methyl-galactitol prepared from authentic 1-β-O-methylgalactofuranoside and did not chromatograph with the partially methylated alditol acetate of β-mannofuranoside.[30] These data are consistent with the designation of peak III as 1,4-di-O-acetyl-2,3,5,6-tetra-O-methylgalactitol or the derivative of a terminal galactofuranose.

That oligosaccharide-VI contained terminal galactofuranoside was consistent with the anomeric configuration data. Treatment of oligosaccharide-VI with α-mannosidase appeared to remove only one mannose residue, since only one mannose residue would be accessible to the glycosidase. The susceptibility of acetylated hexofuranosides to CrO_3 treatment is opposite to the response of acetylated hexopyranosides; acetylated hexofuranosides in β-linkage survive the treatment whereas acetylated hexofuranosides in an α-linkage are oxidized by such treatment.[31] Therefore, the galactofuranose of compound VI is probably in the α-configuration.

The presence of terminal galactofuranoside and mannopyranoside raised the question of the 3-linked or 6-linked substitution of each sugar on the mannose of oligosaccharide-VI. Therefore, oligosaccharide-VI was treated with α-mannosidase for 70 h to remove the terminal mannopyranoside and the products were permethylated and converted to partially methylated alditol acetates. Analysis by chemical ionization mass chromatography revealed that monosubstituted inositol, terminal galactofuranoside, and a monosubstituted hexose have the same retention time as authentic 2,3,4-tri-O-methyl-1,5,6-tri-O-acetylmannitol. Thus, α-mannosidase treatment of oligosaccharide-VI generated a mannitol substituted at the sixth position, indicating that the mannopyranose residue was at the third position in the untreated oligosaccharide. These results suggest that the structure of oligosaccharide-VI is:

$$\begin{array}{l} \mathrm{man_p}(\alpha 1\rightarrow 3) \\ \qquad\qquad\quad | \\ \qquad\qquad\quad \mathrm{man_p}(\alpha 1\rightarrow 2\text{ or }6)\text{-}myo\text{-inositol} \\ \qquad\qquad\quad | \\ \mathrm{gal_f}(\alpha 1\rightarrow 6) \end{array}$$

3. Analysis of Oligosaccharide-VIII

Oligosaccharide-VIII was resistant to treatment with various α-mannosidases, α-galactosidases and β-galactosidases, including single and double enzyme incubation. As an alternate approach, oligosaccharide-VIII was acetylated and subjected to 45 min CrO_3 treatment as described for oligosaccharide-VI.[28,29] No galactose was detected after 45 min of CrO_3 treatment, whereas 95% of the mannose survived the oxidation. These results indicated that both of the mannoses of oligosaccharide-VIII were in an α-linkage and the galactose was in the β-configuration.

The linkages of the hexoses of oligosaccharide-VIII were determined by conversion of the oligosaccharide to partially methylated alditol acetates, as had been done for oligosaccharides V and VI. Analysis by chemical ionization mass spectrometry revealed four peaks in the mass spectrum of oligosaccharide-VIII. Peak I was designated as penta-O-methyl-mono-O-acetyl-inositol. Peaks II and III were identified as 1,5-di-O-acetyl-2,3,4,6-tetra-O-methylmannitol and 1,5-di-O-acetyl-2,3,4,6-tetra-O-methyl-galactitol, respectively, from a comparison with the retention time of authentic standards. Peak IV was 2,6-di-O-methyl-1,3,4,5-tetra-O-acetyl-mannitol. These results indicated the presence of 3,4-disubstituted mannopyranose, terminal mannopyranose, terminal galactopyranose and monosubstituted inositol in oligosaccharide-VIII.

Mass spectrometry did not delineate the position of substitution of the terminal mannose and galactose on the disubstituted mannose. It became necessary to selectively remove one of the terminal hexoses to examine the structure of the resulting trisaccharide by methylation analysis.[27] Since the β-linked galactose was sensitive to 45 min CrO_3 treatment, peracetylated oligosaccharide-VIII was treated with CrO_3 for 0 and 45 min, followed by permethylation, purification by liquid chromatography, conversion to partially methylated alditol acetates and analysis by chemical ionization mass spectrometry. Four peaks were evident at the zero time point, and were identified as monosubstituted inositol, terminal mannose, terminal galactose and disubstituted mannose. After 45 min CrO_3 oxidation, terminal galactose was absent. Terminal mannose had also disappeared and had been replaced by a monosubstituted hexose peak linked at the third position. The retention time of the 3-linked hexopyranoside was identical to the retention time of 3-linked mannopyranose from oligosaccharide-V, demonstrating that the internal mannose was substituted with mannose at the third position and galactose at the fourth position. This established the structure of oligosaccharide-VIII as:

$$
\begin{array}{l}
\text{man}_p(\alpha 1\rightarrow 3) \\
\qquad\qquad | \\
\qquad\qquad \text{man}_p(\alpha 1\rightarrow 2 \text{ or } 6)\text{-}\textit{myo}\text{-inositol} \\
\qquad\qquad | \\
\text{gal}_p(\alpha 1\rightarrow 6)
\end{array}
$$

In summary, five phosphoinositol SPH were purified from the yeast phase of *H. capsulatum*. These compounds are related to previously characterized fungal and plant phosphoinositolceramides. Compounds II and III are variants of inositol SPH purified from *S. cerevisiae*.[21] Compound II was only isolated from large scale cultures, suggesting that under these conditions, there is insufficient hydroxylation due to a slightly anaerobic environment. Compounds V, VI and VIII are novel compounds. Compound V may represent a biosynthetic precursor to compounds VI and VIII since both contain the same trisaccharide core as V but with the addition of a galactofuranose residue at the sixth position of mannose (compound VI) and the addition of galactopyranose at the fourth position of mannose (compound VIII). Compound VIII, or a compound resembling this molecule, is found in both the mycelial and yeast phases. Compounds V and VI appear to be restricted to the yeast phase.[27]

Compound VI represents not only the first reported occurrence of galactofuranose in a glycosphingolipid, but also the first demonstration of the sugar in the yeast phase of *H. capsulatum*. The occurrence of terminal galactofuranose residues in cell wall polymers has been reported, specifically as the constituents of the malonogalactan of *Penicillium citrinium* Thom 1131,[32] the galactomannan of the cell wall of *A. niger*,[33] and polysaccharides isolated from the mycelial forms of *H. capsulatum*, *H. duboisii*, *Paracoccidioides brasiliensis*, and *B. dermatitidis*.[34] It is interesting that compound VI is not found in the mycelial phase.

The *Histoplasma* glyco-SPH, compounds V, VI and VIII, did react with antibodies from histoplasmosis patients. This observation is not surprising since many antigens have been identified as glyco-SPHs.[35] However, it cannot be ruled out that the true immunogen in humans might have been another glycoconjugate, such as a glycoprotein, which could have saccharide structures similar to the glycolipids.

VI. REFERENCES

1. **Pine, L.,** Studies on the growth of *H. capsulatum*. I. Growth of the yeast phase in liquid media, *J. Bacteriol.*, 68, 671, 1954.
2. **Rippon, J.W.,** *Medical Mycology: The Pathogenic Fungi and the Pathogenic Actinomycetes*, W.B. Saunders Company, Pennsylvania, 342, 1982.
3. **Pine, L.,** in *Histoplasmosis*, Sweany, H.C., Ed., C.C. Thomas, Illinois, 1960.
4. **Salvin, S.B.,** Cysteine and related compounds in the growth of the yeast like phase of *H. capsulatum*, *J. Infect. Dis.*, 84, 275, 1949.
5. **Maresca, B., Medoff, G., Schlessinger, D., Kobayashi, G.S. and Medoff J.,** Regulation of dimorphism in the pathogenic fungus *H. capsulatum*, *Nature*, 266, 447, 1977.
6. **Domer, J.E., Hamilton, J.G. and Harkin, J.C.,** Comparative study of the cell walls of the yeast like and mycelial phases of *H. capsulatum*, *J. Bacteriol.*, 94, 466, 1967.
7. **Domer, J.E.,** Monosaccharide and chitin content of cell walls of *Histoplasma capsulatum* and *Blastomyces dermatitidis*, *J. Bacteriol.*, 107, 870, 1971.
8. **Bloch, J.,** Studies on the virulence of *Tubercle bacilli*. Isolation and biological properties of a constituent of virulent organisms, *J. Exptl. Med.*, 91, 197, 1950.
9. **Noll, H., Bloch, H., Assenlineau, J. and Lederer, E.,** The chemical structure of the cord factor of *Mycobacterium tuberculosis*, *Biochim. Biophys. Acta*, 20, 299, 1956.
10. **Al-Doory, Y.,** Free lipids and phospholipid phosphorus of *Histoplasma capsulatum* and other pathogenic fungi, *J. Bacteriol.*, 80, 565, 1960.
11. **Al-Doory, Y. and Larsh, H.W.,** Quantitative studies of total lipids of pathogenic fungi, *Appl. Microbiol.*, 10, 492, 1962.

12. **Nielsen, H.S.,** Variation in lipid content of strains of *Histoplasma capsulatum* exhibiting different virulence properties for mice, *J. Bacteriol.*, 91, 273, 1966.
13. **Hanahan, D.J., Dittmer, J.C. and Warashina, E.,** A column chromatographic separation of classes of phospholipids, *J. Biol. Chem.*, 228, 685, 1957.
14. **Disalvo, A. and Denton, J.F.,** Lipid content of four strains of *Blastomyces dermatitidis* of different mouse virulence, *J. Bacteriol.*, 85, 927, 1963.
15. **Domer, J.E. and Hamilton, J.G.,** The readily extractable lipids of *Histoplasma capsulatum* and *Blastomyces dermatitidis*, *Biochim. Biophys. Acta*, 231, 465, 1971.
16. **Hanson, B.A. and Lester, R.L.,** The extraction of inositol-containing phospholipids and PC from *Saccharomyces cerevisiae* and *Neurospora crassa*, *J. Lipid Res.*, 21, 309, 1980.
17. **Weete, J.D.,** SPH, in *Fungal Lipid Biochemistry, Distribution and Metabolism*, Plenum Press, New York, 1974, chap.9.
18. **Weete, J.D.,** *Lipid Biochemistry of Fungi and Other Organisms*, Plenum Press, New York, 180, 1980.
19. **Prostenik, M.,** *Chemistry and Metabolism of SPH*, Sweeley, C.C., Ed., North Holland Publishing Company, 1, 1970.
20. **Laine, R.A., Hsieh, T.C.Y. and Lester, R.L.,** in *Cell Surface Glycolipids,* Sweeley, C.C., Ed., American Chemical Society Symposium Series 128, Washington, D.C., 65, 1980.
21. **Smith, S.W. and Lester, R.L.,** Inositol phosphorylceramide, a novel substance and the chief member of a major group of yeast SPH containing a single inositol phosphate, *J. Biol. Chem.*, 249, 3395, 1974.
22. **Barr, K. and Lester, R.L.,** Occurrence of novel antigenic phosphoinositol-containing SPH in the pathogenic yeast *Histoplasma capsulatum*, *Biochem.*, 23, 5581, 1984.
23. **Skipski, V.P. and Barclay, M.,** Thin layer chromatography of lipids, *Methods in Enzymol.*, 14, 545, 1969.
24. **Dubois, M., Giles, A.K., Hamilton, J.K., Rebers, P.A. and Smith, F.,** Colorimetric method for determination of sugars and related substances, *Anal. Chem.*, 28, 350, 1956.
25. **Ballou, C.E., Vilkas, E. and Lederer, E.,** Structural studies on the *myo*-inositol phospholipids of *Mycobacterium tuberculosis* (var, *bovis*, strain BCG), *J. Biol. Chem.*, 238, 69, 1963.
26. **Graf, L. and Rapport, M.M.,** Immunochemical studies of organ and tumor lipids. XVI. Gel diffusion analysis of the cytolipid K system, *Int. Arch. Allergy*, 28, 171, 1965.
27. **Barr, K., Laine, R.A. and Lester, R.L.,** Carbohydrate structures of three novel phosphoinositol-containing SPH from the yeast *Histoplasma capsulatum*, *Biochem.*, 23, 5589, 1980.

28. **Hoffman, J., Lindberg, B. and Svensson, S.,** Determination of the anomeric configuration of sugar residues in acetylated oligo- and polysaccharides by oxidation with chromium trioxide in acetic acid, *Acta Chem. Scand.*, 26, 661, 1972.
29. **Laine, R.A. and Renkonen, O.,** Ceramide di- and trihexosides of wheat flour, *J. Lipid Res.*, 16, 102, 1975.
30. **Angyal, S.J., Evans, M.E. and Beveridge, R.J.,** *Methods in Carbohydrate Chemistry*, VIII, 233, 1980.
31. **Oshima, M. and Yamakawa, T.,** Chemical structure of a novel glycolipid from an extreme thermophile, *Flavobacterium thermophilum*, *Biochem.*, 13, 1140, 1974.
32. **Kohama, T., Fujimoto, M., Kuninaka, A. and Yoshina, H.,** Structure of malonogalactan, an acidic polysaccharide of *Penicllium citrinum*, *Agric. Biol. Chem.*, 38, 127, 1974.
33. **Bardallaye, P.C. and Nordin, J.H.,** Chemical structure of the galactomannan from the cell wall of *Aspergillus niger*, *J. Biol. Chem.*, 252, 2584, 1977.
34. **Azumo, I., Kanetsuna, F., Tanaka, Y., Yamamura, Y. and Carbonell, L.M.,** Chemical and immunological properties of galactomannans obtained from *Histoplasma duboisii*, *Histoplasma capsulatum*, *Paracoccidioides brasiliensis* and *Blastomyces dermatitidis*, *Mycopath. Mycol. Appl.*, 54, 111, 1974.
35. **Hakomori, S.I.,** Glyco SPH in cellular interaction, differentiation and oncogenesis, *Ann. Rev. Biochem.*, 50, 733, 1981.

Chapter 5

LIPIDS OF *CANDIDA ALBICANS*

R. Prasad, A. Koul, P. K. Mukherjee and M. A. Ghannoum

CONTENTS

0-8493-4794-7/96/$0.00+$.50

I. INTRODUCTION

Lipids of *Candida albicans* have attracted considerable interest because they are the prime target of antifungals.[1] In addition, lipids of *C. albicans* play an important role in the structure and function of its membrane,[2] in morphogenesis,[3-7] adherence[8,9] and virulence.[10] Recent findings that the lipid composition of *C. albicans* varies with different morphotypes[11] have provided some interesting insights into the pathogenesis of candidosis. Although efforts have been made to understand the role of lipids in various physiological aspects of *Candida*, its metabolism and regulation remain poorly understood. Among the different yeasts, *Saccharomyces cerevisiae* is by far the best studied organism; however, with respect to lipids, *C. albicans* is probably the least understood. Attempts have been made to isolate antibiotic-resistant mutants of *C. albicans*.[12] Because most of these antibiotics interact with sterols of plasma membrane, mutants defective in sterol biosynthesis have been studied in some detail.[12]

Keeping pace with the research, a few articles dealing with the lipids of *Candida* have appeared.[2,13] We, therefore, avoid discussion of those topics covered in earlier reviews. In this chapter we mainly emphasize the biosynthesis of the lipids of *C. albicans*, its mutants defective in lipid biosynthetic pathways, and their lipid composition. A relatively newer role of lipids in adherence and morphogenesis of *C. albicans* is also included in our discussion. For aspects such as extraction, estimation, cellular distribution and composition of lipids, readers are referred to previous articles.[2,13]

II. LIPID COMPOSITION

Total lipid content of *C. albicans* is approximately 3 % of the total dry weight of the cell.[14] The cellular lipids of *C. albicans* consist of two major classes: neutral lipids and phospholipids. Among the neutral lipids, triacylglycerol, fatty acids and sterols are the major constituents. In the whole cell, phospholipids consist of phosphatidylcholine (PC), phosphatidylethanolamine (PE), phosphatidylserine (PS), and phosphatidylinositol (PI), which resembles a typical eukaryotic mixture of phospholipids[15-17] (Table 1). Fatty acids of *C. albicans* show a preponderance of palmitic (16:0) and stearic (18:0) acids. Polyunsaturated fatty acids, e.g., linoleic (18:2) and linolenic (18:3) acids, which are not found in *S. cerevisiae*, were also detected in *C. albicans*.[4,5] Triacylglycerols constitute a major fraction, while mono- and diacylglycerols represent a minor fraction of total acylglycerols.[5] Similar to other yeasts, the major sterol of *C. albicans* is ergosterol; sterols such as zymosterol, fecosterol and other methyl sterols are also detected.[5] The major class of glycolipids present in *C. albicans* are steryl glycosides (both free and esterified forms), while digalactosyldiacylglycerol and monogalactosyldiacyl-

Table 1
Phospholipid Composition of Yeast Form of *C. albicans*

Strain	Source of lipid	Phospholipid composition (%)								Ref.
		PI	PS	PE	PC	CL	SL	PG	Others	
Candida albicans										
6406	P	-	11	70	4	-	15	-	-	5
Clinical isolate	C	10	9	24	37	8	-	7	6	15
ATCC 10231	C[a]	22	20	37	11	-	-	-	10	16
ATCC 10231	C[b]	11	6	28	30	-	-	-	25	16
ATCC 10261	C[a]	18	22	38	12	-	-	-	10	16
ATCC 10261	C[b]	9	7	29	32	-	-	-	23	16
ATCC 46977	C	17	9	24	43	-	-	-	7	17

Note: P = plasma membrane; C = whole cell; PI = phosphatidylinositol; PS = phosphatidylserine; PE = phosphatidylethanolamine; PC = phosphatidylcholine, CL = cardiolipin; SL = sphingolipids; PG = phosphatidylglycerol.
[a] Pulse labeling for 15 min with [^{32}P]; [b] Steady-state labeling with [^{32}P].

glycerol constitute a minor fraction of total lipids.[4] Variations in lipid composition with morphological transition and adherence of *C. albicans* are discussed later in this article. The fluctuations in lipid composition due to environmental stress were described earlier.[13] In the following section, we discuss the lipid composition of mutants of *C. albicans* that are defective in lipid biosynthesis.

In contrast to *S. cerevisiae*, very few lipid mutants of *C. albicans* have been isolated and characterized. The mutants of *C. albicans* defective in ergosterol biosynthesis have been relatively well characterized. These mutants (*erg* mutants) are known to accumulate intermediates of sterol biosynthesis. Since accumulated intermediates can sustain growth and essential membrane functions, they do not pose an auxotrophic requirement of ergosterol. Alterations in the sterol composition of the membrane of these mutants also lead to changes in their phospholipid and fatty acyl composition (Table 2). Because of the overall change in lipid composition, *erg* mutants exhibit a change in their order parameters and a higher phase transition temperature in reconstituted lipids than their parental strain.[18]

A. POLYENE RESISTANT MUTANTS

Polyenes comprise a group of antifungals that exert their action by binding directly to membrane sterols,[19] subsequently forming cross membrane channels that lead to the leakage of ions and metabolites.[20,21]

Reculturing yeasts, including *C. albicans*, on media with successively higher concentrations of polyene antibiotics leads to segregation of resistant cultures.[22-24] The development of such resistance was attributed to the altered ergosterol content in resistant cells. Different levels of total sterol in polyene resistant mutants were observed by various groups.[12,23,25] Because those results were based primarily on the sterol/lipid composition of the whole cell, the reported variations could be partly due to the lack of available information on

Table 2
Lipid Composition of Sterol Mutants of *C. albicans*

Strain	Phospholipid Composition Percentage of total				
	PC	PE	PI	PS	PA
33 erg^{+}	53.9 ± 5.9	16.1 ± 2.8	10.6 ± 3.9	13.2 ± 2.5	6.2 ± 2.0
*erg*2	33.2 ± 5.0	20.7 ± 2.9	25.3 ± 4.0	5.7 ± 0.9	15.3 ± 5.9
*erg*12	34.9 ± 1.3	22.1 ± 3.2	16.1 ± 0.6	8.8 ± 2.5	18.1 ± 2.7
*erg*16	48.0 ± 5.7	14.1 ± 1.3	23.3 ± 2.7	4.3 ± 0.9	10.3 ± 2.9
*erg*20	36.7 ± 4.3	19.1 ± 3.7	14.3 ± 5.9	10.3 ± 6.0	19.6 ± 5.4
*erg*37	44.1 ± 2.5	14.4 ± 1.8	21.7 ± 0.5	10.2 ± 3.6	9.6 ± 5.6
*erg*40	40.4 ± 6.1	31.5 ± 2.5	11.9 ± 0.9	6.4 ± 1.2	9.8 ± 2.7
*erg*41	40.4 ± 4.9	14.6 ± 4.9	17.2 ± 6.8	11.7 ± 6.1	16.1 ± 7.5

Strain	Fatty Acid Composition							
	16:0	16:1	16:2	16:3	18:0	18:1	18:2	18:3
33 erg^{+}	19.1	10.5	0.9	2.2	4.3	35.0	25.7	2.3
*erg*2	19.1	9.4	1.7	3.7	6.0	35.7	20.2	4.2
*erg*12	16.1	12.6	1.2	6.3	8.1	26.5	24.2	5.0
*erg*16	18.0	10.8	1.3	3.7	4.1	40.9	19.3	1.9
*erg*20	20.6	13.8	0.8	4.5	7.7	25.5	21.2	5.9
*erg*37	19.0	11.2	2.1	6.2	7.6	22.6	27.8	3.5
*erg*40	20.1	9.9	1.0	2.8	8.6	17.4	32.8	7.4
*erg*41	15.3	12.7	1.6	5.9	6.1	25.7	25.3	7.4

Compiled from Pesti, M., Horvath, L., Vigh, L., and Farakas, T., *Acta Microbiol. Hungary*, 32, 305, 1985. With permission.

plasma membrane lipid composition.[12]

A series of polyene resistant mutants (C7, E4, C4 and D10), which were isolated by chemical mutagenesis, were also found to be cross resistant to other polyenes.[25] The lipid composition of C7 was the same as that of sensitive parental strain. Mutant E4 contained lichesterol and other Δ^8 sterols indicating a block at Δ^7-Δ^8-isomerase, and most of the sterols consisted of 4,14-desmethylsterols. C4 and D10, which were most resistant to polyenes, also had altered phospholipid and fatty acid contents and were marked by the absence of 4,14-desmethylsterols. Among the sterols of C4 and D10, 24-methylenelanosterol was found to be the major component and obtusifoliol and lanosterol were the minor components. It was interesting to note that the resistance of these mutants increased in the presence of Δ^8- and 4, 14-desmethylsterols.[25,26]

Pesti and co-workers analyzed the lipid composition of a number of ergosterol mutants,[18] which were initially isolated by the same group by chemical mutagenesis.[27] Except for *erg12*, which accumulates ergosterol as a major sterol, all other mutants accumulate various intermediates of sterol biosynthesis. Based on the accumulated sterol intermediates, the enzymic defects were defined as follows:

1. *erg12*, which accumulates cholesta-7,24-dien-3-β-ol and ergosterol, is assigned as a consequence of the leaky mutation of 5,6-dehydrogenase.
2. *erg2* and *erg40* accumulate Δ^8-sterols where a block at Δ^8-Δ^7-isomerase is suggested.
3. *erg20* and *erg37* predominantly contain ergosta-7-en-3-β-ol, which is the last product of ergosterol biosynthesis due to a defect in either 5,6-dehydrogenase or 22,23-dehydrogenase.
4. *erg16* and *erg41* accumulate ergosta-7,22-dien-3-β-ol, which may be due to a block at the 5,6-dehydrogenase level.

The fatty acid composition of these mutants showed a slight decrease in the average chain lengths and relative proportion of unsaturation. Despite some variations, the level of hexadecatrienoic (16:3), stearic (18:0) and γ-linolenic (18:3) acids increased at the expense of oleic acid (18:1) (Table 2). Changes were also observed in the phospholipid composition of these mutants. In general, PC and PS were decreased whereas the levels of PI and PA were significantly increased (Table 2). These changes in lipid composition led to changes in membrane fluidity.[18,25,27]

B. AZOLE-SENSITIVE/RESISTANT STRAINS

Azoles and their derivatives constitute another class of antifungals. The lipids of both azole-sensitive (A, B2630) and azole-resistant (AD, KB) strains of *C. albicans* were analyzed.[28] Although the total lipid content of these two types of mutant strains was similar, it varied with the growth status of the cells. The phospholipid composition of azole-sensitive strains varied markedly. Large proportions of PC, PE and PS were found in *C. albicans* A as compared to B2630. In contrast, the phospholipid composition of the two imidazole-resistant strains (AD, KB) was not different.

No significant differences in triacylglycerol and nonesterified fatty acid fractions were detected in azole-sensitive and resistant strains. The azole resistant strains (AD and KB) contained more nonesterified sterols and, thus, the phospholipid:sterol ratio was approximately half that of azole- sensitive strains. The ratio correlated well with the permeability of triazoles.[28] Resistant strain Darlington showed a higher uptake of triazole as compared to sensitive strain A. Recently, Hitchcock et al. observed that the mycelial form of *C. albicans* containing higher phospholipid/nonesterified sterols also had an enhanced ability to take up triazole as compared to the yeast form, which had a lower ratio.[29]

The total lipid content of *C. albicans* 6.4 (resistant to both polyenes and azoles) was approximately three times higher than azole-sensitive and -resistant strains grown under similar conditions.[30] The higher lipid content of *C. albicans* 6.4 was attributed to an increase in both neutral and polar lipids; however, it was primarily due to an increase in neutral lipids. Thus, the ratio of polar:neutral lipids in strain 6.4 was low (1.49) as compared to other azole-sensitive/resistant strains (3.31 to 3.97). In addition, *C. albicans* 6.4 has relatively higher proportions of PE and PC, while no sphingomyelin was detected. Similar to other strains, PC was the most unsaturated and PI was the most saturated phospholipid. The neutral lipid composition of *C. albicans* 6.4 was distinct because it had large proportions of triacylglycerol. The most remarkable feature of *C. albicans* 6.4 was the absence of ergosterol, which is the only nonesterified sterol in other strains.[28] Instead, ergosterol is replaced by methylated sterols, mainly lanosterols, 24-methylene-24,25-dihydro lanosterol and 4-methyl-ergosta-diene-3-ol. Thus, *C. albicans* 6.4 resembles a sensitive strain treated with azoles, where such methylsterols are known to accumulate.[31,32] *C. albicans* 6.4 does not synthesize ergosterol and does not contain 14-demethylase activity, indicating that it lacks a key target site for azole interaction.[30]

C. ACULEACIN A-RESISTANT MUTANT

Aculeacin A is an antifungal that inhibits membrane-bound β-1,3-glucan synthase.[33] The *C. albicans* mutant resistant to aculeacin A develops cross resistance to other glucan synthase inhibitors.[34] Lipid analyses revealed that the aculeacin A-resistant mutant, designated as Acl_{R-1}, contained about two fold higher total lipid content as compared to the parental strain.[35] The free fatty acid contents were higher, while the mono-, di- and triacylglycerol contents were lower in the mutant. The resistant strain has higher amounts of PC and low levels of PS and PE. Lysophosphatidylserine, which was undetectable in the parental strain, was present in the mutant. Among the fatty acids analyzed, saturated fatty acids (16:0 and 18:0) and monounsaturated fatty acids (18:1) were significantly higher as compared to polyunsaturated fatty acids (18:2 and 18:3). Probably, the saturated nature of lipids plays a major role in aculeacin A resistance of *C. albicans*.[35]

III. METABOLISM OF LIPIDS

A. PHOSPHOLIPID BIOSYNTHESIS

Phospholipid metabolism and its regulation have been studied in great detail in *S. cerevisiae*.[36-39] However, such studies in *C. albicans* have been rather limited.[16,40,41] The biosynthetic pathways of major phospholipids in yeasts are summarized in Figure 1. The enzymes of biosynthesis have been localized in various subcellular fractions of *S. cerevisiae* and *C. albicans*.[42] CDP-diacylglycerol, which is an important branch point intermediate in the biosynthesis of PC, PE, PI and PS,[37,43] plays an important regulatory role in PI and PC synthesis.[38,44] A number of enzymes, e.g., CDP-DG synthase, phospholipid methyltransferase, PS synthase and PS decarboxylase, are repressed when choline and *myo*-inositol are exogenously added to the growth media of *S. cerevisiae*.[45-48]

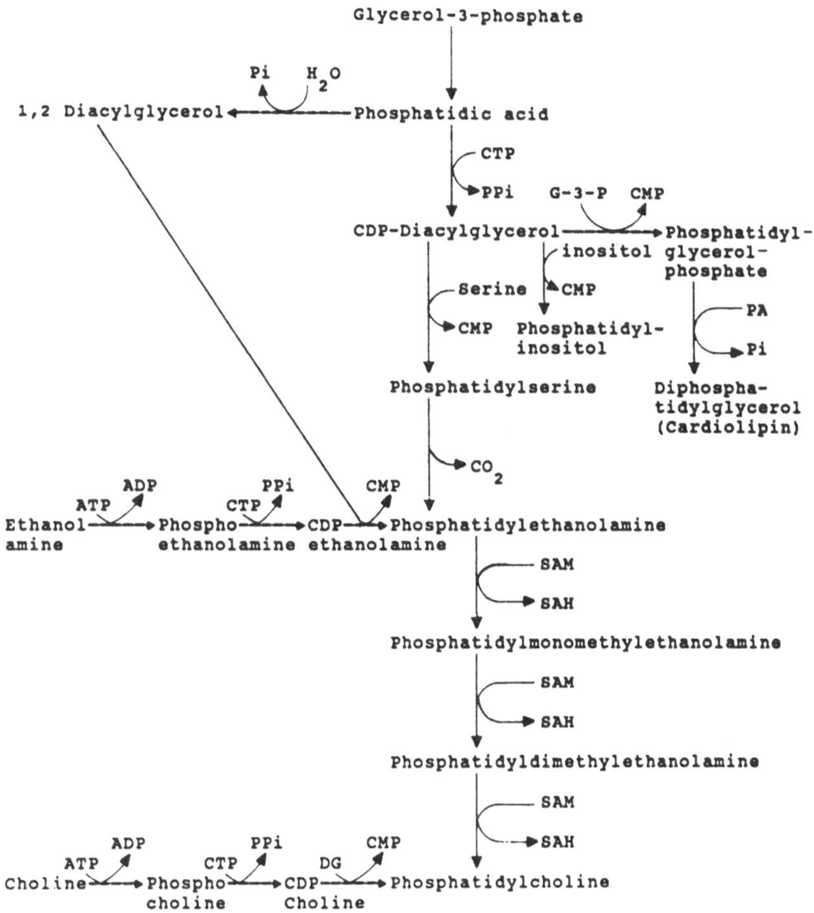

Figure 1. Biosynthetic Pathways of major phospholipids in yeasts,
SAM = S-adenosylmethionine; SAH = S-adenosylhomocysteine; DG = diacylglycerol; CMP = cytosinemonophosphate; CTP = cytosine triphosphate; PA = phosphatidic acid; PPi = pyrophosphate; G-3-P = glycerol-3-phosphate.

Recently, the existence of a CDP pathway was confirmed by pulse labeling of polar head groups. It became apparent that the phospholipid biosynthetic pathway of *C. albicans* was similar to that of *S. cerevisiae*. Mago and Khuller[41] also observed that the CDP pathway was probably the main pathway for the synthesis of PC and PE, but not for the synthesis of PS in *C. albicans*. The difference in the CDP pathway between *C. albicans* and *S. cerevisiae* observed earlier could be attributed to strain variations.[49] That the strain variations were considerable was evident from the work of Klig et al.,[16] wherein incorporation of [^{14}C]-ethanolamine and [^{14}C]-choline was observed to be different not only between *S. cerevisiae* and *C. albicans* but also between *C. albicans* ATCC

10231 and *C. albicans* ATCC 10261.[16] The presence of an active methylation pathway for the synthesis of PC in *C. albicans* was further substantiated by using a potent inhibitor of the methylation pathway, 2-hydroxyethylhydrazine (HEH).[50] Thus, similar to *S. cerevisiae*, *C. albicans* can synthesize PC by a *de novo* pathway and by successive methylation of PE. PE, on the other hand, can be synthesized either from CDP-ethanolamine or by decarboxylation of PS and base exchange reaction (Figure 1). PS is synthesized by PS-synthase or by base exchange reaction.

A coordinated regulation of phospholipid biosynthesis has been found in *S. cerevisiae*.[38,44,46,47] Phospholipid methyltransferases, PS-synthase and PS-decarboxylase are repressed by the addition of choline to the medium containing inositol.[45-48] The addition of choline alone does not repress the above mentioned enzymes of *S. cerevisiae*.[38,47,51] The synthesis of PC, a major neutral, dipolar, ionic phospholipid, and that of PI, a major anionic phospholipid, are well coordinated. For example, inositol with choline represses PC synthesis[38,47,48] as well as inositol-1-phosphate synthase.[52,53] Thus, this regulation balances the overall charge of the membrane. The net charge(s) of phospholipid is largely found to be balanced in mutants defective in phospholipid biosynthesis. *opi*3, a mutant of *S. cerevisiae*, has very little PC but is capable of maintaining its net charge by increasing the contents of phosphatidylmonomethyl ethanolamine (PMME) and phosphatidyldimethyl ethanolamine (PDME).[54] Similarly, *cho*1 mutant, which contains no PS, has a higher PI content.[55,56] Furthermore, during inositol starvation in *S. cerevisiae*, although phospholipid composition changes dramatically, its net charge remains unaffected.[57] Similar charge compensation has also been observed in inositol and choline auxotrophs of *Neurospora crassa*.[58] Klig et al.[16] confirmed that the phospholipid biosynthetic pathways of *C. albicans* are similar to those of *S. cerevisiae*; however, regulation of the pathway in the presence of exogenously supplemented precursors, inositol and choline of *C. albicans* markedly differs from that of *S. cerevisiae*. *C. albicans* has evolved a regulatory mechanism in response to choline, while *S. cerevisiae* has evolved a mechanism for maximum utilization of inositol whenever available. Higher levels of inositol synthesis in *C. albicans*, as compared to *S. cerevisiae*, may constitute a regulatory difference between the two yeasts. In view of the difference in the base utilization for phospholipid biosynthesis in *C. albicans* and *S. cerevisiae*, a detailed study of methyltransferases in *C. albicans* is required.

B. STEROL BIOSYNTHESIS

As pointed out earlier, ergosterol is a specific fungal sterol that provides the basis of action of many antifungals. Accordingly, intensive studies have been carried out to understand the biosynthesis of sterols in *C. albicans*. Most of the studies are based on the isolation of antifungal resistant mutants.[18,21,22,25,28,30] These mutants accumulate various intermediates of sterol biosynthesis, and thus, reveal information about their biosynthetic pathway (Table 3). Another

Table 3
Mutants of *C. albicans* Defective in Sterol Metabolism

Parent Strain	Mutant Strain designated/ genotype	Method of Isolation	Phenotype and lipid content	Ref.
Candida albicans	Res A4	Mutagenesis	Amphotericin B resistant; decreased total sterol content; lower content of ergosterol	123
	Res P2	Mutagenesis	Pimaricin resistant; lacks ergosterol	
	Res P4 Res P5			
799-S	799-XL 799-XY 799-YL 799-YS	Chemical mutagenesis	Nystatin resistant; no auxotrophic requirement; increased ergosterol content	21
Isolate	C7 E4 C4 D10	Chemical mutagenesis	Nystatin resistant; cross-resistant to other polyenes; increased total sterol content	25
ATCC 752	*Nys*-30 *Nys*-1000 *Luc*-12 *Luc*-18 *Luc*-24	Growth selection	Nystatin resistant; absence of ergosterol; total sterol unchanged; Lucensomycin resistant; decrease in ergosterol with increase in resistance	23
NCPF 3302	AD		Azole resistant	28
NCPF 3303	KB		Increased total sterol	
	Darlington		Azole resistant; total sterol unchanged	
6406	6.4	Spontaneous	Azole and polyene resistant; increased total sterol	30
33 erg^{+}	*erg* 2 *erg* 12 *erg* 16 *erg* 20 *erg* 37 *erg* 40 *erg* 41	Chemical mutagenesis	Nystatin resistant; no auxotrophic requirement of ergosterol	27
KD 14	KD 4700	UV-radiation	Amphotericin B resistant; cross resistant to other polyenes; hyphae minus; unchanged total sterol content; decreased ergosterol and increased 14-methylsterol content	69

approach has been to use inhibitors of sterol biosynthesis, which also leads to the accumulation of sterol intermediates.[59]

It seems that the biosynthetic pathways of sterols share a commonality, except the biosynthesis of episterol in both *S. cerevisiae* and *C. albicans*.[23,60] Episterol is converted to ergosterol through different pathways in *S. cerevisiae* and *C. albicans*. In *C. albicans*, the intermediates formed between episterol and ergosterol are ergosta-7-en-β-ol and ergosta-5,7-diene-3-β-ol, whereas in *S. cerevisiae* ergosta-7,22,24(28)-triene-3-α-ol and ergosta-5,7,22,24(28)-tetraene-3-β-ol are formed.[23]

Although the pathway of sterol biosynthesis was proposed two decades ago in yeasts, the enzymes involved have not been fully characterized. Some of the enzymes of *C. albicans* that are potential targets of antifungal drugs have, however, been studied in some detail.

2,3-Oxidosqualene cyclase is one such target enzyme. The genes coding for their enzyme were cloned by complementation of 2,3-oxidosqualene cyclase (*erg*7) mutation in *S. cerevisiae* strain SGY 163.[61] The level of 2,3-oxidosqualene cyclase produced from the expression of a single copy of the *Candida ERG*7 sequence was sufficient to allow growth of *S. cerevisiae erg*7 mutants in the absence of exogenous ergosterol. Based on *in vivo* incorporation studies, it was suggested that the gene product of *C. albicans ERG*7 resulted in the same phenotype in *S. cerevisiae* as the squalene cyclase inhibitor U18666A[3-β-(2-diethylamino ethoxy)-androst-5-en-17-one]. When *erg*7 mutants were transformed with *Candida ERG*7 DNA, unlike wild-type *S. cerevisiae*, they accumulated lanosterol. The accumulation of lanosterol was also observed in wild-type *S. cerevisiae* if U18666A was present. Probably, 2,3-oxidosqualene exerts a negative regulatory effect on lanosterol demethylase. *ERG*7 gene of *Candida* could not restore the wild-type level of 2,3-oxidosqualene cyclase activity because of the poor utilization of the *C. albicans* promoter by *S. cerevisiae*. In addition, U18666A was shown to inhibit 2,3-squalene cyclase of *C. albicans* both *in vivo* and *in vitro* to a greater extent than the corresponding *S. cerevisiae* enzyme.[61] This difference was attributed to the structural difference in these two enzymes. It is also likely that the regulation of ergosterol biosynthesis differs in the two yeasts.[62]

C-14 demethylase is another enzyme that has been widely studied in *C. albicans* because it is the main target of azoles.[59] These antifungals inhibit cytochrome P450-dependent 14α-demethylase, a key enzyme in ergosterol and cholesterol biosynthesis.[59-63] The accumulation of methylated sterols as a result of azole action is known to disrupt membrane structure and function. In order to understand the molecular basis of interaction between azoles and C-14 demethylase, it was recently purified from *C. albicans*.[64] The molecular weight of the purified enzyme is estimated to be 51 kDa. Its reconstitution in a model membrane system of dilauroylphosphatidylcholine with NADH and O_2 catalyzed the complete 14α-demethylation of lanosterol, which was inhibited by carbon monoxide. The enzyme demonstrates a high degree of specificity towards oxidation of lanosterol.[64] A wide array of compounds, including azoles, pyridine, pyrimidine and piperazine derivatives, are known inhibitors of C-14 demethylase.[65]

A cytochrome P450-deficient mutant of *C. albicans* strain, D10, has been characterized.[66] This mutant accumulates 14α-methylsterols and thus mimics the sterol profile of azole-treated, wild-type strain. Further studies revealed that substitution of 14α-methyl sterol for ergosterol resulted in defective hyphal formation and the cells became more susceptible to azole antifungals,[67] probably, because of increased membrane rigidity.[68] Another mutant of *C. albicans*, KD4700, was isolated by Shimokawa et al.,[69] which also accumulated 14-α-methyl sterol and had a defect in hyphal formation. This mutant had increased susceptibility to a number of compounds, suggesting an altered membrane permeability and, thereby, an increased susceptibility to membrane active agents.[70]

Using the cloned gene sequence of lanosterol 14α-demethylase (*ERG16*) from *S. cerevisiae* as a heterologous probe, a similar gene from *C. albicans* was isolated.[71] The cloned gene of *C. albicans* produces imidazole resistance comparable to that produced by the *S. cerevisiae* gene when cloned into *S. cerevisiae* on a 2 μm-based vector.[71] However, cytochrome P450 measurements revealed differences; while the *S. cerevisiae* gene increased cytochrome P450 levels by three to six folds, it was much less by the *Candida* sequence. Based on this and other observations, it is suggested that the *Candida ERG16* gene is inefficiently expressed in *S. cerevisiae*.

Another series of yeast mutants that require ergosterol for their growth has a defect in heme biosynthesis.[72,73] A number of such *hem* mutants are already known in *C. albicans*.[37,72,73] Recently, a *hem* mutant in *C. albicans* was also isolated by sequential gene disruption.[74] This mutant is defective in the enzyme involved in the synthesis of the tetrapyrrole ring (uroporphyrinogen I synthase, E.C. 4.3.1.8). The *C. albicans HEM3* gene was cloned by the complementation of the heme deficiency in the *hem*3 mutant of *S. cerevisiae*. Based on experimental evidence, the cloned *HEM3* gene of *C. albicans* was similar to the *S. cerevisiae HEM3* gene.[74] This conclusion was based on the fact that *S. cerevisiae hem*3 transformants carrying the *Candida* gene could express the enzyme activity. In addition, disruption of both copies of the *HEM* 3 gene produces a heme-dependent *Candida* mutant with no enzymatic activity. Lastly, gene prototrophy and enzyme activity could be restored in this mutant by transformation with vectors carrying the *HEM3* gene, which suggests that this requirement was due to gene disruption.[74]

C. FATTY ACID BIOSYNTHESIS

The first step in the biosynthesis of fatty acids in yeast involves generation of acetyl CoA, which is produced during metabolism, e.g., from acetate utilizing acetyl CoA synthase, from acetyl-carnitine using acetyl carnitine transferase, from citrate using ATP citrate lyase, or from acetyl phosphate using phosphotransacetylase. The supply of acetyl units from the mitochondria to cytosol of *C. albicans*, however, appears to be dependent only on the activity of carnitine acetyl transferase because the enzyme ATP citrate lyase, the major source of acetyl units in oleaginous yeast, is absent in *C. albicans.*[75] In general, the fatty acid metabolism of *C. albicans* is poorly understood. That cerulenin can inhibit fatty acid synthesis of *C. albicans*[76] suggests the presence of a *FAS* gene similar to *S. cerevisiae* that is probably involved in the synthesis of fatty acids. Isolation of mutants defective in desaturase enzyme and auxotrophic for unsaturated fatty acid[77] suggests the involvement of a desaturase enzyme.

Originally the A44 strain of *C. albicans* was isolated as an oleic acid-requiring mutant; however, later it was found to grow on other unsaturated fatty acids, e.g., palmitoleic, linoleic and linolenic acid. Saturated fatty acids were unable to support growth of this mutant and it was suggested that it was a Δ^9-desaturase defective mutant because supplemented 16:1 fatty acid is not converted to longer chain fatty acids. However, supplemented 18:1 or 18:2 got converted to 18:2 and 18:3, respectively, suggesting that the capability of this mutant to desaturate 18:1 and 18:2 was intact.[77] While fatty acid auxotrophs of *S. cerevisiae* have been extensively employed to study their role in membrane functions,[78-80] for such studies *C. albicans* mutants remain to be exploited.[77]

IV. LIPIDS IN MEMBRANE STRUCTURE

Although the role of lipids in membrane structure is well established in other microorganisms,[81,82] relatively little is known about this aspect in yeast.[83,84] There are even fewer reports on the structural role of lipids of *C. albicans*. One of the approaches to obtain information about the role of lipids in the structure of membrane involves the selective degradation of lipids followed by studying its effect on the structure. Using this approach, Prasad and co-workers[85] demonstrated that phospholipase A treatment of spheroplasts prepared from exponentially growing *C. albicans* resulted in the selective cleavage of PE (40%) and PS + PI (50%) while most of the PC remained inaccessible to the enzyme. The structural alterations, subsequent to phospholipid cleavage, were assessed by a fluorescent probe, 1-anilino-8-naphthalene sulfonate (ANS).[85] Compared to untreated spheroplasts, the relative fluorescence intensity of ANS was reduced in phospholipase A-treated spheroplasts. The decrease in relative fluorescence intensity was found to be associated with an increase in the apparent dissociation constant and reduced number of binding sites.[85]

Using a polyene antibiotic, filipin, as an *in situ* probe for membrane sterols, Takeo[86] inferred that the structural organization of the plasma membrane of stationary phase cells of *C. albicans* differs from that of exponential phase cells. The plasma membrane of exponential phase cells revealed numerous drug-induced deformations while most of the areas of the plasma membrane of stationary phase cells resisted such deformations.[86] Plasma membrane

Table 4
Lipid-Dependent Membrane Functions of *C. albicans*

Methods to alter lipid composition	Alteration in lipid composition	Affected function	Ref.
	1. Media Supplementation		
a. Alkanes of different chain length	Gross lipid composition	Amino acid transport,	124
		Polyene sensitivity,	125
		Radiosensitivity	126
b. Hydroxylamine hydrochloride	Accumulation of PS	Amino acid transport	17
c. Hydroquinone	Increased level of ergosterol	Amino acid transport,	125
		Polyene sensitivity	127
d. Ascorbic acid	Decreased level of ergosterol	Amino acid transport,	125
		Polyene sensitivity	127
e. Ergosterol	Level of ergosterol	Germ tube formation, Chitin synthesis	128
f. Cerulenin	Blocks fatty acid synthesis	Germ tube formation, Chitin synthesis	76
g. Aqueous garlic extract	Lowers PS and blocks fatty acid synthesis	Growth and viability	129
h. Polyene antibiotics	Level of sterols	Amino acid transport and K^+ release	125
i. Alcohol	PI levels	Blocks morphogenesis	130
j. Miconazole	Accumulation of ergosterol biosynthesis intermediates	Reduced uptake of amino acid and increased membrane order parameter	135
	2. Growth Status		
Growth up to prolonged stationary phase	Alteration in phospholipid composition	Cell cycle	131
	3. Genetic Manipulations		
a. *erg16*	Accumulates zymosterol, episterol, ergosta-7-en-β-ol, ergosta-7,22 diene-3-ol	Miconazole sensitivity	88
		Amino acid transport	134
b. *erg2*	Accumulates zymosterol and fecosterol	Membrane permeability; glycerol uptake	132
		Amino acid transport	

contd.

Table 4 (contd.)
Lipid-Dependent Membrane Functions of *C. albicans*

Methods to alter lipid composition	Alteration in lipid composition	Affected function	Ref.
c. *erg20* and *erg37*	Accumulates zymosterols, episterols, ergosta-7-en-3-β-ol	Growth and respiration; assimilation of nutrients	27, 132
d. *erg40*	Accumulates zymosterol and fecosterol		
e. *erg41*	Accumulates zymosterol, episterol, ergosta-7-en-3-β-ol, ergosta-7,22-diene-3-β-ol		
f. Cerulenin resistant mutants	No change in [^{3}H] incorporation into lipids	Adherence and virulence	10, 76, 133
g. Fatty acid auxotroph A'44	Altered fatty acid composition	Growth and polyene sensitivity	77

structures of *S. cerevisiae* and *C. albicans* were studied during growth by freeze-fracturing and it was observed that undifferentiated regions of the plasma membrane of *C. albicans* were severely deformed by filipin whereas the plasma membrane of small buds were mildly deformed, suggesting high and low levels of ergosterol, respectively. The bottom of evagination and the plasma membrane of the neck between the mother and daughter cells usually lacked such deformations.[87]

Using a similar approach, the effect of miconazole on the plasma membrane structure of *C. albicans* 33*erg*$^+$ and of an ergosterol-less mutant *erg16* was studied.[88] Scanning electron microscopy (SEM) showed that *33erg*$^+$ was wrinkled and displayed randomly distributed bud scars, while its mutant resulted in pronounced alterations in its cell surface, e.g., some of the cells had deep depressions in the cell wall and most of the cells were deformed and had thread-like structures among them. Such pronounced alterations of miconazole-treated cells of *erg16* were probably due to an altered and more rigid plasma membrane.[88]

When grown in the presence of ketoconazole and miconazole (10^{-6}M) for 6 h, *C. albicans* shows a shift from mono- to di-unsaturated fatty acids in the phospholipids.[33,89] This shift reflects the adaptation that yeast cells make changes and how the cells alter their membrane fluidity and thus compensate for azole-induced alterations.[59] On the contrary, the presence of 10^{-6}M clotrimazole, econazole, miconazole or ketoconazole for 16 h resulted in an increased content of palmitate,[90] which suggests that azole-induced depletion of ergosterol and the concomitant accumulation of 14α-methylsterols alter the membrane fluidity in such a way that enzyme desaturase is affected.[59] Other effects of various antifungal agents on membrane integrity of *C. albicans* plasma membrane have also been studied.[90]

Pesti and co-workers employed ergosterol-less mutants of *C. albicans* to study the structural role of ergosterol and its precursors.[18] Ergosterol-less mutants displayed no ultrastructural alterations as revealed by freeze-etch electron microscopy.[91] However, using an ESR probe 5-DS (5′-doxylstearic acid), the differences in membrane order parameters between wild-type and sterol mutant strains of *C. albicans* were evident.[18] The mutants exhibited a higher plasma membrane order parameter than their ergosterol-producing parental strain (33*erg*$^+$) suggesting that the parental strain had a more fluid membrane. These results were further confirmed by another recent study employing fluorescence polarization measurements. The decrease in membrane fluidity of *erg* mutants also led to reduced uptake of amino acids in *C. albicans* cells.[134,135]

Lees and co-workers[68] have studied the membrane fluidity of a cytochrome P450-deficient mutant of *C. albicans* (D10) by electron paramagnetic resonance. Using nitroxyl spin labels, the degree of fluidity (order parameter) was examined at three nitroxyl label positions, i.e., 5-DS, 7-DS and 10-DS, in both exponential and stationary phase *C. albicans* cells. Mutant strain D10 was found to be uniformly more rigid than wild-type cells. Furthermore, membrane rigidity increased with culture age.[68] However, this trend was not evident in the mutant when 7-DS and 10-DS were used. The microviscosity measurements based on the 16-DS label also showed that the mutant had a more rigid membrane both in exponential and stationary phase cells. However, using 16-DS, the increase in rigidity of the membrane with culture age was observed neither in the mutant nor in its parent. The work related to the structural role of membrane lipids of *Candida* is not yet sufficient to envisage the structural role of individual membrane lipid components of *C. albicans*.

The emerging role of *C. albicans* lipids in membrane functions is being realized.[2] Some of the lipid-dependent membrane functions in *C. albicans* are summarized in Table 4; a detailed discussion on the role of lipids in adherence and morphogenesis is presented in following sections.

V. LIPIDS IN ADHERENCE

A. ADHERENCE IN GENERAL

Adherence of microorganisms to host cell surfaces is now recognized as an important initial step in the pathogenesis of microbial infections.[92-95] Indications suggesting the involvement of adherence in this colonization process of *Candida* came from studies of Lilijemark and Gibbons.[96] Subsequently, King et al.[97] gave evidence that adherence may be the initial step in mucocutaneous candidosis. The realization that adherence may be a key determinant of the capacity of microorganisms, including *Candida*, to cause disease led to extensive research into this phenomenon. This research

Table 5
Effect of Lipid Extracts from *Candida* Cells on Their Adherence to BEC

Organism	Control	Mean Adherence (±SE) in[a]	
		Acetone Extract	Chloroform-Methanol Extract
		(% reduction; p)	(% reduction; p)
C. albicans			
ATCC 10231	523 ± 25	373 ± 25 (28.7; < 0.001)	353 ± 31 (32.5; < 0.001)
KCCC 13878	570 ± 25	330 ± 27 (42.1; < 0.001)	350 ± 27 (38.6; < 0.001)
KCCC 14172	565 ± 26	364 ± 25 (35.6; < 0.001)	381 ± 25 (32.6; < 0.001)
C. tropicalis			
KCCC 13605	421 ± 25	352 ± 22 (16.4; < 0.001)	340 ± 24 (19.2; < 0.005)
C. pseudotropicalis			
KCCC 13709	277 ± 23	284 ± 21 (NS)	276 ± 20 (NS)

[a]Data are expressed as mean numbers of yeast cells adhering to 100 BEC.
NS = Not significant.
From Ghannoum, M.A., Burns, G.R., Abu Elteen, K., and Radwan, S.S., *Infect. Immun.*, 54, 189, 1986. With permission.

addressed various related aspects including the mechanism(s) of *Candida* adherence, specific chemicals involved in this process, and the responses to environmental variables, to cite a few examples. Earlier investigations showed that the initial contact between the microorganism and host tissues is most probably a random and an instantaneous process that depends on general, rather than specific surface characteristics.[98] This type of adhesion is reversible and serves to stabilize the microbial cell at that site and prepares for the more permanent association between the microorganism and the epithelial cell. The permanent interaction, known as irreversible adhesion,[99] is a specific interaction mediated by the macromolecule on the microbial surface (adhesins) that combines with complementary structures on the epithelial cell surfaces (receptors).[100] The chemical nature of *Candida* adhesin(s) is unknown, with some reports proposing that *C. albicans* produces more than one adhesin.[101,102] Based on different experimental approaches, a number of compounds, e.g., mannan, mannoproteins, glucans, protein, chitin and lipids, have been suggested as possible *Candida* adhesins.[94,95] In the following section, we particularly discuss the role of lipids in adherence of *Candida* species.

B. LIPIDS AS ADHESINS

Involvement of lipids in the form of lipoteichoic acids (LTA) and glycosphingolipids in the adherence of bacteria to host cells has been reported.[103-105] The fatty acid moieties of LTA are believed to bind

Table 6
Effects of Constituent Lipid Classes of Total Lipid from *C. albicans* on Its Adherence to BECs

Lipid classs	Control	Mean Adherence (±SE) in[a]	
		Acetone Extract	Chloroform-Methanol Extract
		(% reduction; p)	(% reduction; p)
Steryl esters	515 ± 28	423 ± 27 (17.9; < 0.02)	435 ± 25 (15.5; < 0.05)
Triacylglycerols	515 ± 28	510 ± 28 (NS)	522 ± 30 (NS)
Fatty acids	515 ± 28	503 ± 25 (NS)	498 ± 22 (NS)
Sterols	515 ± 28	416 ± 28 (19.2; < 0.005)	413 ± 251 (9.8; < 0.01)
Diacylglycero-phospho-ethanolamines	515 ± 28		277 ± 24 (46.2; < 0.001)
Diacylglycero-phospho-glycerols	515 ± 28		315 ± 26 (38.8; < 0.001)
Diacylglycero-phospho-cholines	515 ± 28		309 ± 25 (40.0; < 0.001)
Diacylglycero-phospho-inositols	515 ± 28		228 ± 25 (55.7; < 0.001)
Diacylglycero-phospho-serines	515 ± 28		292 ± 24 (43.3; < 0.001)
Other non-phospholipid compounds	515 ± 28	298 ± 25 (42.1; < 0.001)	

[a]Data are expressed as mean numbers of yeast cells adhering to 100 BECs.
NS = Not significant.
From Ghannoum, M.A., Burns, G.R., Abu Elteen, K. and Radwan, S.S., *Infect. Immun.*, 54, 189, 1986. With permission.

spontaneously to the molecules of the animal cells.[105] It is known that acylated LTA are present mainly in association with the plasma membrane and such compounds become exposed to the surfaces of intact bacteria.[106] Hakomori[104] reported that glycosphingolipids are present in small amounts in the plasma membranes of animal cells, to which the fimbriae of Gram-negative bacteria adhere.

Ghannoum and co-workers[8] investigated the role of various lipid classes in the adherence of *Candida* cells to buccal epithelial cells (BECs). BECs lack cell walls and, therefore, their plasma membrane represents the outermost surface. In contrast, the plasma membranes of *Candida* cells are coated with the cell wall. Whether lipid moieties of the plasma membrane of BECs penetrate through the cell wall and become exposed to the cell surface, as in the case of LTA and bacteria,[103] is yet to be established.

Table 7
Effects of Total Lipid from Whole Cells, Cell Walls, and Cell Membranes of *C. albicans* and of Whole Epithelial Cells on the Yeast Adherence to BEC

Source of lipids (175 μg/ml of assay mixture)	Mean Adherence(±SE)[a] Without lipids	With lipids	(%reduction; p)
Whole yeast forms	495 ± 23	324 ± 18	(35; < 0.001)
Whole mycelial forms	495 ± 23	309 ± 21	(38; < 0.001)
Yeast cell walls	495 ± 23	214 ± 19	(57; < 0.001)
Yeast cell membrane	495 ± 23	298 ± 23	(40; < 0.001)
Epithelial cells	495 ± 23	232 ± 23	(53; < 0.001)

[a]Data are expressed as mean number of yeast cells adhering to 100 BECs.
From Ghannoum, M.A., Abu Elteen, K., and Radwan, S.S., *Mykosen*, 30, 371, 1987. With permission.

Table 8
Effects of Individual Glycolipids on the Adherence of Yeast Cells to BEC

Glycolipid (μg/ml of assay mixture)	Mean Adherence(±SE)[a] Without lipids	With lipids	(%reduction; p)
Ceramide monohexosides	529±29	276±24	(48; < 0.001)
Ceramide dihexosides	529±29	268±21	(49; < 0.001)
Steryl glycosides	529±29	243±20	(54; < 0.001)
Monogalactosyl diacylglycerols	529±29	579±25	NS
Digalactosyl diacylglycerols	529±29	510±28	NS
Sulfoquinovosyl diacylglycerols	529±29	524±24	NS

[a]Data are expressed as mean number of yeast cells adhering to 100 BEC.
NS = Not Significant.
From Ghannoum, M.A., Abu Elteen, K., and Radwan, S.S, *Mykosen*, 30, 371, 1987. With permission.

Lipids extracted from *C. albicans* and *C. tropicalis* blocked *in vitro* the adherence of various *Candida* species to BECs (Table 5). Individual lipid classes, making up the total lipid extracts isolated by preparative thin layer chromatography (TLC) and tested for their effects on the adherence of *C. albicans*, showed that phospholipids, sterols and steryl esters significantly blocked adherence. In contrast, triacylglycerols and fatty acids did not have any significant effect (Table 6). It seems that lipid classes found to be efficient in blocking adherence of *Candida* are also constituents of its plasma membrane.[9]

Irrespective of the morphological form of *Candida*, its total lipids dramatically block adherence (Table 7). Furthermore, not only sterols and phospholipids, but also glycolipids (ceramide monohexosides, ceramide dihexosides and steryl glycosides) can block the process of adherence (Table 8). The latter lipid classes were also suggested to have a role in adherence of *Escherichia coli* to epithelial cells.[107] It is relevant to recognize ceramide hexosides as surface agents with a capacity to mediate interactions of the cell with the environment.[104]

Involvement of lipids, particularly fatty acids, in the adhesion and virulence of *C. albicans* has also been suggested by using cerulenin-resistant mutants.[108] These mutants were spontaneous mutants and adhered less readily *in vitro* to human vaginal epithelial cells than the parent strain. They were also found to be less virulent in a mouse model of vaginal candidosis. [109] The mutant resembled the wild-type strain in growth rate, hyphae formation, and proteinase secretion (neither strain secreted proteinase). Biochemical characterization of these mutants suggested that the resistance in at least two mutants was due to the result of a mutation in one of the genes encoding fatty acid synthase, which differed in the two strains.[10] A detailed study of the similar mutants is essential to establish a direct role of lipids in adherence and virulence processes.

What has emerged so far is that the surface lipids not only of the pathogen but also of the host are involved in adherence.[9] By virtue of the lipid distribution on the outer envelope of both, the microorganism and the host, the role of lipids needs to be more closely scrutinized in the infection process.

VI. LIPIDS IN MORPHOGENESIS

A. DIMORPHISM

C. albicans is a dimorphic yeast that is capable of growing in culture as an ellipsoidal bud referred to as the blastospore or yeast form, as a pseudohypha of elongated yeast-like cells, or as true septate hyphae, referred to as the filamentous or mycelial form. Both the true hyphae and pseudohypha may regenerate the yeast phase by producing clusters of yeast like blastoconidia.[110] The growth phenotype depends on environmental conditions and the growth history of the cells.[111]

The ability of *C. albicans* to exist in the two forms *in vivo* has received much attention, particularly since the presence of one form or the other was correlated with pathogenicity.[112] Many reports claim that there is a relationship between mycelia formation and infection.[113,114] This is based on the assumption that hyphae penetrate tissue more readily than yeast cells and are more difficult to ingest and form emboli. It appears, therefore, that both forms may be involved in the pathogenicity of this yeast because both dimorphic forms of *C. albicans* are found in infected tissues and have been implicated in the pathogenicity of this fungus.[115] An understanding of the molecular basis of dimorphism is essential to the development of new strategies of therapy. Complete characterization of both the yeast and mycelial forms of *C. albicans* will allow the development of appropriate antifungal and preventive treatments.

In order to understand the dimorphic phenomenon, mainly the cell wall composition, the protein disulfide reductase enzyme assay and mechanical model for "explosive extrusion" between different forms have been

Table 9
Chemical Composition of Purified Plasma Membranes of *C. albicans*

Composition	Composition (%)	
	Yeast form	Mycelial form
Protein	52 ± 2.0	45 ± 3.0
Lipid	43 ± 3.0	31 ± 3.0
Carbohydrates	9 ± 0.5	25 ± 4.0
Nucleic acid	0.3 ± 0.1	0.5 ± 0.1
Phospholipids		
Phosphatidylethanolamine	70 ± 4.0	50 ± 4.0
Phosphatidylserine	11 ± 2.0	--
Phosphatidylcholine	4 ± 1.0	50 ± 4.0
Sphingolipids	15 ± 4.0	--
Neutral lipids		
Steryl esters	40 ± 4.0	28 ± 1.0
Triacylglycerols	24 ± 3.0	36 ± 2.0
Free fatty acids	17 ± 3.0	27 ± 4.0
Free sterols	19 ± 2.0	9 ± 1.0

From Marriott, M.S., *J. Gen. Microbiol.*, 86, 115, 1975. With permission.

analyzed.[116] Since the concept of the close association of membrane fluidity with certain functions is well accepted for a variety of membranes, it is reasonable to postulate that the plasma membrane may participate in cell wall formation and that significant differences in membrane composition/properties would be present in yeast and mycelial forms. A comparison of lipids of purified plasma membranes of *C. albicans* was made by Marriott.[5] He observed that the composition of the plasma membrane of *C. albicans*, which consists mainly of proteins and lipids, resembles that of other eukaryotes. Major differences in the composition of phospholipids and the free fatty acids of lipids were observed in membranes of yeast and mycelial forms. Moreover, PS and sphingolipids were present in yeast but were lacking in mycelial membranes (Table 9). Irrespective of the morphological forms, the predominant free sterol in the plasma membranes of *C. albicans* was ergosterol.

Yamada[116] showed that there is a large increase in PC and PE with a corresponding decrease in PI/PS within 3 to 5 h after the initiation of yeast to mycelial conversion of *C. albicans*. In addition, a great increase in linoleic acid in all phospholipids, in accordance with elongation of germ tubes, was observed.[116]

Sadamori[6] compared the lipid composition of whole yeast cells and mycelia rather than cellular fractions. He reported that the contents of fatty acids, phospholipids, sterols (except zymosterol) and squalene of the mycelial form of the fungus were significantly higher than those in the yeast form. Furthermore, in the latter form, the relative composition of the fatty acids, which are considered to be further metabolites of the nascent palmitic acid, e.g., $C_{18:0}$, $C_{18:1}$, $C_{18:2}$ and $C_{18:3}$, was higher than those in the mycelial form, whereas fatty acids such as $C_{16:0}$ and $C_{16:1}$ were higher in the mycelial form.

Based on these results, Sadamori[6] suggested that there exists an immaturity in lipid composition of the mycelial form as compared to the yeast form.[6]

In most of the studies where comparisons of lipids between yeast and the mycelial form of *C. albicans* were undertaken, the two morphological forms were grown in separate cultures at various temperatures and/or using media of different composition.[5,15,117,118] Such factors dramatically alter the lipid composition of various organisms.[119] To eliminate this possibility, Ghannoum et al.[4] analyzed the lipids from yeast and mycelial forms grown in the same cultures and showed that mycelial lipids are poorer in sterols than the yeast lipids, but are much richer in complex lipids that contain sterols such as sterylglycosides and esterified steryl glycosides. The two morphological forms, however, had similar phospholipid patterns.[4] As reported earlier, higher levels of polyunsaturated fatty acids were observed in mycelial lipids than in yeast lipids.[118]

The observed fluctuation of fatty acids, sterols and sterol-containing complex lipids in the yeast and mycelial forms do suggest a possible role of cellular lipids in the dimorphic behavior of *C. albicans*. The data are, however, not yet sufficient to relate specifically any lipid class(es) to morphogenesis.

B. WHITE-OPAQUE TRANSITION

Slutsky et al.[120] reported a new "phenotypic switching" mechanism in *C. albicans* in which smooth "white opaque" colonies arose as variants among the progeny of a clinical isolate. The white-opaque transition of *C. albicans*, which occurs frequently and reversibly, involves fundamental changes in both the colonial and cellular phenotypes. In the white phenotype, cells form colonies that are white and smooth. In the opaque phenotype, cells form colonies that are larger, flatter and opaque or grey in color. The differences observed in colony characteristics appear to be due to dramatic differences at the cellular level.[121] White cells are round to ellipsoidal and exhibit a budding pattern similar to most strains of *C. albicans*; in contrast, opaque cells are elongated or bean shaped and exhibit a different budding pattern. While white cells undergo bud-to-hypha transition under the standard regimen of pH-regulated dimorphism, opaque cells exhibit quite different requirements for the bud-to-hypha transition.[121] White cells exhibit the standard, continuous wall morphology like other *Candida* as well as *Saccharomyces* strains. In contrast, opaque cells exhibit unique pimples on the mature wall with complex morphology. Under SEM, a central pimple pore is observed and, in many cases, a bleb emerges from the central pore.[122] It appears that membrane-bound vesicles traverse the opaque pimple pore. It has been demonstrated that an opaque, specific 14.5 kDa antigen either resides in the pimple channel or is a component of the vesicle emerging from pimple pore.[122]

Because it was demonstrated that variations in lipid and sterol composition are probably associated with the yeast-mycelial transition of *C. albicans*,[3,4] Ghannoum et al.[11] undertook a study to examine whether variations in lipid and sterol patterns were also associated with the white opaque transition of

Table 10
Lipid Composition of White and Opaque Phenotypes of *C. albicans*

Lipids	Mid-exponential cultures		Stationary cultures	
	White	Opaque	White	Opaque
Total lipid (% dry wt.)	7.9 ± 0.05	5.9 ± 0.03	4.1 ± 0.01	3.9 ± 0.04
Sterols (% dry wt.)	1.2 ± 0.002	0.3 ± 0.001	1.0 ± 0.001	0.1 ± 0.001
Apolar lipids[a]				
Steryl esters	10.0 ± 0.7	14.8 ± 1.1	6.8 ± 0.5	15.3 ± 1.3
Alkyl esters	12.0 ± 0.9	10.3 ± 0.7	9.8 ± 0.6	2.1 ± 0.05
Triacylglycerols	10.5 ± 0.8	7.2 ± 0.3	Tr	3.8 ± 0.1
Fatty acids	10.8 ± 0.6	4.5 ± 0.2	6.4 ± 0.3	1.8 ± 0.02
Diacylglycerols	7.1 ± 0.2	1.1 ± 0.07	8.4 ± 0.5	6.7 ± 0.4
Sterols	12.2 ± 1.0	6.3 ± 0.4	10.5 ± 0.9	7.5 ± 0.6
Monoacylglycerols	9.0 ± 0.5	7.8 ± 0.5	6.5 ± 0.2	9.2 ± 0.7
Polar Lipids[a]				
Monogalactosyl-diacylglycerol	Tr	Tr	6.0 ± 0.09	4.8 ± 0.1
Steryl glycosides	1.4 ± 0.02	15.5 ± 1.3	6.0 ± 0.3	15.6 ± 1.1
Ceramide-monohexoside	1.6 ± 0.01	Tr	4.2 ± 0.1	6.4 ± 0.5
Phosphatidyl-ethanolamine	8.6 ± 0.8	12.1 ± 0.9	9.6 ± 0.6	7.4 ± 0.5
Phosphatidylglycerol	12.8 ± 1.4	Tr	Tr	5.6 ± 0.1
Phosphatidylcholine	4.0 ± 0.09	13.5 ± 1.1	16.5 ± 1.5	11.8 ± 1.0
Digalactosyl-diacylglycerol	ND	ND	2.3 ± 0.08	ND
Phosphatidylinositol	Tr	Tr	2.9 ± 0.06	Tr
Phosphatidylserine	Tr	Tr	1.4 ± 0.01	Tr
Phosphatidic acid	Tr	Tr	Tr	2.0 ± 0.06
Cardiolipin	Tr	6.9 ± 0.3	2.7 ± 0.02	Tr
Sterols				
Squalene	2.8 ± 0.02	17.5 ± 1.1	7.4 ± 0.6	2.8 ± 0.03
Calciferol	ND	Tr	ND	1.5 ± 0.01
Zymosterol	1.6 ± 0.01	ND	ND	Tr
Ergosterol	14.6 ± 1.00	49.3 ± 5.2	64.6 ± 6.1	76.1 ± 6.3
Stigmasterol	6.1 ± 0.09	Tr	9.8 ± 0.7	9.1 ± 0.8
Lanosterol	48.7 ± 4.90	33.2 ± 4.1	13.1 ± 1.3	10.5 ± 0.9
24-Methylene-dihydrolanosterol	26.2 ± 2.7	ND	5.1 ± 0.09	ND

[a]Values are expressed as percentage (w/w) of total lipids; each value is the mean ± SD of three determinations. Variation was <10%. Tr = Trace; ND = Not detected.

Compiled from Ghannoum, M.A., Swairjo, I. and Soll, D.R., *J. Med. Vet. Mycol.*, 28, 103, 1990. With permission.

C. albicans strain WO-1. It was observed that white cells were higher in lipid and sterol contents in both mid-exponential and stationary phase cultures (Table 10). In mid-exponential phase cultures, the lipids of white cells accumulated a substantial amount of apolar compounds, including steryl esters, alkyl esters, triacylglycerols, fatty acids, free sterols and mono- and diacylglycerols, while opaque cells accumulated nearly equal portions of apolar and polar lipids (Table 10). In stationary phase cultures, both white and opaque cells had slightly higher proportions of polar lipids. In addition, white cells had higher proportions of free sterols than the opaque cells.[11] Steryl glycosides and steryl esters, however, were higher in white cells (Table 10). A comparison of the sterols between white and opaque cells by UV, TLC and GLC showed that a qualitative as well as a quantitative difference exists between the two phenotypes (Table 10). Fatty acid analyses of white and opaque cells revealed that C_{16} and C_{18} were the most abundant fatty acids in both phenotypes. White and opaque cells differ in their fatty acid composition. The former had higher proportions of palmitoleic (16:1) and stearic (18:0) and lower proportions of linoleic (18:2) acid than the opaque cells.[11] Analysis of fatty acids of major lipid classes present in both forms showed that the fatty acid pattern varied dramatically.[11] The lipid composition, particularly of sterol and polyunsaturated fatty acids, of the opaque phenotype resembles that of mycelial cultures.[4,11]

VII. CONCLUSION

From the foregoing discussions, it is evident that lipids of *C. albicans* have acquired considerable importance in recent years because lipids provide information about the mechanism of adherence and morphogenesis of the organism and also offer target sites for antifungals. Unfortunately, limited studies on the lipid metabolism of *C. albicans* have been conducted so far. One of the main reasons for this has been the poorly understood genetics of *C. albicans*. The organism is a diploid in which classical methods of genetics are not applicable. However, with the advent of parasexual genetics and application of new molecular biological techniques, the genetic analyses of *C. albicans* have become amenable and as a result a few genes of lipid-metabolizing enzymes have been cloned and sequenced. Unlike *S. cerevisiae*, lipid mutants of *C. albicans* are also limited. Few sterol mutants have indeed been isolated that have shed some light regarding the sterol biosynthesis of *C. albicans*. What emerges from studies conducted so far is that the basic mechanisms of various lipid biosynthesis and degradation in *C. albicans* are not very different from *S. cerevisiae*. There are, however, some interesting differences that have been observed between the two yeasts with regard to ergosterol, fatty acids and phospholipid biosynthesis. At this point, it is too early to comment on the regulation of lipid metabolism of this pathogenic yeast. Further efforts are needed to resolve the biochemistry, genetics and metabolism of lipids so that their role in pathogenesis can be defined.

ACKNOWLEDGMENTS

The work from our laboratories included in this review was partially supported by the Indo-French Centre For Promotion of Advanced Research, Department of Science and Technology, Pfizer Pharmaceuticals, USA and NIH (grant AI 31696-01).

VIII. REFERENCES

1. **Vanden Bossche, H., Willemsens, G. and Marichal, P.,** Anti-*Candida* drugs — the biochemical basis for their activity, *Crit. Rev. Microbiol.*, 15, 57, 1987.
2. **Mishra, P. and Prasad, R.,** Lipids of *Candida albicans*, in *Candida albicans: Cellular and Molecular Biology,* Prasad, R., Ed., Springer-Verlag, Heidelberg, 128, 1991.
3. **Cannon, R.D. and Kerridge, D.,** Correlation between the sterol composition of membrane and morphology in *Candida albicans*, *J. Med. Vet. Mycol.*, 26, 57, 1988.
4. **Ghannoum, M.A., Janini, G., Khamis, L. and Radwan, S.S.,** Dimorphism-associated variations in the lipid composition of *Candida albicans*, *J. Gen. Microbiol.*, 132, 2367, 1986.
5. **Marriott, M.S.,** Isolation and chemical characterization of plasma membranes from the yeast and mycelial forms of *Candida albicans*, *J. Gen. Microbiol.*, 86, 115, 1975.
6. **Sadamori, S.,** Comparative study of lipid composition of *Candida albicans* in the yeast and mycelial forms, *Hiroshima J. Med. Sci.*, 36, 53, 1987.
7. **Sundaram, S., Sullivan, P.A. and Shepherd, M.G.,** Changes in lipid composition during starvation and germ tube formation in *Candida albicans*, *Exp. Mycol.*, 5, 140, 1981.
8. **Ghannoum, M.A., Burns, G.R., Abu Elteen, K. and Radwan, S.S.,** Experimental evidence for the role of lipids in adherence of *Candida* spp. to human buccal epithelial cells, *Infect. Immun.*, 54, 189, 1986.
9. **Ghannoum, M.A., Abu Elteen, K. and Radwan, S.S.,** Blocking adherence of *Candida albicans* to buccal epithelial cells by yeast glycolipids, yeast wall lipids and lipids from epithelial cells, *Mykosen*, 30, 371, 1987.

10. **Cihlar, R.L.**, in *2nd Conf. on Candida and Candidosis: Biology, Pathogenesis and Management*, Abs. No. 58, Philadelphia, PA, 1990.
11. **Ghannoum, M.A., Swairjo, I. and Soll, D.R.**, Variation in lipid and sterol contents in *Candida albicans* white and opaque phenotypes, *J. Med. Vet. Mycol.*, 28, 103, 1990.
12. **Hitchcock, C.A., Barrett-Bee, K.J. and Russell, N.J.**, Sterols in *Candida albicans* mutants resistant to polyenes or azole antifungals, and of a double mutant *Candida albicans* 6.4, *Crit. Rev. Microbiol.*, 15, 111, 1987.
13. **Mishra, P. and Prasad, R.**, An overview of lipids of *Candida albicans*, *Prog. Lipid Res.*, 29, 65, 1991.
14. **Nishi, K., Ichikawa, H., Tomochika, K., Okabe, A., Kanechica, K. and Kanemasa, Y.**, Lipid composition of *Candida albicans* and effect of growth temperature on it, *Acta Med. Okayama*, 27, 73, 1973.
15. **Ballmann, G.E. and Chaffin, W.L.**, Lipid synthesis during reinitiation of growth from stationary phase cultures of *Candida albicans*, *Mycopathol.*, 67, 39, 1979.
16. **Klig, L.S., Friedli, L. and Schmidt, E.**, Phospholipid biosynthesis in *Candida albicans*: regulation by the precursors inositol and choline, *Biochim. Biophys. Acta*, 172, 4407, 1990.
17. **Trivedi, A., Singhal, G.S. and Prasad, R.**, Effect of phosphatidylserine enrichment on amino acid transport, *Biochim. Biophys. Acta*, 729, 85, 1983.
18. **Pesti, M., Horvath, L., Vigh, L. and Farakas, T.**, Lipid contents and ESR determination of plasma membrane order parameter in *Candida albicans* sterol mutants, *Acta Microbiol. Hungary*, 32, 305, 1985.
19. **Hammond, S.H.**, Biological activity of polyene antibiotics, in *Progress in Medical Chemistry*, Ellis, G.P. and West, G.B., Ed., vol. 14, Elsevier, Amsterdam, 105, 1977.
20. **Hamilton-Miller, J.M.T.**, Chemistry and biology of the polyene macrolide antibiotics, *Bacteriol. Rev.*, 37, 166, 1973.
21. **Hamilton-Miller, J.M.T.**, Sterols from polyene resistant mutants of *Candida albicans*, *J. Gen. Microbiol.*, 73, 201, 1973.
22. **Fryberg, M., Oehlschlager, A.C. and Unrau, A.M.**, Sterol biosynthesis in antibiotic resistant yeast strains, *Arch. Biochem. Biophys.*, 160, 83, 1974.
23. **Fryberg, M., Oehlschlager, A.C. and Unrau, A.M.**, Sterol biosynthesis in antibiotic sensitive and resistant *Candida*, *Arch. Biochem. Biophys.*, 173, 171, 1975.
24. **Lomb, H., Fryberg, M., Oehlschlager, A.C. and Unrau, M.**, Sterols and fatty acid composition of polyene macrolide resistant *Torulopsis glabrata*, *Can. J. Biochem.*, 53, 1309, 1975.
25. **Pierce, A.M., Jr., Pierce, H.D., Unrau, A.M. and Oehlschlager, A.C.**, Lipid composition and polyene antibiotic resistance of *Candida albicans* mutants, *Can. J. Biochem.*, 56, 135, 1978.

26. **Chen, C., Kalb, V.F., Turi, T.G. and Loper, J.C.,** Primary structure of the cytochrome P 450 lanosterol 14-α-demethylase gene from *Candida tropicalis*, *DNA*, 7, 617, 1988.
27. **Pesti, M., Paku, S. and Novak, E.K.,** Some characteristics of nystatin-resistant sterol mutants of *Candida albicans*, *Acta Microbiol. Hungary*, 29, 55, 1982.
28. **Hitchcock, C.A., Barrett-Bee, K.J. and Russell, N.J.,** The lipid composition of azole-sensitive and azole-resistant strains of *Candida albicans*, *J. Gen. Microbiol.*, 132, 2421, 1986.
29. **Hitchcock, C.A., Barrett-Bee, K.J. and Russell, N.J.,** The lipid composition and permeability to the triazole antibiotic [C] 153066 of serum grown mycelial cultures of *Candida albicans*, *J. Gen. Microbiol.*, 135, 1949, 1989.
30. **Hitchcock, C.A., Barrett-Bee, K.J. and Russell, N.J.,** The lipid composition and permeability to azole of an azole- and polyene-resistant mutant of *Candida albicans*, *J. Med. Vet. Mycol.*, 25, 29, 1987.
31. **Borgers, M.,** Antifungal azole derivatives, in *The Scientific Basis of Antimicrobial Chemotherapy*, GreenWood, D. and O'Grady, F., Eds., Cambridge University Press, Cambridge, 133, 1985.
32. **Vanden Bossche, H., Willemsens, G., Cools, W., Marichal, P. and Lauwers, W.,** Hypothesis on molecular basis of the antifungal activity of N-substituted imidazoles and triazoles, *Biochem. Soc. Trans. U.K.*, 11, 665, 1983.
33. **Mizoguchi, J., Saito, T., Mizuno, K. and Hayano, K.,** On mode of action of a new antifungal antibiotic aculeacin A. Inhibition of cell wall synthesis in *Saccharomyces cerevisiae*, *J. Antibiotics*, 30, 308, 1977.
34. **Mehta, R.J., Boyer, J.M. and Nash, C.H.,** Aculeacin A resistant mutants of *Candida albicans*: alterations in cellular lipids, *Microbiol. Lett.*, 27, 25, 1984.
35. **Mehta, R.J., Nash, C.H., Grappel, S.F. and Actor, P.,** Aculeacin A resistant mutant of *Candida albicans*, *J.Antibiotics*, 35, 707, 1982.
36. **Carman, G.M. and Henry, S.A.,** Phospholipid biosynthesis in yeast, *Ann. Rev. Biochem.*, 58, 636, 1989.
37. **Henry, S.A.,** Membrane lipids of yeast: biochemical and genetic studies, in *The Molecular Biology of the Yeast Saccharomyces: Metabolism and Gene Expression*, Strathern, J. N., Jones, E. W. and Broach, J. R., Eds., Cold Spring Harbor Laboratory, Cold Spring Harbor, New York, 101, 1982.
38. **Henry, S.A., Klig, L.S. and Loewy, B.S.,** The genetic regulation and coordination of biosynthetic pathways in yeast: amino acid and phospholipid synthesis, *Ann. Rev. Genet.*, 18, 207, 1984.
39. **Hill, J.E., Chung, C., McGraw, P., Summers, E. and Henry, S.A.,** Biosynthesis and role of phospholipids in yeast membranes, in *Biochemistry of Cell Walls and Membranes in Fungi*, Khun, P.J.,

Trinci, A.P.J., Jung, M.L., Goosen, M.W. and Copping, L.G., Eds., Springer-Verlag, Berlin, 245, 1990.
40. **Mago, N. and Khuller, G.K.,** Lipids of *Candida albicans* subcellular distribution and biosynthesis, *J. Gen. Microbiol.*, 136, 993, 1990.
41. **Mago, N. and Khuller, G.K.,** Biosynthesis of major phospholipids in *Candida albicans*, *Curr. Microbiol.*, 20, 369, 1990.
42. **Mago, N. and Khuller, G.K.,** Subcellular localisation of enzymes of phospholipid metabolism in *Candida albicans*, *J. Med. Vet. Mycol.*, 28, 355, 1990.
43. **Steiner, M.R. and Lester, R.L.,** *In vitro* studies of phospholipid biosynthesis in *Saccharomyces cerevisiae*, *Biochim. Biophys. Acta*, 260, 222, 1972.
44. **Letts, V.A. and Henry, S.A.,** Regulation of phospholipid synthesis in phosphatidylserine synthase deficient (*cho*1) mutants of *S. cerevisiae*, *J. Bacteriol.*, 163, 560, 1985.
45. **Carson, M., Atkinson, K.D. and Waecheter, C.J.,** Properties of particulate and solubilized phosphatidylserine synthase activity from *S. cerevisiae*: inhibitory effect of choline in growth medicine, *J. Biol. Chem.*, 257, 8115, 1982.
46. **Carson, M., Emala, M., Hogsten, P. and Waecheter, C.J.,** Coordinate regulation of phosphatidylserine synthase and decarboxylase activity and phospholipid methylation in yeast, *J. Biol. Chem.*, 259, 6267, 1984.
47. **Klig, L.S., Homann, M.J., Carman, G.M. and Henry, S.A.,** Coordinate regulation of phospholipid biosynthesis in *Saccharomyces cerevisiae*: pleiotropically constitutive *opi* 1 mutant, *J. Bacteriol.*, 162, 1135, 1985.
48. **Waecheter, C.J. and Lester, R.L.,** Differential regulation of the N-methyltransferases responsible for phosphatidylcholine synthesis in *Saccharomyces cerevisiae*, *Arch. Biochem. Biophys.*, 158, 401, 1973.
49. **Trivedi, A., Dudani, A.K. and Prasad, R.,** Why choline supplementation did not enhance phosphatidylcholine level in *Candida albicans*, *Biochem. Int.*, 6, 119, 1983.
50. **Nikawa, J. and Yamashita, S.,** 2-hydroxyethyl-hydrazine as a potent inhibitor of phospholipid methylation in yeast, *Biochim. Biophys. Acta*, 751, 201, 1983.
51. **Yamashita, S., Oshima, J., Nikawa, J. and Hosaka, K.,** Regulation of phosphatidylethanolamine methylation pathway in *Saccharomyces cerevisiae*. *Eur. J. Biochem.*, 128, 589, 1982.
52. **Culbertson, M.R., Donahue, T.F. and Henry, S.A.,** Control of inositol biosynthesis in *Saccharomyces cerevisiae*: properties of a repressible enzyme system in extracts of wild type *Ino*$^+$ cells, *J. Bacteriol.*, 126, 232, 1976.

53. **Donahue, T.F. and Henry, S.A.,** *Myo*-inositol-1-phosphate synthase: characteristic of enzyme and identification of its structural gene in yeast, *J. Biol. Chem.*, 256, 7077, 1981.
54. **Greenberg, H.L., Klig, L.S., Letts, V.A., Loewy, B.S. and Henry, S.A.,** Yeast mutants defective in phosphatidylcholine synthesis, *J. Bacteriol.*, 153, 791, 1983.
55. **Atkinson, K.D., Fogel, S. and Henry, S.A.,** Yeast mutant defective in phosphatidylserine synthesis, *J. Biol. Chem.*, 255, 6653, 1980.
56. **Atkinson, K.D., Jensen, B., Kilat, A.I., Storm, E.M., Henry, S.A. and Fogel, S.,** Yeast mutants auxotrophic for choline or ethanolamine, *J. Bacteriol.*, 141, 558, 1980.
57. **Becker, G.W. and Lester, R.L.,** Changes in phospholipids of *Saccharomyces cerevisiae* associated with inositol-less death, *J. Biol. Chem.*, 252, 8684, 1977.
58. **Hubbard, S.C. and Brody, S.,** Glycerophospholipid variation in choline and inositol auxotroph of *Neurospora crassa*, *J. Biol. Chem.*, 250, 7173, 1975.
59. **Vanden Bossche, H.,** Ergosterol biosynthesis inhibitors, in *Candida albicans, Cellular and Molecular Biology*, Prasad, R., Ed., Springer-Verlag, Heidelberg, 239, 1991.
60. **Ratledge, C. and Evans, C.T.,** Lipids and their metabolism, in *Yeasts*, Rose, A.H. and Harrison, J.S., Eds., Academic Press, London, vol. 3, 2nd ed., 367, 1989.
61. **Kelly, R., Miller, S.M., Lai, M.H. and Kirsch, D.R.,** Cloning and characterisation of 2,3-oxidosqualene cyclase coding gene of *Candida albicans*, *Gene*, 87, 177, 1990.
62. **Kurtz, M.B., Kelly, R. and Kirsch, D.R.,** Molecular genetics of *Candida albicans*, in *The Genetics of Candida*, Kirsch, D.R., Kelly, R. and Kurtz, M.B., Eds., CRC Press, Boca Raton, FL, 21, 1990.
63. **Vanden Bossche, H.,** Biochemical targets for antifungal azole derivatives: hypothesis on the mode of action, *Curr. Top. Med. Mycol.*, 1, 313, 1985.
64. **Hitchcock, C.A., Dickinson, K., Brown, S.B., Evans, E.G.V. and Adams, D.J.,** Purification and properties of cytochrome P450-dependent 14-α-sterol demethylase from *Candida albicans*, *Biochem. J.*, 263, 573, 1989.
65. **Vanden Bossche, H.,** Mode of action of pyridine, pyrimidine and azole antifungals, in *Sterol Biosynthesis Inhibitors*, Berg, D. and Plempel, M., Eds., Ellis Horwood, Chichester, England, 79, 1988.
66. **Bard, M., Lees, N.D., Barbuch, R.J. and Sanglard, D.,** Characterization of a cytochrome P450 deficient mutant of *Candida albicans*, *Biochem. Biophys. Res. Commun.*, 147, 794, 1987.

67. **Lees, N.D., Broughton, M.C., Ganglard, D. and Bard, M.,** Azole susceptibility and hyphal formation in cytochrome P450-deficient mutant of *Candida albicans*, *Antimicrob. Agents Chemother.*, 34, 831, 1990.
68. **Lees, N.D., Kleinhans, F.W., Broughton, M.C., Pennington, D.E., Ricker, V.A. and Bard, M.,** Membrane fluidity alterations in cytochrome P450 deficient mutant of *Candida albicans*, *Steroids*, 53, 567, 1989.
69. **Shimokawa, O., Kato, Y. and Nakayama, H.,** Accumulation of 14-methyl sterols and defective hyphal growth in *Candida albicans*, *J. Med. Vet. Mycol.*, 24, 327, 1986.
70. **Shimokawa, O., Kato, Y. and Nakayama, H.,** Increased drug sensitivity in *Candida albicans*, *J. Med. Vet. Mycol.*, 24, 481, 1986.
71. **Kirsch, D.R., Lai, M.H. and O'Sullivan, J.,** Isolation of gene for cytochrome P450 L1A1 (lanosterol 14α-demethylase), *Gene*, 68, 229, 1988.
72. **Bard, M., Woods, R. and Haslam, J.,** Porphyrin mutants of *Saccharomyces cerevisiae* correlated lesions in sterol and fatty acid biosynthesis, *Biochem. Biophys. Res. Commun.*, 56, 324, 1974.
73. **Golloub, E., Lin, K., Doyan, J., Adlersberg, M. and Sprinson, D.,** Yeast mutant deficient in heme biosynthesis and a heme mutant additionally blocked in cyclization of 2,3-oxidosqualene, *J. Biol. Chem.*, 252, 2854, 1977.
74. **Kurtz, M.B. and Marrinam, J.,** Isolation of *Hem3* mutants from *Candida albicans* by sequential gene disruption, *Mol. Gen. Genet.*, 217, 47, 1989.
75. **Sheridan, R.C., Ratledge, C. and Chalk, P.A.,** Pathway to acetyl CoA formation in *Candida albicans*, *FEMS Microbiol. Lett.*, 69, 165, 1990.
76. **Hoberg, K.A., Cihlar, R.L. and Calderone, R.A.,** Inhibitory effect of cerulenin and sodium butyrate on germination of *Candida albicans*, *Antimicrob. Agents Chemother.*, 24, 401, 1983.
77. **Koh, T.Y., Marriott, M.S., Taylor, J. and Gale, E.F.,** Growth characteristics and polyene sensitivity of a fatty acid, *J. Gen. Microbiol.*, 102, 105, 1977.
78. **Mishra, P. and Prasad, R.,** Alterations in fatty acyl composition can selectively affect amino acid transport in *Saccharomyces cerevisiae*, *Biochem. Int.*, 15, 499, 1987.
79. **Mishra, P. and Prasad, R.,** Relationship between fluidity and L-alanine transport in a fatty acid auxotroph of *Saccharomyces cerevisiae*, *Biochem. Int.*, 19, 1019, 1989.
80. **Mishra, P. and Prasad, R.,** Relationship between ethanol and fatty acyl composition of *Saccharomyces cerevisiae*, *Appl. Microbiol. Biotechnol.*, 30, 294, 1989.

81. **Cronan, J.E., Jr. and Vagelos, P.R.,** Metabolism and function of the membrane phospholipids of *Escherichia coli*, *Biochim. Biophys. Acta*, 265, 25, 1972.
82. **McElhaney, R.N.,** Membrane lipid fluidity, phase state and membrane function in prokaryotic microorganisms, in *Membrane Fluidity in Biology, Cellular Aspects*, Aloia, R.C. and Boggs, J.M., Eds., Academic Press, Orlando, FL, vol. 4, 147, 1985.
83. **Prasad, R.,** Lipids in the structure and function of yeast membrane, *Adv. Lipid Res.*, 21, 187, 1986.
84. **Prasad, R. and Rose, A.H.,** Involvement of lipids in solute transport in yeasts, *Yeast*, 3, 205, 1986.
85. **Trivedi, A., Singh, M., Khare, S., Singhal, G.S. and Prasad, R.,** Effect of phospholipase A on the structure and function of yeast, *Candida albicans*, *Ind. J. Biochem. Biophys.*, 19, 336, 1982.
86. **Takeo, K.,** Plasma membrane of *Saccharomyces cerevisiae* and *Candida albicans* as revealed by freeze-fracturing before and after treatment with filipin, *FEMS Microbiol. Lett.*, 26, 71, 1985.
87. **Takeo, K.,** Resistance of the stationary phase plasma membrane of *Candida albicans* to filipin-induced deformation, *FEMS Microbiol. Lett.*, 27, 73, 1985.
88. **Pesti, M., Becher, D. and Bartsch, G.,** The effect of miconazole on ergosterol-less mutant of *Candida albicans*, *Acta Microbiol. Hungary*, 30, 25, 1983.
89. **Vanden Bossche, H., Willemsens, G., Cools, W. and Lauwers, W.F.,** Effect of miconazole on the fatty acid pattern in *Candida albicans*, *Acta Int. Physiol. Biochim.*, 89:B, 131, 1981.
90. **Georgopapadakou, N.H., Dix, B.A., Smith, S.A., Freudenberger, J. and Funke, P.T.,** Effects of antifungal agents on lipid biosynthesis and membrane integrity in *Candida albicans*, *Antimicrob. Agents Chemother.*, 31, 46, 1987.
91. **Pesti, M., Novak, E.K., Ferenczy, L. and Svoda, A.,** Freeze-fracture electron microscopical investigation on *Candida albicans* cells sensitive and resistant to nystatin, *Sabouraudia*, 19, 17, 1981.
92. **Calderone, R.A. and Braun, P.C.,** Adherence and receptor relationships of *Candida albicans*, *Microbiol.*, 55, 1, 1991.
93. **Douglas, L.J.,** Adhesion to surfaces, in *The Yeasts*, Rose, A.H. and Harrison, J.S., Eds., vol. 2, 2nd ed., Academic Press, London, 239, 1987.
94. **Ghannoum, M.A. and Radwan, S.S,** *Candida Adherence to Epithelial Cells*, CRC Press, Boca Raton, FL, 1990.
95. **Kennedy, M.J.,** Adhesion and association mechanisms of *Candida albicans*, in *Current Topics in Medical Mycology*, McGinnis, M.R., Ed., vol. 2, Springer-Verlag, New York, 73, 1988.

96. **Lilijemark, W.F. and Gibbons, R.J.,** Suppression of *Candida albicans* by human oral streptococci in gnotobiotic mice, *Infect. Immun.*, 8, 846, 1973.
97. **King, R.D., Lee, J.C. and Morris, A.L.,** Adherence of *Candida albicans* and other *Candida* species to mucosal epithelial cells, *Infect. Immun.*, 27, 667, 1980.
98. **Jones, G.W.,** The attachment of bacteria to the surfaces of animal cells, in *Microbial Interactions, Receptors and Recognition*, Reissig, J.L., Ed., Series B, vol. 3, Chapman & Hill, London, 139, 1977.
99. **Marshall, K.C., Stout, R. and Mitchell, R.,** Mechanism of initial events in the sorption of marine bacteria to surfaces, *J. Gen. Microbiol.*, 68, 337, 1971.
100. **Ofek, I., Lis, H. and Sharon, N.,** Animal cell surface membranes, in *Bacterial Adhesion: Mechanism and Physiological Significance*, Savage, D.C. and Fletcher, M., Eds., Plenum Press, New York, 71, 1985.
101. **Kennedy, M.J. and Sandin, R.L.,** Influence of growth conditions on *Candida albicans* adhesion, hydrophobicity and cell wall ultrastructure, *J. Med. Vet. Mycol.*, 26, 79, 1988.
102. **Sandin, R.L.,** Studies on cell adhesion and concanavalin A induced agglutination of *Candida albicans* after mannan extraction, *J. Med. Vet. Mycol.*, 24, 145, 1987.
103. **Christensen, G.D., Simpson, A. and Beachey, E.H.,** Adhesion of bacteria to animal tissue-complex mechanism, in *Bacterial Adhesion: Mechanism and Physiological Significance*, Savage, D.C. and Fletcher, M., Eds., Plenum Press, New York, 279, 1985.
104. **Hakomori, S.,** Glycosphingolipids, *Sci. Amer.*, 254, 44, 1986.
105. **Ofek, I., Beachy, E.H., Jefferson, W. and Campbell, G.L.,** Cell membrane binding properties of group A streptococcal lipoteichoic acids, *J. Exp. Med.*, 141, 990, 1975.
106. **Wicken, A.J. and Knox, K.W.,** Immunological properties of teichoic acids, *Bacteriol. Rev.*, 37, 215, 1980.
107. **Davis, C.P., Avots-Avotins, A.E. and Fader, R.C.,** Evidence for a bladder cell glycolipid receptor for *Escherichia coli* and the effect of neuraminic acid and colominic acid on adherence, *Infect. Immun.*, 34, 994, 1981.
108. **Calderone, R.A., Cihlar, R.L., Lee, D., Hoberg, K. and Scheld, W.M.,** Yeast adhesion in the pathogenesis of *Candida albicans* encarditis: studies with adherence negative mutants, *J. Infect. Dis.*, 154, 710, 1985.
109. **Lehrer, N., Segal, E., Cihlar, R.L. and Calderone, R.,** Pathogenicity of vaginal candidiasis: studies with a mutant has reduced ability to adhere *in vitro*, *J. Med. Vet. Mycol.*, 24, 127, 1986.

110. **Szaniszlo, P.J., Lacobs, C.W. and Geis, P.A.,** Dimorphism: morphological and biochemical aspects, in *Fungi Pathogenic for Humans and Animals*, Part A, *Biology*, Howard, D.H., Ed., Marcel Dekker, New York, 323, 1988.
111. **Soll, D.R.,** *Candida albicans*, in *Fungal Dimorphism, with Emphasis on Fungi Pathogenic for Humans*, Szaniszlo, J., Ed., chap. 7, Plenum Press, New York, 167, 1985.
112. **MacKenzie, D.W.,** Morphogenesis of *Candida albicans in vivo*, *Sabouraudia*, 2, 225, 1964.
113. **Gresham, G.A. and Whittle, C.H.,** Studies on the invasive mycelial form of *Candida albicans*, *Sabouraudia*, 1, 30, 1961.
114. **Young, G.,** The process of invasion and the persistence of *Candida albicans* injected intraperitoneally into mice, *J. Infect. Dis.*, 102, 114, 1964.
115. **Evans, Z.A.,** Tissue responses to the blastospores and hypha of *Candida albicans* in the mouse, *J. Med. Vet. Mycol.*, 14, 307, 1980.
116. **Yamada, T.,** Biochemical studies on the mechanism for dimorphism of fungal cell membrane alterations during conversion of yeast to mycelium in *Candida albicans*, *Acta. Sch. Med. Univ. Gifu.*, 34, 15, 1986.
117. **Bianchi, D.E.,** The lipid content of cell walls obtained from juvenile yeast-like and filamentous cells of *Candida albicans*, *Antonie Van Leeuwenhoek J. Microbiol. Serol.*, 23, 159, 1982.
118. **Yano, K.T., Yamada, Y., Banno, T., Sekiye, T. and Nozawa, Y.,** Modification of lipid composition in a dimorphic fungus, *Candida albicans* during yeast cell to hypha transformation, *Jpn. J. Med. Mycol.*, 15, 159, 1982.
119. **Wassef, M.K.,** Fungal lipids, *Adv. Lipid Res.*, 15, 159, 1977.
120. **Slutsky, B.M., Staebell, M., Pfaller, M. and Soll, D.R.,** The " White-Opaque transition": a second switching system in *Candida albicans*, *J. Bacteriol.*, 169, 189, 1987.
121. **Anderson, J.M., Cundiff, L., Schnars, B., Gao, M., MacKenzie, I. and Soll, D.R.,** Hypha formation in the white-opaque transition of *Candida albicans*, *Infect. Immun.*, 57, 458, 1989.
122. **Anderson, J.M., Mihalik, R. and Soll, D.R.,** Ultra-structure and antigenicity of the unique cell wall pimple of the *Candida* opaque phenotype, *J. Bacteriol.*, 172, 224, 1990.
123. **Subden, R.E., Safe, L., Morris, D.C., Brown, R.G. and Safe, S.,** Eburicol, lichesterol, ergosterol and obtusifoliol from polyene antibiotic resistant mutants of *Candida albicans*, *Can. J. Microbiol.*, 23, 751, 1977.
124. **Singh, M., Jayakumar, A. and Prasad, R.,** The effect of altered lipid composition on the transport of various amino acids in *Candida albicans*, *Arch. Biochem. Biophys.*, 190, 680, 1978.

125. **Singh, M., Jayakumar, A. and Prasad, R.,** Lipid composition and polyene sensitivity in isolates of *Candida albicans*, *Microbiol.*, 24, 7, 1979.
126. **Khare, S., Jayakumar, A., Trivedi, A., Kesavan, P.C. and Prasad, R.,** Radiation effects on membrane. I. Cellular permeability and cell survival, *Radiat. Res.*, 90, 233, 1982.
127. **Singh, M., Jayakumar, A. and Prasad, R.,** The effect of altered ergosterol content on the transport of various amino acids in *Candida albicans*, *Biochim. Biophys. Acta*, 555, 42, 1979.
128. **Chiew, Y.Y., Sullivan, P.A. and Shepherd, M.G.,** The effects of ergosterol and alcohols on germ-tube formation and chitin synthesis in *Candida albicans*, *Can. J. Microbiol.*, 60, 15, 1982.
129. **Ghannoum, M.A.,** Studies on the anti-candidal mode of action of *Allium sativum* (garlic), *J. Gen. Microbiol.*, 134, 2917, 1988.
130. **Lingappa, B.T., Prasad, M., Lingappa, Y., Hunt, D.F. and Biemann, K.,** Phenylalcohol and tryptophol: Autoantibiotics produced by fungus *Candida albicans*, *Science*, 163, 192, 1969.
131. **Dudani, A.K. and Prasad, R.,** Differences in amino acid transport and phospholipid content during the cell cycle of *Candida albicans*, *Folia Microbiol.*, 30, 493, 1985.
132. **Pesti, M. and Novak, R.,** Decreased permeability of glycerol in an ergosterol-less mutant of *Candida albicans*, *Acta Microbiol. Hungary*, 31, 81, 1984.
133. **Hoberg, K.A., Cihlar, R.L. and Calderone, R.A.,** Characterization of cerulenin-resistant mutants of *Candida albicans*, *Infect. Immun.*, 51, 102, 1986.
134. **Ansari, S., Gupta, P., Mahanty, S. K. and Prasad, R.,** The uptake of amino acids by *erg* mutants of *Candida albicans*, *J. Med. Vet. Mycol.*, 31, 377,1993.
135. **Ansari, S. and Prasad, R.,** Effect of miconazole on the structure and function of plasma membrane of *Candida albicans*, *FEMS Microbiol. Lett.*, 114, 93, 1993.

Chapter 6

LIPIDS OF *ASPERGILLUS*

P. Chakrabarti, M. Kundu and J. Basu

CONTENTS

0-8493-4794-7/96/$0.00+$.50

I. INTRODUCTION

The genus *Aspergillus* belongs to a large group of filamentous fungi, the Ascomycetes. Members of this genus are relatively common in air with distribution varying from 0.1 to 22% of total aerospore samples. Some species are capable of causing disease either by tissue invasion or colonization or by means of reactions involving the immune system. Since the spores of these organisms are airborne, mostly the lung and respiratory tract are affected. The study of membrane lipids of *Aspergillus* has drawn some attention during the past few decades since drugs directed against these lipids are often used to cure aspergillosis, a disease caused by an *Aspergillus* species. Moreover, there are increasing reports on the possible involvement of fungal lipids in virulence, allergenicity and pathogenesis of fungi. *Aspergillus* is also used as a tool in several industrial processes by growing them on media which are poorly utilized by other organisms. Dramatic changes in the amount and organization of lipids occur in the membrane of fungal cells during their developmental cycle. This article is an attempt to review different aspects of the lipids of *Aspergillus*. This will include the structure and function relationship of membrane lipids, their role in the growth cycle, as well as their variation under different conditions of growth and their involvement in the pathogenesis of the species.

II. LIPID COMPOSITION

A. PHOSPHOLIPIDS

Lipid analysis is a useful tool for taxonomic classification of microorganisms. Major lipids of membranes of eukaryotic cells are glycerophospholipids, sphingolipids, glycolipids and sterols. The total content of lipid may vary between 30% and 50%.[1] The content of phospholipid, sterol and carbohydrate differs markedly in the plasma membrane and also in the membranes of the cellular organelles.

Detailed phospholipid compositions have been studied in *A. niger*,[2] *A. fischeri*,[3] *A. flavus*,[4] *A. ochraceus*[4] and *A. vesicolor*.[5] Phosphatidylcholine (PC) and phosphatidylethanolamine (PE) are the major phospholipids of *A. niger*.[2] Phosphatidylinositol (PI) (3 to 15%), phosphatidylserine (PS) (1 to 24%) and cardiolipin (CL) (2 to 6%) are also present in variable amounts. Both neutral lipids and phospholipids contain O-alkyl and O-alkenyl analogs. As in other eukaryotes, CL occurs specifically in mitochondria of *A. niger*.[6]

B. GLYCOLIPIDS

The most abundant glycosylacylglycerols in nature are the galactolipids, monogalactosyldiacylglycerols (MGDG) and digalactosyldiacylglycerols (DGDG). Generally glycolipids are the least studied fungal lipids, although they are widely distributed in fungi. In *A. niger*, the major glycolipids are MGDG and DGDG with traces of the corresponding mannose analogs.[2] Diacylglycerols are considered to be the precursors of galactolipids.[7]

The formation of MGDG involves the transfer of galactose from the uridine diphosphate (UDP-gal) derivative to a 1,2-diacylglycerol. DGDG is formed by galactosylation of MGDG in a similar manner. The two galactosylation reactions are catalyzed by different enzymes. The presence of a water-soluble phosphorylated glycosphingolipid has been reported in *A. niger.*[8] Glucosyltransferase activity has been detected in *A. niger*, which catalyzes the transfer of glucose from UDP-glucose to 2-hydroxyoctadec-trans-3-enoic acid.[9] Monoglucosyl-diacylglycerol and diglucosyl-diacylglycerol are formed in the microsomes of *A. niger* from diacylglycerol and UDP-glucose.[10]

C. FATTY ACIDS

Long-chain fatty acids, usually from C_{12} to C_{18}, are present in saturated and unsaturated forms with one to three double bonds. Like other Ascomycetes, mycelial lipids of *Aspergillus* generally contain palmitic, oleic and linoleic acids.[11] Ascomycetes (higher fungi) synthesize mainly α-linolenic acid (octadeca-9,12,15-trienoic acid) while lower fungi synthesize γ-linolenic acid. Minor amounts of fatty acids with chain length >18 carbon atoms are present in *A. nidulans*. The presence of lignocerate (24:0) has been reported in *A. nidulans.*[12] The fatty acid composition of the conidia and mycelia of *A. nidulans* was analyzed. Several quantitative and qualitative variations were observed. Most notable was 15-fold increase in linoleate observed during the first day of incubation and its subsequent total disappearance by the fourth day.

D. STEROLS

Sterols are constituents of biomembranes of eukaryotic cells. They maintain membrane stability by modulating membrane fluidity. Ergosterol has been reported as the major sterol of *Aspergillus* species.[13] However, in *A. oryzae*, brassicasterol is the main sterol.[14] Lanosterol is also found in *A. nidulans.*[15] Unlike other yeasts and fungi, *A. nidulans* contains sterol mostly in the free form and not as the steryl esters.[16] In addition to ergosterol, 14-dehydroergosterol, ergosterol peroxide and cervisterol are also present in *A. niger.*[17] Sterol esters formed from sterol and fatty acids are the storage forms of sterols in fungi.

E. ACYLGLYCEROLS

Glycerol may be fully acylated (triacylglycerols) or only partially acylated (mono- and diacylglycerols). Alkyl and alkenyl analogs are sometimes present. Triacylglycerols may comprise over 90% of the lipid in fungi, varying considerably according to the species, stage of development and growth conditions. A nystatin-resistant mutant of *A. niger* has been found to contain more triacylglycerols in mycelia (20% of the total lipid) and conidia (60% of the total lipid) than the nystatin-sensitive parent strain.[18,19] The fungal cells are known to contain a large number of lipid particles that accumulate during various growth phases and also when the organism is growing under stress conditions. These lipid particles contain about 95% lipids consisting of triacylglycerols (50%) and sterol ester (~45%) along with some phospholipids and fatty acids. Small amounts of diacylglycerols may also be present but monoacylglycerols are absent.[20]

III. LIPID METABOLISM

Although mutants defective in lipid biosynthesis have been successfully employed to elucidate the structure-function relationship in microorganisms,[21-32] such studies in *Aspergillus* are scanty.

The effect of temperature on the composition of fatty acids has been studied using temperature-sensitive mutants. In *A. nidulans*, total lipids of mycelia showed an increase in the ratio of unsaturated to saturated fatty acids with an increase in the growth temperature of wild type strain,[33] whereas the ratio of short-chain to long-chain fatty acids changed inversely with temperature. When the temperature-sensitive mutant was grown at a restrictive temperature, it physiologically adapted itself by decreasing the ratio of unsaturated to saturated fatty acids.

A nystatin-resistant mutant of *A. niger* has been shown to contain lower amounts of sterol compared to the wild type strain.[18,19] This is accompanied by qualitative changes in the phospholipid composition of the mutant. The observed increase in the PC content and simultaneous decrease in the PE content in the mutant as compared to the wild type is one of the adaptive responses of *A. niger* to the presence of nystatin. Even more significant than the change in phospholipid composition is the altered fatty acid composition of the mutant. The proportion of linoleic acid in phospholipid was elevated in the mutant from 38 to 56% and the proportion of oleic acid was lowered from 28 to 12%.[34] Overall alteration of membrane microviscosity was also observed in this mutant. The modulation effect of exogenous sterol on membrane phospholipid and fatty acid composition and on physical properties has also been shown in *Saccharomyces cerevisiae*.[33]

There are three different permeases responsible for the uptake of different amino acids in *Aspergillus* species: a general amino acid permease, an acidic amino acid permease [35,36] and a basic amino acid permease.[37] Mazumder et al.[18] have demonstrated that different permeases are affected to different extent due to the alteration in sterol and phospholipid composition in a nystatin-resistant mutant of *A. niger*, suggesting the regulation of membrane-bound enzymes by the sterol levels of the membrane, as has been reported in the case of other microorganisms.[38,39]

Some unsaturated fatty acid (UFA) requiring mutants designated as *ufa1*, *ufa2*, *ufa3*, and *ufa4* have been isolated[40,41] from wild type *A. niger* V35 by chemical mutagenesis. The mutants can grow only in the presence of a suitable fatty acid supplement having at least one Δ^9-*cis*-double bond. The fatty acids,viz., palmitoleic, oleic, vaccenic, linoleic and linolenic acid, support growth of the mutant. Fatty acid profiles of the mutants[41] grown in presence of these fatty acids suggest that the mutants may be defective in Δ^9-desaturase activity . This has been confirmed by Δ^9-desaturase assay with the microsome using palmitoyl-CoA as substrate. In *ufa4*, the sterol content is significantly lower and the accumulation of supplemented fatty acid is more than that in the other cases. The wild type contains linoleic, oleic and palmitic acid as the major fatty acids and linolenic acid as the minor one.[41]

A comparative study[42,43] of the mitochondrial lipid composition of *A. nidulans* and a respiratory-deficient mutant (*rd3*) of *A. niger* grown at different temperatures (25°, 30°, 35°, and 40°C) has been performed. The lipid spectrum, though it is qualitatively similar, differs quantitatively in both cases. At optimum growth temperatures (30°C) depletion in ergosterol (40%) and cardiolipin (52%) was observed. This probably indicates the formation of defective mitochondria in the mutant with impaired respiratory system.

Complete depletion of cardiolipin species containing the fatty acid (20:5 ω-3) in the respiratory deficient mutant suggests a possible role of this lipid in mitochondriogenesis, at least in *A. niger*. The effect of temperature is predominantly on the degree of unsaturation and sterol ester formation. The linoleic acid (18:2, ω-6) content decreases with a concomitant increase in oleic acid (18:1, ω-9) content as the growth temperature increases for both cell types. Some morphological changes and effects on the vegetative life cycle have been observed with variation in growth temperature in the wild type and also in the mutant strain.[43]

A respiratory-deficient acetate non-utilizing mutant, *acu*10,[44] with increased cyanide sensitivity has been isolated from the wild type strain V35 of *A. niger* by chemical mutagenesis.[43,44] Respiration is less in the presence of glucose in the whole cells and cellular extract of the mutant, *acu*10. Hexokinase activity remains unaltered in the mutant. Cyanide-sensitive respiration is 3-fold greater in whole cells of *acu*10 than that in V35, but almost similar in their cellular extracts. Activities of some respiratory enzymes, viz. succinate dehydrogenase and cytochrome oxidase, are lower in *acu*10 than that in V35 after 18 h of growth but almost the same at 41 h of growth. Mitochondrial lipid profile is almost unaltered in *acu*10. Salicylhydroxamic acid, a specific inhibitor of cyanide-insensitive respiration, inhibits the respiration of both strains. The extent of inhibition is lower in *acu*10 revealing that *A. niger* possesses two respiratory pathways: one sensitive to cyanide and the other to salicylhydroxamic acid. [^{14}C]-cyanide uptake is much higher in *acu*10. This may be due to changes in cyanide permeability due to altered mycelial lipid composition of the mutant. Cyanide permeability and cyanide-insensitive respiration are affected in the mutant, *acu*10 of *A. niger*.

Under pyridoxine deficiency, pyridoxine-requiring mutants[45] of *A. nidulans* have been reported to contain increased levels of total lipid, sterols, phospholipids and triacylglycerols. Although total fatty acid levels decrease, there is an increase in the level of saturated fatty acids, specially the lower carbon chain fatty acids and a decrease in unsaturated fatty acid levels. A possible involvement of pyridoxine in the elongation and desaturation of fatty acid chains in *A. nidulans* has been suggested.

In eukaryotic microorganisms, probably three desaturases, Δ^9-, Δ^{12}- and Δ^{15}-, are involved in the biosynthesis of unsaturated fatty acids. Since the mutant (*ufa2*) can desaturate the supplemented 18:1 Δ^9-*cis* fatty acid to 18:2 and 18:3 acids to the same extent as the wild type, it may be assumed that the biosynthesis of 18:2 and 18:3 fatty acids and, consequently, the enzymes Δ^{12}-desaturase and Δ^{15}-desaturase, which are responsible for the conversion of 18:1 to 18:2 and 18:2 to 18:3 acids, are not affected by mutation present in *ufa2*, while only Δ^9-desaturase activity is probably affected. From the above analyses it may be assumed that the biosynthesis of unsaturated fatty acids in *Aspergillus* follows the same pathway as that of yeast,[46] *Neurospora*[47] and *Penicillium*.[48]

Aflatoxins are secondary metabolites produced primarily by some strains of *A. flavus* and *A. parasiticus*. Because of their pronounced toxicity and extreme carcinogenicity in many animal species, aflatoxins continue to be the subject of intense investigation. Culture conditions have been found to control the biosynthesis of lipids and aflatoxin by *A. parasiticus*. These include cultures grown at different optimal growth incubation temperatures. Incubations at 35°C and 45°C favour lipid synthesis and suppress production of aflatoxin. These observations suggest that formation of both the products share some similar biosynthetic steps. Perhaps, key enzymes responsible for synthesis of either lipid or aflatoxin are regulated by strain and culture conditions, and these conditions determine the amount of the final products formed.[49]

IV. FUNCTION OF LIPIDS

Phospholipids are exclusively present in cellular membranes. They regulate many membrane-bound enzymatic activities. The spatial organization and composition of phospholipids in the membrane modulates membrane functions such as transport of solutes, amino acids, metabolic flux within intracellular compartments, etc. Although there are several reports on the study of the role of phospholipids on membrane functions in the case of yeast and several fungi like *Neurospora*, *Penicillium*, etc., such studies on *Aspergillus* are insufficient. A mutant of *A. niger*, resistant to a polyene antibiotic, shows overall changes in phospholipid composition as compared to the wild type. As a result of such alterations, the mutant shows significant changes in the transport of amino acids.[18,19]

Studies on the influence of external factors on the membrane lipid composition in fungi suggest that the ability to respond rapidly towards stress by altering the length as well as extent of unsaturation of fatty acids is important in maintaining the membrane functions. Fatty acid compositional differences between different stages of growth of *A. nidulans* cultures had been reported.[34,50] The changes in total conidial fatty acid composition results in alteration of sterol and phospholipid composition of *A. niger*, presumably to maintain membrane fluidity and membrane functions.[18,19]

Glycolipids play an important role in many cellular processes, such as cell-cell interaction and recognition and cell growth regulation.[51] They are also involved in the synthesis of various cell wall polysaccharides. Glycolipid content has been reported to increase when the diacylglycerol content decreases in *A. niger*[2] supporting the earlier hypothesis that diacylglycerols are the precursors of galactolipids.[7]

Sterols to phospholipid ratio directs the microviscosity of the plasma membrane. Polyene antibiotics have been shown to interact with membrane sterols of fungal cells.[52] This interaction results in the leakage of cellular constituents, initially small ions (like potassium) followed by macromolecules like proteins and nucleic acids[53] (for details see chapter 12 of this volume). Polyene antibiotics have extensively been used to study the biosynthesis[54] and importance of sterols in fungal membranes.[55] Various polyene antibiotics such as nystatin and amphotericin B have been used to isolate resistant strains of fungi containing low levels of sterols.[56,57] There are now reports to suggest that besides sterols, other lipids also interact with polyene antibiotics.[58] The regulation of sterol production is a complex process in eukaryotes. *A. nidulans* produces ergosterol as its major 4-desmethyl sterol and lanosterol as its initial 4,4-dimethyl sterol. The sterol composition of *A. nidulans* has been reported to be consistent throughout the entire 7-day growth period, with the exception that the 4,4-dimethyl sterols, which are early cyclized precursors of ergosterol, cease to accumulate after 2 days of growth. All other 4-desmethyl components accumulate in parallel with the growth of the culture, indicating that a homeostatic regulatory process is involved in sterol biosynthesis.[16] Many fungi, including *Aspergillus*, accumulate specific metabolites upon entering the stationary phase of the life cycle. Sterols may be secondary metabolites in some fungal systems.[59,60]

In comparison to cholesterol, ergosterol has a greater disordering effect due to the presence of a bulky methyl substitution at C-24. It can weaken the van der Waals interactions when introduced into PC vesicles. Weakening of such interactions may contribute to greater mobility of phospholipids and permits cells to grow at low temperature. However, like other fungi, *Aspergillus* species contain dienoic and trienoic fatty acids.[12] These fatty acids also help in the maintenance of membrane fluidity.

A. fumigatus produces a soluble extracellular inhibitor of the alternative complement pathway, called *Aspergillus* complement inhibitor or CI. Phospholipids from *A. fumigatus*, specially those which co-migrate with PS, and PI on thin-layer chromatography, have been found to possess significant complement inhibitory activity.[61] Fungizone (amphotericin B) action on *A. niger* is manifested by the decrease in total lipids, free sterols, sterol esters and triacylglycerols.[62]

De novo synthesis and turnover of membrane components are the two distinct properties that contribute to the dynamic state of biological membranes. Several factors may be responsible for phospholipid turnover. Phospholipases also influence phospholipid turnover; however, this aspect of lipid metabolism has not been investigated in *Aspergillus*.[63]

V. FACTORS THAT AFFECT LIPID COMPOSITION

Dramatic changes in the membranes of fungal cells occur during the developmental cycle. However, most of the studies have been done in batch or continuous cultures which cannot reflect the temporal events that occur during the division of the daughter cells. A synchronized population of cells offers a more valuable tool for studying the biochemical events of the cell cycle. Evans et al.[12] have reported both qualitative and quantitative differences in the fatty acid content of exponentially growing and stationary phase cultures of *A. nidulans*. Farag et al.[64] have shown that oleate is the principal unsaturated fatty acid, but this result varies from one strain to another. Variations of fatty acid composition have been reported in *A. parasiticus*.[65] In *A. nidulans*, linolenate is the principal fatty acid at the early stage of growth, whereas the content of linoleate is less during this period and increases only during conidiation. Chavant et al.[66] have reported the relationship of fatty acid composition with the growth of four fungi, including *A. ochraceus*. The lipids from *A. ochraceus*[67] contain palmitic, stearic, oleic, linoleic and linolenic acids. Changes in lipids of the conidia of *A. fumigatus*[68] during maturation and germination have been observed. *A. vesicolor*, cultivated in a synthetic medium for 22 days, yields maximum dry weight, neutral lipids and sterigmatocystin appear, respectively on the 4th, 7th and the 20th day.[69]

The culture conditions affect the biosynthesis of lipids in *Aspergillus*.[49] Lipid composition has been studied in *A. niger* grown on various carbon sources, namely glucose, xylose, avicel (microcrystalline cellulose) and bagasse (a lignocellulosic substrate).[70] The amount of lipid accumulated ranged from 13.6-16.6 %. Neutral lipids, phospholipids and glycolipids of the mycelia varied from 41.0-46.2 %, 34.9-38.4% and 18.7-22.6 % of total lipid, respectively. Unsaturated fatty acids comprise around 80% of the total fatty acids with linoleic and oleic acid as predominant components. Of the four nitrogen sources tested, the optimum temperature range for growth and lipid synthesis was 25°-30°C. The lipid composition of a toxigenic and non-toxigenic strain of *A. flavus* grown in stationary and shake cultures in glucose-salt (CAM) medium and in sucrose-yeast extract (YES) medium has been analyzed. The levels of total lipids and sterols are higher in YES medium shake cultures which increased with time. The toxigenic strain has higher total lipid content than the non-toxigenic strain. YES medium yielded two to three times more phospholipids than CAM medium. PC and PE are the major phospholipids formed. The total phospholipid content and the relative amounts of individual phospholipids are similar in both strains.[4] Fatty acid composition of the lipids of the genus *Aspergillus* cultured on media with different nitrogen sources showed variations.[71] Wool-colonizing keratinolytic fungi, *A. fumigatus* and *A. flavus*, are capable of utilizing wool- lipids and fatty acids as sole sources of carbon and energy. These grow well in inorganic medium supplemented with total wool-lipids of different animals. The predominant fatty acyl moiety in wool-lipids is linolenic acid (18:2). They can also utilize palmitic and linolenic acids and cholesterol.[72]

The organophosphorus fungicides, iprobenfos and edifenphos, inhibit the germination of spores of *A. nidulans*[73]. In *A. nidulans*, iprobenfos caused a 70% increase in PE concentration and a 90% inhibition of chitin synthetase activity, accompanied by abnormal hyphae formation. Fungizone treatment[74] also leads to a decrease in total lipids, free sterols, sterol esters and triacylglycerols.

Growth inhibition of various pathogenic fungi correlates to some extent with the degree of inhibition of ergosterol biosynthesis. The mechanism of cell death appears to be related to accumulation of very high levels of intracellular squalene.[75] In *A. fumigatus*, terbinafine at a level equal to its minimum inhibitory concentration (MIC) inhibits sterol biosynthesis by 95%. Some antimycotics such as miconazole, ketoconazole and tolnaftate have been reported to inhibit lipid synthesis and consequently affect biosynthesis of cellular macromolecules.[76] UV irradiation has no effect on lipid synthesis in *A. flavus*.[77]

A. carbonarius is able to grow in Harrold's media amended with concentrations of cadmium chloride upto 2.5% (w/v). A considerable amount of cadmium is absorbed by the organism. In the presence of cadmium, the cellular lipid content has been found to be extraordinarily high and lipase activity is inhibited. Conidiophore formation has been affected by cadmium.[78,79] An inverse relationship[79] between aflatoxin and lipid production by two strains of *A. flavus* has been observed when aluminium is included in a sucrose-asparagine-salt growth medium. The growth rate, the fatty acid unsaturation and the lipid microviscosity have been measured in *A. ochraceus* grown at different temperatures. A significant deviation from the commonly followed line, i.e., a correlation between cold temperature acclimation and an increase in the fatty acid unsaturation, has been observed.[66]

VI. CONCLUSION

The foregoing discussion reveals that information on *Aspergillus* lipids is inadequate. Though several mutants defective in fatty acid and sterol biosynthesis are available, very little is known about the metabolism of *Aspergillus* lipids. Lipid catabolizing enzymes have been suggested to assist the fungus in host tissue invasion. Therefore, more intensive studies are needed on the metabolism of *Aspergillus* lipids. Development of mutants with defects in synthesis and catabolism of lipids would be very useful in understanding the functions and probable role of *Aspergillus* lipids in pathogenesis.

ACKNOWLEDGMENTS

The work reported from our laboratory was supported in part by grants from the Department of Science and Technology and the Department of Atomic Energy, Government of India. The authors are grateful to Mr. P.T. Srinivasan for the preparation of this manuscript.

VII. REFERENCES

1. **Weete, J.D.,** *Lipid Biochemistry of Fungi and Other Organisms,* Plenum Press, New York, 1980.
2. **Chattopadhyay, P., Banerjee, S.K., Sen, K. and Chakrabarti, P.,** Lipid profiles of *Aspergillus niger* and its unsaturated fatty acid auxotroph, *ufa2*, *Can. J. Microbiol.*, 31, 352, 1985.
3. **El-Nockrasky, A.S., Hassan, N.M., Sadek, M. and Gad, A.M.,** Mold phosphatides. I. Thin layer and column chromatography investigations of *Aspergillus fischeri*, *Grasas, Aceites*, 16, 113, 1965.

4. **Gupta, S.R., Viswanathan, L. and Venkitosubramaniam, T.A.**, A comparative study of the lipids of a toxigenic and non-toxigenic strain of *A. flavus*, *Ind. J. Biochem.*, 7, 108, 1970.
5. **Chavant, L., Le Bars, J. and Sancholle, M.**, Lipid metabolism of *A. vesicolor*. Relation to the biogenesis of sterigmatocystin, *Mycopathologia*, 60, 151, 1977.
6. **Letoublon, R., Mayet, B., Frot-coutaz, J., Nicolan, C. and Got, R.**, Subcellular distribution of phospholipids and of polyprenol phosphate in *Aspergillus niger* Van Tieghem, *Biochim. Biophys. Acta*, 711, 509, 1982.
7. **Kates, M.**, Biosynthesis of lipids in microorganisms, *Ann. Rev. Microbiol.*, 20, 13, 1966.
8. **Brennan, P.J. and Roe, J.**, The occurrence of a phosphorylated glycosphingolipid in *Aspergillus niger*, *Biochem. J.*, 147, 179, 1975.
9. **Brennan, P.J. and Losel, D.M.**, Physiology of fungal lipids: selected topics, *Adv. Microb. Physiol.*, 17, 47, 1978.
10. **Sastry, P.S.**, Glycosylglycerides, *Adv. Lipid Res.*, 12, 251, 1974.
11. **Bhatia, I.S., Raheja, R.K. and Chahal, D.S.**, Fungal lipids. I. Effect of different N-sources on chemical composition, *J. Sci. Food Agric.*, 23, 1197, 1972.
12. **Evans, J.L., Moclock, M.A. and Gealt, M.A.**, The fatty acid composition of the conidia and mycelia of the fungus *Aspergillus nidulans*, *Can. J. Microbiol.*, 32, 179, 1986.
13. **Elliot, C.G.**, Sterols in fungi, their functions in growth and reproduction, *Adv. Microb. Physiol.*, 15, 121, 1977.
14. **Fujino, Y. and Ohnishi, M.**, Characterization and composition of sterols in the free and esterified sterol fractions of *Aspergillus niger*, *Lipids*, 14, 663, 1979.
15. **Shapiro, B.F. and Gealt, M.A.**, Ergosterol and lanosterol from *A. nidulans*, *J. Gen. Microbiol.*, 128, 1053, 1982.
16. **Evans, J.L. and Gealt, M.A.**, The sterols of growth and stationary phases of *Aspergillus nidulans* cultures, *J. Gen. Microbiol.*, 131, 279, 1985.
17. **Goodwin, T.W.**, Comparative biochemistry of sterols, in eukaryotic microorganisms, in *Lipid and Biomembranes of Eukaryotic Microorganisms*, Erwin, J.A., Ed., Academic Press, New York and London, 15, 1973.
18. **Mazumder, C., Kundu, M., Basu, J. and Chakrabarti, P.**, Lipid composition and amino acid transport in a nystatin-resistant mutant of *Aspergillus niger*, *Lipids*, 22, 609, 1987.
19. **Mazumder, C., Basu, J., Kundu, M. and Chakrabarti, P.**, Changes in membrane lipids and amino acid transport in a nystatin-resistant *Aspergillus niger*, *Can. J. Microbiol.*, 36, 435, 1990.
20. **Henry, H.A. and Halvorson, H.O.**, Lipid synthesis during sporulation of *Saccharomyces cerevisiae*, *J. Bacteriol.*, 114, 1158, 1973.

21. **Keith, A., Resnick, M. and Haley, A.,** Fatty acid desaturase mutants of *Saccharomyces cerevisiae*, *J. Bacteriol.*, 98, 415, 1969.
22. **Atkinson, K.D., Fogel, S. and Henrys, S.A.,** Yeast mutant defective in phosphatidylserine synthesis, *J. Biol. Chem.*, 255, 6653, 1980.
23. **Atkinson, K.D., Jensen, B., Starm, E., Kolat, A., Henry, S.A., and Fogel, S.,** Yeast mutants auxotrophic for choline or ethanolamine, *J. Bacteriol.*, 141, 558, 1980.
24. **Rodriguez, R.J., Taylor, F.R. and Parks, L.W.,** A requirement for ergosterol to permit growth of yeast sterol auxotrophs on cholesterol, *Biochem. Biophys. Res. Commun.*, 106, 435, 1982.
25. **Rodriguez, R.J. and Parks, L.W.,** Structural and physiological features of sterols necessary to satisfy bulk membrane and sparking requirements in yeast sterol auxotrophs, *Arch. Biochem. Biophys.*, 225, 861, 1983.
26. **Pinto, W.J., Lozano, R., Sekula, B.C. and Nes, W.R.,** Stereochemically distinct roles for sterol in *Saccharomyces cerevisiae*, *Biochem. Biophys. Res. Commun.*, 112, 47, 1983.
27. **Pinto, W.J. and Nes, W.R.,** Stereochemically distinct roles for sterol in *Saccharomyces cerevisiae*, *J. Biol. Chem.*, 258, 4472, 1983.
28. **Höfer, M., Huh, H. and Kunemund, A.,** Membrane potential and cation permeability: A study with a nystatin-resistant mutant of *Rhodotorula gracilis* (*Rhodosporidium toruloides*), *Biochim. Biophys. Acta*, 735, 211, 1983.
29. **Gollub, E., Lin, K., Doyan, J., Adlersherg, M. and Sprinson, D.,** Yeast mutant deficient in heme biosynthesis and heme mutant additionally blocked in cyclization of 2,3-oxidosqualene, *J.Biol. Chem.*, 252, 2846, 1977.
30. **Kovac, L., Gbelska, V.T., Poliachova, V., Subix, J. and Kovacova, V.,** Membrane mutants: A yeast mutant with a lesion in PS biosynthesis, *Eur. J. Biochem.*, 111, 491, 1980.
31. **Chopra, A. and Khuller, G.K.,** Lipid metabolism in fungi, *CRC Crit. Rev. Microbiol.*, 11, 209, 1984.
32. **Roshanara, B. and Shanmugasundaram, E.R.,** Lipid components variation among the temperature-sensitive mutants and wild strain of *A. nidulans*, *Ital. J. Biochem.*, 30, 127, 1981.
33. **Low, C., Rodriguez, R.J. and Parks, L.W.,** Modulation of yeast plasma membrane composition of a yeast sterol auxotroph as a function of exogenous sterol, *Arch. Biochem. Biophys.*, 240, 530, 1985.
34. **Singh, J. and Walker, T.,** Changes in the composition of the fat of *Aspergillus nidulans* with age of culture, *Biochem. J.*, 62, 286, 1956.
35. **Pateman, J.A., Kinghorn, J.R. and Duran, E.,** Regulatory aspects of L-glutamate transport in *A . nidulans*, *J. Bacteriol.*, 119, 534, 1974.
36. **Robinson, J.H., Anthony, C. and Drabble, W.T.,** The acidic amino acid permease of *Aspergillus nidulans*, *J. Gen. Microbiol.*, 79, 53, 1973.

37. **Piotrowska, M., Stephen, P.P., Bartnik, E. and Zakrewska, E.**, Basic and neutral amino acid transport in *Aspergillus nidulans*, *J. Gen. Microbiol.*, 92, 890, 1976.
38. **Höfer, M., Thiele, O.W., Huh, H., Hummemna, D.H. and Mracek, K.**, A nystatin-resistant mutant of *Rhodotorulla gracilis*. Transport properties and sterol content, *Arch. Microbiol.*, 132, 313, 1982.
39. **Chattopadhyay, P., Banerjee, S.K., Sen, K. and Chakrabarti, P.**, An unsaturated fatty acid mutant of *Aspergillus niger* with partially defective Δ^9-desaturase, *Can. J. Microbiol.*, 31, 246, 1985.
40. **Chattopadhyay, P., Banerjee, S.K., Sen, K. and Chakrabarti, P.**, Isolation and characterization of an unsaturated fatty acid requiring mutant (*ufa4*) of *Aspergillus niger*, *Ind. J. Exp. Biol.*, 24, 421, 1986.
41. **Chattopadhyay, P., Banerjee, S.K., Sen, K. and Chakrabarti, P.**, Lipid profiles of conidia of *Aspergillus niger* and a fatty acid auxotroph, *Can. J. Microbiol.*, 33, 1116, 1987.
42. **Mandal, S.B., Sen, P.C., Chakrabarti, P. and Sen, K.**, Effect of respiratory deficiency and temperature on the mitochondrial lipid metabolism of *Aspergillus niger*, *Can. J. Microbiol.*, 24, 586, 1978.
43. **Banerjee, S.K., Chattopadhyay, P., Chakrabarti, P. and Sen, K.**, A respiratory-deficient (*rd*) mutant of *Aspergillus niger* with increased cyanide sensitivity, *Ind. J. Exp. Biol.*, 23, 466, 1985.
44. **Chattopadhyay, P., Banerjee, S.K., Chakrabarti, P. and Sen, K.**, Biochemical studies on acetate non-utilizing (*acu*) mutants of *Aspergillus niger*, *Ind. J. Exp. Biol.*, 18, 764, 1980.
45. **Mohana, K. and Shanmugasundaram, E.R.**, Pyridoxine and its relation to lipids. Studies with pyridoxineless mutants of *Aspergillus nidulans*, *J. Nutr. Sci. Vitaminol. (Tokyo)*, 24, 397, 1978.
46. **Rattray, J.B.M., Schibeci, A. and Kidby, D.K.**, Lipids of yeast, *Bacteriol. Rev.*, 39, 197, 1975.
47. **Scott, W.A.**, Mutations resulting in an unsaturated fatty acid requirement in *Neurospora*. Evidence for Δ^9-desaturase defects, *Biochemistry*, 16, 5274, 1977.
48. **Bennet, A.S. and Quakenbush, P.W.**, Synthesis of unsaturated fatty acids by *Penicillium chrysogenum*, *Arch. Biochem. Biophys.*, 130, 567, 1969.
49. **Shih, C.H. and Marth, E.H.**, Some cultural conditions that control biosynthesis of lipid and aflatoxin by *Aspergillus parasiticus*, *Appl. Microbiol.*, 27, 452, 1974.
50. **Singh, J., Walker, T. and Meara, M.**, The component fatty acids of the fat of *Aspergillus nidulans*, *Biochem. J.*, 61, 85, 1955.
51. **Hakomori, S.I.**, Glycosphingolipids in cellular interaction, differentiation, and oncogenesis, *Ann. Rev. Biochem.*, 50, 733, 1987.
52. **Kobayashi, G.S. and Medoff, G.**, Antifungal agents: recent developments, *Ann. Rev. Microbiol.*, 31, 291, 1977.

53. **Hammond, S.M., Lambert, P.A. and Klinger, B.N.,** The mode of action of polyene antibiotics; induced potassium leakage in *Candida albicans*, *J. Gen. Microbiol.*, 81, 325, 1974.
54. **Pierce, A.M.H.D., Jr., Unrau, A.M. and Ochlschlager, A.C.,** Lipid composition and polyene antibiotic resistance of *Candida albicans* mutants, *Can. J. Biochem.*, 56, 135, 1978.
55. **Hamilton-Miller, M.T.,** Fungal sterols and mode of action of the polyene antibiotics, *Adv. Appl. Microbiol.*, 17, 109, 1974.
56. **Singh, M., Jayakumar, A. and Prasad, R.,** The effect of altered lipid composition on the transport of various amino acids in *Candida albicans*, *Arch. Biochem. Biophys.*, 191, 680, 1978.
57. **Woods, R.A.,** Nystatin-resistant mutants of yeast: Alterations in sterol content, *J. Bacteriol.*, 108, 69, 1971.
58. **Singh, M., Jayakumar, A. and Prasad, R.,** Lipid composition and polyene antibiotic sensitivity in isolates of *Candida albicans*, *Microbios*, 24, 7, 1979.
59. **Demain, A.L.,** Industrial microbiology, *Science*, 214, 987, 1981.
60. **Malik, V.S.,** Microbial secondary metabolism, *TIBS*, 5, 68, 1980.
61. **Washburn, R.G., Dehart, D.J., Agwn, D.E., Bryant-Varela, B.J. and Julian, N.C.,** *Aspergillus fumigatus* complement inhibitor: production, characterization and purification by hydrophobic interaction and thin-layer chromatography, *Infect. Immun.*, 58, 3508, 1990.
62. **Elwan, S.H., Radwan, S.S. and Assad, A.A.,** Relation of lipids to the action of streptomycin on *Bacillus subtilis* and *Escherichia coli* and griseoflavia and fungizone on *Aspergillus niger*, *Zentralbl Bakeriol Parasitenkd Infektionskr Hyg*, 132, 109, 1977.
63. **Zaikina, N.A. and Robakidze, T.N.,** Phospholipase C of fungi and *Staphylococci*, *Mikrobiologiya*, 45, 466, 1976.
64. **Farag, R., Youseff, A., Khalil, F. and Taha, R.,** The lipids of various fungi grown on an artificial medium, *J. Am. Oil Chem. Soc.*, 58, 765, 1987.
65. **Rambo, G. and Bean, G.,** Sterols and fatty acids of aflatoxin and non-aflatoxin producing isolates of *Aspergillus*, *Phytochemistry*, 13, 195, 1974.
66. **Chavant, L., Wolf, C., Fonvieille, J.L. and Dargent, R.,** Deviation from the usual relationships between the temperature, the growth rate, of the fatty acid composition and the lipid microviscosity of four different fungi, *Biochem. Biophys. Res. Commun.*, 101, 912, 1981.
67. **Sood, M.G. and Singh, J.,** The component fatty acids of *Aspergillus ochraceus* fat, *J. Sci. Food Agric.*, 24, 1171, 1973.
68. **Tsukahara, T.,** Changes in chemical composition of conidia of *Aspergillus fumigatus* during maturation and germination, *Microbiol. Immunol.*, 24, 747, 1980.

69. **Chavant, L.**, Lipid metabolism in *Aspergillus vesicolor*, relation to biogenesis of sterigmatocystin, *Mycopathologia*, 60, 151, 1977.
70. **Sing, A.**, Lipid accumulation by a cellulolytic strain of *Aspergillus niger*, *Experimentia*, 48, 234, 1992.
71. **Kolesnikova, T.G. and Tolstikova, G.V.**, Fatty acid composition of the lipids of fungi of the genus *Aspergillus* cultured on media with various nitrogen sources, *Mikrobiologiya*, 53, 826, 1985.
72. **Musallam, A.A. and Radwan, S.S.**, Wool-colonizing micro-organisms capable of utilizing wool-lipids and fatty acids as sole sources of carbon and energy, *J. Appl. Bacteriol.*, 69, 806, 1990.
73. **Craig, G.D. and Peberdy, J.F.**, The effect of S-benzyl O,O-di-isopropylphosphorothioate (Kitazin) and dicloran on the total lipid, sterol and phospholipids in *Aspergillus nidulans*, *FEMS Microbiol. Lett.*, 18, 11, 1983.
74. **Ryder, N.S., Seidl, G., Petranyi, G. and Suez, A.**, Mechanism of the fungicidal action of SF 86-327, a new allylamine antimycotic agent, in *Recent Advances in Chemotherapy*, Ishigami, J., Ed., University of Tokyo Press , Tokyo, 1985, 2558.
75. **Ryder, N.S., Frank, I. and Dupont M.C.**, Ergosterol biosynthesis inhibition by the thiocarbonate antifungal agents tolnaftate and tolciclate, *Antimicrob. Agents Chemother.*, 29, 858, 1986.
76. **Oh, K., Matsuoka, H., Teraoka, T., Sumita., O., Takatori, K. and Kurata, H.**, Effects of antimycotics on the biosynthesis of cellular macromolecules in *Aspergillus niger* protoplasts, *Mycopathalogia (Netherlands)*, 122, 135, 1993.
77. **Osman, M., Mohamed, Y.A., El-Sayed, M.A. and Abo-Zeid, A.**, Effect of UV radiation on some aspects of metabolic activities in *Aspergillus flavus* and *Penicillium notatum*, *Microbios*, 56, 79, 1988.
78. **Ramadan, S.E., Razak, A.A. and Saliman, H.G.**, Influence of cadmium on certain biological activities in a cadmium tolerant fungi, *Biol. Trace Elem. Res.*, 18, 179, 1988.
79. **Malini, R., Mukherji, K.G. and Venkitasubramaniam, T.A.**, Effect of aluminium and nickle on aflatoxin production by *Aspergillus flavus*, *Folia Microbiol. (Praha)*, 29, 104, 1984.

Chapter 7

LIPIDS OF *CRYPTOCOCCUS NEOFORMANS*

A. S. Ibrahim, H. Sanati and M. A. Ghannoum

CONTENTS

0-8493-4794-7/96/$0.00+$.50

I. INTRODUCTION

The encapsulated *Cryptococcus neoformans* is one of the major yeast pathogens in patients with impaired cell-mediated immunity, and those diagnosed with the acquired immunodeficiency syndrome (AIDS).[1,2] This organism is widely distributed in nature and has been isolated from milk, soil, and excrement of a variety of birds.[3-8] Furthermore, the organism has also been isolated from materials of *Eucalyptus camaldulensis* (river red gum) trees.[9,10] Since lipids are major constituents of the microbial cell membrane, they are considered as a promising target for drug discovery. Thus, characterization of the cryptococcal lipids may help to identify potential therapeutic targets. Little is known about the different classes of lipid in *C. neoformans*. Two studies from the same laboratory examined the lipid composition of this important pathogen. The first examined the lipid content of a strain of *C. neoformans*,[11] while the second investigated the interrelationship between the capsule size, lipid composition and virulence of three strains of *C. neoformans*.[12] In a recent study our laboratory analyzed the sterol composition of different strains of *C. neoformans*.[13,14] This chapter will review the lipid and sterol composition of *C. neoformans* and discuss the effect of triazole antifungals on the sterol composition of this yeast.

II. LIPID COMPOSITION

The total lipid content of *C. neoformans* varies among different strains. Using one strain of *C. neoformans*, Rawat et al.[11] reported that lipids comprise 0.96% of the cell dry weight. Another study by the same group showed that the lipid content of three strains of *C. neoformans* varied from 0.9% to 3.4% of the cell dry weight.[12] Our group found higher levels of lipids in three cryptococcal strains (lipid content ranged from 3.8% to 8.6% of the dry weight).[14] These differences may represent either innate variation between strains studied or due to growth conditions employed for growing the yeast.

A. NON-POLAR LIPIDS

Unlike *Candida albicans*, which has almost equal levels of polar and non-polar lipids,[15,16] the majority of the lipids extracted from *C. neoformans* are non-polar in nature. Almost 88% of the lipid content belongs to the non-polar lipid fractions. Steryl esters (SE) and alkyl esters (AE) are the predominant non-polar lipid classes obtained, followed by triacylglycerols (TG) and diacylglycerols (DG). Minor quantities of monoacylglycerols (MG), sterols (S) and fatty acids (FA) were also detected[14] (Table 1).

Table 1
Composition of Lipid Classes Extracted from Different Strains of *C. neoformans*

Lipid Class	Strains			
	SP 66	1253	T1	T20
Non-polar				
SE + AE	50.2	39.0 ± 2.9	49.8 ± 3.7	37.1 ± 3.1
TG	20.3	22.3 ± 5.8	22.4 ± 6.3	22.0 ± 5.9
FA	6.1	0.3 ± 0.4	0.7 ± 1.0	0.4 ± 0.6
DG	8.3	17.4 ± 8.1	1.6 ± 2.2	21.4 ± 4.2
S	5.2	3.3 ± 1.0	Tr	1.5 ± 2.2
MG	2.5	5.1 ± 0.5	8.6 ± 1.4	4.1 ± 3.2
Unidentified	3.8	ND	ND	ND
Polar				
SG	ND	0.4 ± 0.3	1.2 ± 1.0	Tr
PI	0.4	Tr	0.1 ± 0.0	0.4 ± 0.2
PS	ND	Tr	0.1 ± 0.1	0.2 ± 0.0
PC	1.1	4.6 ± 0.1	3.2 ± 0.6	3.9 ± 0.2
LPC	0.2	ND	ND	ND
PE	0.5	3.44 ± 0.4	4.53 ± 0.9	4.9 ± 0.3
LPE	0.4	ND	ND	ND
PG	ND	0.7 ± 1.1	1.8 ± 1.8	1.5 ± 0.2
GL	0.4	2.9 ± 2.1	4.3 ± 1.0	2.3 ± 0.3
CMH	ND	0.6 ± 0.4	1.8 ± 0.5	0.4 ± 0.3
PA	Tr	Tr	Tr	0.1 ± 0.1
MGD	ND	Tr	Tr	Tr
CL	0.4	Tr	Tr	Tr
Unidentified	0.2	ND	ND	ND

Values are expressed as the percentage (w/w) of the total amount of lipids. Tr, trace < 0.05%; ND, not detected. Unpublished data from Ibrahim, A.S., Sanati, H., Edwards, J.E., Jr. and Ghannoum, M.A., *J. Med. Vet. Mycol.* (submitted).

The finding that most of the cryptococcal lipids are non-polar in nature is very interesting. The presence of high levels of non-polar lipids would have considerable effect on membrane fluidity which would affect the structure and function of cell membrane. In this regard, it is worth mentioning that sterols[17] and membrane lipid fatty acyl composition [18,19] have been shown to affect the solute transport in yeast.

B. POLAR LIPIDS

The major polar lipids of *C. neoformans* are phosphatidylcholines (PC), phosphatidylethanolamines (PE) and tentatively identified glycolipid fraction (GL) (based on spraying with α-naphthol reagent).[14] Phosphatidylglycerols (PG), phosphatidylinositols (PI), ceramide monohexosides (CMH), steryl glycosides (SG), phosphatidylserines (PS), lysophosphatidylcholine (LPC), lysophosphatidylethanolamine (LPE), phosphatidic acid (PA), monogalactosyl diacylglycerols (MGD), and

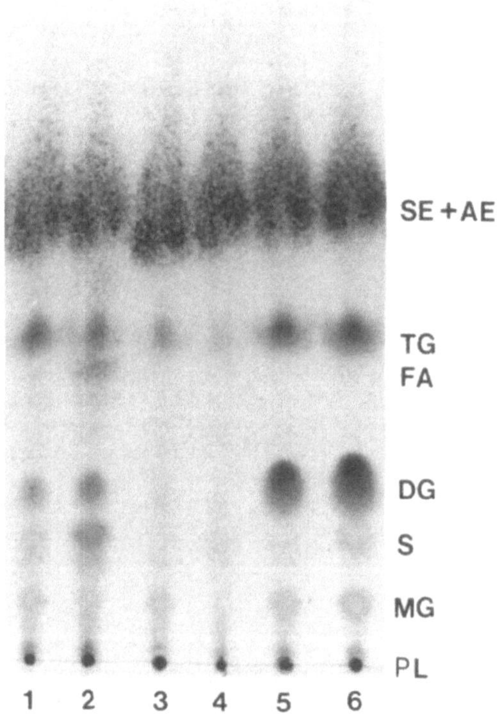

Figure 1. Typical TLC plate showing non-polar lipids of three isolates of *C. neoformans*. The adsorbent was Silica Gel G; the solvent was hexane:diethyl ether:acetic acid (75:25:1, v/v). The lipids were visualized by charring. 1,2 *C. neoformans* 1253; 3,4 *C. neoformans* T1; 5,6 *C. neoformans* T20. PL, polar lipids; MG, monoacylglycerols; S, sterols; DG, diacylglycerols; FA, fatty acids; TG, triacylglycerols; SE + AE, steryl esters and alkyl esters.

cardiolipins (CL) are also present in variable amounts among *C. neoformans* isolates (Table 1).

Figures 1 and 2 represent the separation of the non-polar and the polar cryptococcal lipids by thin layer chromatography, respectively. In general, qualitatively the lipid classes of *C. neoformans* resemble those reported in *C. albicans*.[15,16] However, significant quantitative differences exist between these two yeasts. For example, alkyl esters combined with steryl esters comprised 42.0% ± 5.5% of the total lipids in the studied *C. neoformans* strains. This amount is almost twice as much as that of lipid classes reported in *C. albicans*.[16] In addition to the lower levels of phospholipid, *C. neoformans* had a small amount of sterols and free fatty acids when

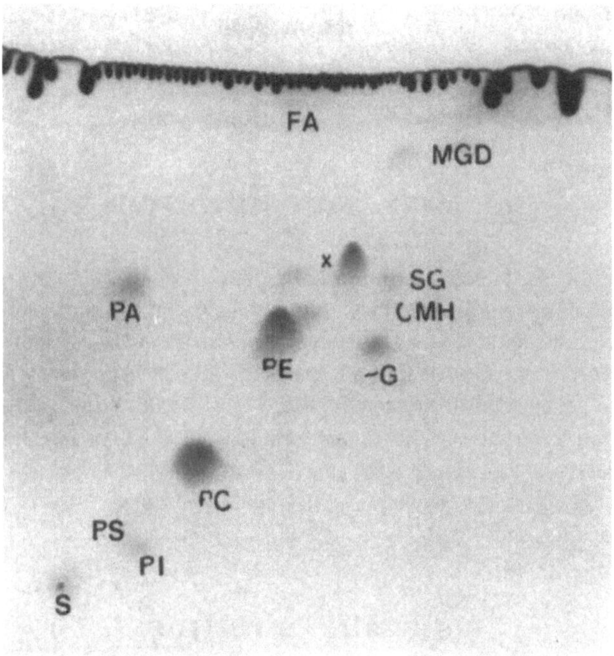

Figure 2. Typical TLC plate showing polar lipids of *C. neoformans*. The adsorbent was Silica Gel G; the solvent was (I) chloroform:methanol:7 M ammonium hydroxide (65:30:4, v/v) and (II) Chloroform:methanol:acetic acid:water (170:25:25:4, v/v). The lipids were visualized by charring. S, start; PI, phosphatidylinositols; PS, phosphatidylserines; PC, phosphatidylcholines; PE, phosphatidylethanolamines; PG, phosphatidylglycerols; X, glycolipids; PA, phosphatidic acids; CMH, ceramide monohexosides; SG, steryl glycosides; MGD, monogalactosyl diacylglycerols; FA, fatty acids.

Table 2

Constituent Fatty Acids of Total Lipids Extracted from Three Clinical Strains of *C. neoformans*.

Fatty acid	Strains		
	1253	T1	T20
$C_{14:0}$	10.9 ± 0.3	2.7 ± 3.4	0.6 ± 0.1
$C_{14:1}$	5.9 ± 0.8	1.1 ± 1.0	0.1 ± 0.1
$C_{16:0}$	19.3 ± 2.7	20.6 ± 5.8	19.6 ± 4.4
$C_{16:1}$	4.2 ± 0.4	5.9 ± 5.5	2.2 ± 0.7
$C_{16:2}$	11.3 ± 3.8	12.3 ± 2.7	11.8 ± 3.4
$C_{18:0}$	17.6 ± 0.3	27.1 ± 8.2	19.4 ± 4.3
$C_{18:1}$	2.3 ± 0.3	2.5 ± 3.5	3.4 ± 2.0
$C_{18:2}$	28.3 ± 0.3	25.8 ± 7.5	42.5 ± 4.0
$>C_{18:2}$	0.2 ± 0.0	2.1 ± 2.6	0.3 ± 0.0

Values are expressed as the percentage (w/w) of the total amount of fatty acids and are the mean ± SD of two separate experiments. Unpublished data from Ibrahim, A.S., Sanati, H., Edwards, J.E., Jr. and Ghannoum, M.A., *J. Med. Vet. Mycol.* (submitted).

compared with *C. albicans*.[16] The differences observed in the lipid composition between *C. neoformans* and *C. albicans* may account for the differences observed in these two clinically important yeasts in antifungal susceptibility and resistance pattern.

III. FATTY ACID COMPOSITION

The fatty acids detected in the total lipids of three *C. neoformans* strains are predominantly palmitic (16:0), stearic (18:0) and linoleic (18:2) acids. Hexadecadienoic acids (16:2) are present to a lesser extent.[14] Minor levels of myristic (14:0), myristoleic (14:1), palmitoleic (16:1), oleic (18:1), and $>C_{18:2}$ fatty acids are also detected (Table 2). The fatty acid composition resembles that reported in other fungi. In addition, *C. neoformans* is similar to other yeasts such as *C. albicans* and *S. cerevisiae* in the presence of trace amounts of C_{14} and the lack of C_{20} fatty acids (for a review of fungal fatty acid composition, see Weete[20]).

IV. STEROL COMPOSITION

While only a few studies dealt with the lipid composition of *C. neoformans*, sterols were investigated with more details. Kim et al. was the first group to study the sterol composition of *C. neoformans*.[21] These investigators characterized the sterol composition of six mutants of *C. neoformans* resistant to nystatin and pimaricin, three mutants resistant to amphotericin B and the wild type strains. These authors found that the major sterol in the wild type and the amphotericin B resistant strain was ergosta-7-en-3β-ol (fungisterol) rather than ergosterol. They also came to the conclusion that the development of polyene resistance in *C. neoformans* does not appear to be totally dependent on the sterol changes in the cells. This conclusion was supported by the fact that the mutants resistant to amphotericin B showed no substantial changes in sterol composition as compared to that of the wild type. Furthermore, they found that sterols of the mutants which were resistant to nystatin and pimaricin had no ergosterol and instead they accumulated ergosterol precursors such as ergosta-7,22-dien-3-ol, ergosta-8-en-3β-ol (in the mutants that lacked the ability to desaturate the C_5-C_6 bond in the ergosterol biosynthesis) and ergosta-5,8,22-trien-3β-ol (in the mutants that lost the enzyme responsible for the isomerization of the $\Delta^{8(9)}$ system to a Δ^7 system in the ergosterol biosynthesis). In 1993, Vanden Bossche et al.[22] reported that ergosterol was the major sterol extracted from a *C. neoformans* strain. They also detected the presence of obtusifoliol, 24-methylenedihydrolanosterol, 14-methylfecosterol, fungisterol, obtusifolione, ergosta-7,22-dien-3-ol, and ergosta-5, 7-dien-3-ol in lower levels.

Table 3
Sterol Content of *C. neoformans* Clinical Strains

Sterol	R_{erg}	Strains 1253	T1	T20
Squalene	0.31	0.8 ± 1.1	0.1 ± 0.2	1.6 ± 0.6
Calciferol	0.81	3.1 ± 1.4	4.5 ± 0.2	3.8 ± 0.7
Zymosterol	0.90	2.7 ± 1.0	3.7 ± 0.7	3.2 ± 1.0
Ergosterol	1.00	35.6± 0.3	30.7± 0.3	60.0± 2.5
4,14-dimethylzymosterol	1.09	Tr	Tr	ND
Obtusifoliol	1.21	48.8± 6.4	55.1± 0.9	27.5± 0.4
Lanosterol	1.30	Tr	ND	ND
24-methylenedihydrolanosterol	1.45	9.1± 5.4	5.9± 0.5	3.8 ± 1.0

Values are expressed as the percentage (w/w) of the total amount of sterols and are the means ± SD of two separate experiments. Tr, trace < 0.1%; ND, not detected. R_{erg}, retention time relative to ergosterol. Unpublished data from Ibrahim, A.S., Sanati, H., Edwards, J.E., Jr. and Ghannoum, M.A., *J. Med. Vet. Mycol.* (submitted).

The observed differences in sterol composition of *C. neoformans* between studies undertaken by Kim et al.[21] and Vanden Bossche et al.[22] may reflect strain differences, or differences in growth conditions employed for growing the fungus. Furthermore, *C. neoformans* clinical isolates are characterized by extensive genetic diversity,[23,24] which may also be associated with variation in cell membrane sterol composition. Keeping that in mind, recently our group undertook the characterization of several clinical isolates of *C. neoformans* that represent the initial and relapse isolates from patients with cryptococcal meningitis.[24,25] The results showed that, unlike pathogenic fungi which have a majority of ergosterol, the primary sterols of *C. neoformans* are obtusifoliol (range 21.1% to 68.2%) and ergosterol (range 0.0% to 46.5%). Squalene, lanosterol, 24-methylenedihydrolanosterol, 4, 14-dimethylzymosterol, zymosterol and calciferol were also present in smaller quantities (Table 3). There was considerable variation in sterol content among these isolates with total sterol contents ranging from 0.31 to 5.5% of dry weight. For the five patients who relapsed, the isolates differed in total sterol content and composition compared to the pretreatment isolates, indicating that the sterols had either been changed by therapy or the patients were infected with new isolates with different sterol composition. However, the latter possibility is highly unlikely since the relapse isolates had the same restriction fragment length polymorphism (RFLP) of the initial isolates.[24,25]

In another recent study we investigated the effect of host factors on the sterol composition and susceptibility to fluconazole and amphotericin B.[26] Similar to the sterol composition of the clinical isolates tested in our previous study,[13] we observed that five environmental isolates of *C. neoformans* had a majority of obtusifoliol and ergosterol as their sterols (comprising between 70% to 90% of the total sterols present). Mouse passage of the five environmental isolates resulted in a significant increase in amphotericin B MIC values. The average ergosterol increased (from 39.2 to

Table 4
Effect of Subinhibitory Concentration of Fluconazole on Sterol Composition of 13 Cryptococcal Isolates

Sterol	% of total sterols	
	Control	Fluconazole
Squalene	7.9 ± 5.1	12.0± 8.8
24-Methylenedihydrolanosterol	2.0 ± 3.3	ND
Lanosterol	4.6 ± 8.0	10.7± 8.2
4,14-Dimethylzymosterol*	5.2 ± 6.0	34.5± 15.4
Zymosterol	4.4 ± 9.3	0.4± 0.82
Obtusifoliol*	44.2 ± 14.1	16.1± 8.9
Calciferol	5.7 ± 4.6	9.3± 4.5
Ergosterol	27.4 ± 15.6	18.6±13.4

Data are the mean ± SD for 13 isolates of *C. neoformans*. *P* values were calculated by the paired *t*- test with the Bonferroni correction for multiple comparisons. **P* value = 0.001.

56.4%) and the average obtusifoliol content decreased (from 42.1% to 23.7%) in four out of five isolates after the passage. No detectable changes in squalene, calciferol, zymosterol, lanosterol and 24-methylene-dihydrolanosterol content occurred in all strains after the passage. This study confirms the unique sterol composition of *C. neoformans* compared to other fungi and also suggests that amphotericin B-resistant variants may arise *in vivo* without exposure to this antifungal agent.

V. EFFECT OF TRIAZOLES ON CRYPTOCOCCAL STEROLS

In an attempt to elucidate the mode of action of itraconazole against *C. neoformans* Vanden Bossche et al. extracted sterols from cryptococcal cells treated with different concentrations of the drug.[22] Upon treating with itraconazole, *C. neoformans* cells accumulated both 24-methylene-dihydrolanosterol and obtusifoliol. Accumulation of these two sterols (C-4 methyl sterols) proves that itraconazole is an inhibitor of the cytochrome P450-dependent 14α-demethylase.[22] Accumulation of 3-ketosteroid indicates that the drug also interferes with the last step in the demethylation at C-4.[22]

Our group investigated the effect of subinhibitory concentrations of fluconazole on the sterol content of 13 clinical isolates.[13] Growth of the cryptococcal isolates in the presence of subinhibitory concentrations of fluconazole (0.25×MIC) significantly altered the sterol content and pattern. Total sterol content decreased in 9 isolates and increased in 4 isolates in response to pretreatment with fluconazole. Fluconazole caused a decrease in obtusifoliol and an increase in 4,14-dimethylzymosterol in all isolates (Table 4). In contrast, this antifungal had no consistent effect on ergosterol levels. The increase in 4,14-dimethylzymosterol levels is consistent with the

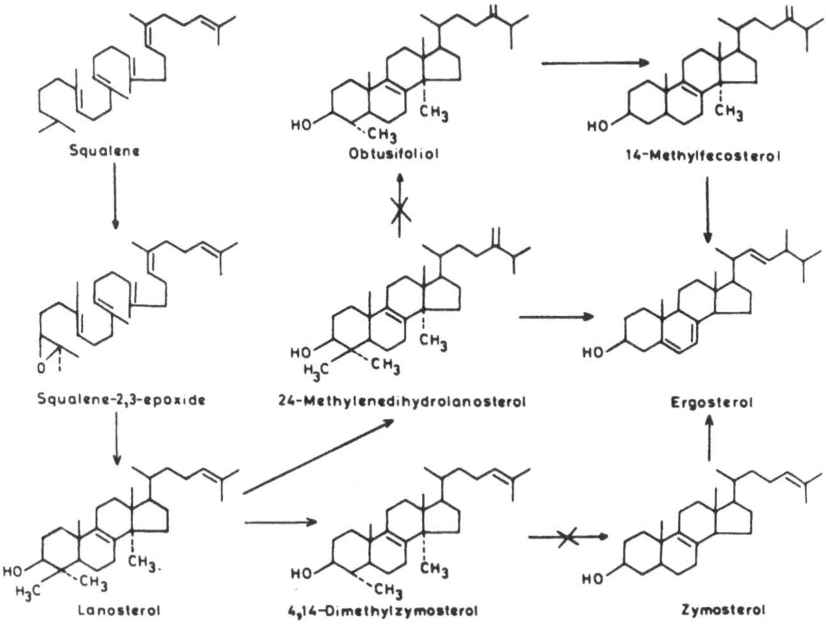

Figure 3. Pathways of metabolic conversion of sterols in *C. neoformans*. The X's indicate possible sites where fluconazole affects ergosterol synthesis.

study of Vanden Bossche et al.[22] showing that triazole antifungal agents inhibit cytochrome P450-dependent 14 α-demethylase.

It is well recognized that sterols serve as bioregulators of membrane fluidity, asymmetry, and consequently affect membrane integrity in eukaryotic cells.[27] Sterols which lack C-4 methyl group are best suited to maintain membrane integrity.[22] Our data demonstrate that fluconazole mediates the blockage of the conversion of 4,14-dimethylzymosterol to zymosterol, resulting in the accumulation of the former (Figure 3). Thus like itraconazole,[22] one possible fungistatic effect of fluconazole may be due to the accumulation of 4,14-dimethylzymosterol. The decrease in obtusifoliol in the presence of fluconazole is not readily explained. Since the synthesis of this compound does not involve a 14α-demethylation step, there is probably an additional site of fluconazole inhibition in the cryptococcal sterol biosynthetic pathway. Fluconazole may have a mode of action similar to that of itraconazole, which was shown to inhibit the 14α-demethylase activity as well as the reduction of obtusifoliol in *C. neoformans*.[22]

VI. VIRULENCE AND LIPID COMPOSITION

Limited information is available about the role of cryptococcal lipids in the virulence of *C. neoformans*. One study that tried to address the role of cryptococcal lipids in virulence was carried out by Upreti et al.[12] This group investigated the relationship between the virulence of *C. neoformans* and their capsule size and lipid composition. Virulence was evaluated by injecting *C. neoformans* intraperitoneally into mice. They found that even though the least virulent strain contained the least amount of total lipid and phospholipids, none of the lipids showed any quantitative relationship with virulence. In addition, fungal cells with bigger capsules had a lower lipid content. The polar lipids detected by Upreti et al.[12] resembled the ones mentioned in Table 1. This study did not elucidate the role of cryptococcal lipids in virulence and it is hard to interpret their data since they had used three unrelated cryptococcal strains. Furthermore, in addition to the differences in lipids, the strains studied differed in capsule size which is considered to be a virulence factor in *C. neoformans*.[28-30] The difference in the capsule size makes conclusions regarding the role of lipids in cryptococcal virulence equivocal.

In an earlier study Kim et al.[21] also investigated the role of cryptococcal sterols in virulence. UV derived mutants that differed in their sterol pattern were injected into mice intravenously. Their survival was compared with that of mice injected with wild type strain. The amphotericin B resistant mutants showed slightly less virulence as compared to the wild type, whereas the nystatin and the pimaricin mutants did not show any virulence in mice and failed to grow at 37°C. However, all the animal data were compared on gross observations and were not statistically analyzed. Additionally, since the mutants were derived by UV irradiation, the differences observed in virulence could be due to changes not related to the observed alterations in sterols.

More extensive studies should employ isogenic strains (obtained by targeted mutagenesis) that differ in a specific lipid constituent. Comparison of virulence of these isogenic strains in appropriate animal models may determine the role of lipids in virulence.

VII. CONCLUSION

Cryptococcal lipids have not been investigated in detail. However, few studies that have investigated the lipid composition of *C. neoformans* seem to agree on the unique lipid composition of this yeast which is manifested by the presence of almost 88% of the total lipids as non-polar lipids. Furthermore, the presence of high levels of obtusifoliol as compared to ergosterol, the major sterol of other fungi, is supportive of the distinct lipid composition, the significance of which in the overall physiology of *C. neoformans* is not known. Whether this could have an effect on the organism's antifungal susceptibility and resistance pattern also needs to be addressed. More studies should be directed towards investigating the effect of non-polar lipids on the transport of compounds across the cell membrane and trying to understand the role of lipid classes in the virulence of this medically important yeast.

ACKNOWLEDGMENTS

We would like to thank Roerig-Pfizer Pharmaceuticals for their financial support for part of the work reported in this chapter.

VIII. REFERENCES

1. **Levitz, S.M.**, Activation of human peripheral blood mononuclear cells by interleukin-2 and granulocyte-macrophage colony-stimulating factors to inhibit *Cryptococcus neoformans*, *Infect. Immun.*, 59, 3393, 1991.
2. **Mukherjee, S., Lee, S.C. and Casadevall, A.**, Antibodies to *Cryptococcus neoformans* glucuronoxylomannan enhance antifungal activity of murine macrophages, *Infect. Immun.*, 63, 573, 1995.
3. **Sanfelice, F.**, Uber einen neuen pathogenen Blastomyceten, welcher innerhalb de Gewebe unter Bidung Kalkartig aussenhender Massendegerenert, *Int. J. Med. Microbiol.*, 18, 521, 1985.
4. **Klein, E.**, Pathogenic microbes in milk, *J. Hyg. (Cambridge)*, 1, 78, 1901.
5. **Emmons, C.W.**, Isolation of *Cryptococcus neoformans* from soil, *J. Bacteriol.*, 62, 685, 1951.
6. **Emmons, C.W.**, Saprophytic sources of *Cryptococcus neoformans* associated with the pigeon (*Columba livia*), *Am. J. Hyg.*, 62, 227, 1955.
7. **Littman, M.L. and Walter, J.E.**, Cryptococcosis: current status, *Am. J. Med.*, 45, 922, 1968.
8. **Bauwens, L., Swinne, D., DeVroey, C. and DeMeurichy, W.**, Isolation of *Cryptococcus neoformans* var. *neoformans* in the aviaries of the Antwerp Zoological Gardens, *Mykosen*, 29, 291, 1986.
9. **Ellis, D.H. and Pfeiffer, T.J.**, Natural habitat of *Cryptococcus neoformans* var. *gattii*, *J. Clin. Microbiol.*, 28, 1642, 1990.
10. **Ellis, D.H. and Pfeiffer, T.J.**, Ecology, life cycle, and infectious propagule of *Cryptococcus neoformans*, *Lancet*, 336, 923, 1990.
11. **Rawat, D.S., Upreti, H.B. and Das, S.K.**, Lipid composition of *Cryptococcus neoformans*, *Microbiologica*, 7, 299, 1984.

12. **Upreti, H.B., Rawat, D.S. and Das, S.K.,** Virulence, capsule size and lipid composition interrelation of *Cryptococcus neoformans, Microbiologica,* 7, 371, 1984.
13. **Ghannoum, M.A., Spellberg, B.J., Ibrahim, A.S., Ritchie, J.A., Currie, B., Spitzer, E., Edwards, J.E., Jr. and Casadevall, A.,** Sterol composition of *Cryptococcus neoformans* in the presence and absence of fluconazole, *Antimicrob. Agents Chemother.*, 38, 2029, 1994.
14. **Ibrahim, A.S., Sanati, H., Edwards, J.E., Jr. and Ghannoum, M.A.,** Characterization of polar and non-polar lipid classes of *Cryptococcus neoformans* isolated from AIDS patients, *J. Med. Vet. Mycol.* (submitted).
15. **Ghannoum, M.A., Burns, G.R., Abu Elteen, K. and Radwan, S.S.,** Experimental evidence for the role of lipids in adherence of *Candida* spp. to human buccal epithelial cells, *Infect. Immun.*, 54, 189, 1986.
16. **Ghannoum, M.A., Swairjo, I. and Soll, D.R.,** Variation in lipid and sterol contents in *Candida albicans* white and opaque phenotypes, *J. Med. Vet. Mycol.*, 28, 103, 1990.
17. **Singh, M., Jayakumar, A. and Prasad, R.,** The effect of altered lipid composition on the transport of various amino acids in *Candida albicans, Arch. Biochem. Biophys.*, 191, 680, 1978.
18. **Mishra, P. and Prasad, R.,** Alterations in fatty acyl composition can selectively affect amino acid transport in *Saccharomyces cerevisiae, Biochem. Intl.*, 15, 499, 1987.
19. **Mishra, P. and Prasad, R.,** Relationship between fluidity and L-alanine transport in a fatty acid auxotroph of *Saccharomyces cerevisiae, Biochem. Intl.*, 19, 1019, 1989.
20. **Weete, J.D.,** Ed., *Lipid Biochemistry of Fungi and Other Organisms,* Plenum Press, New York, 1980.
21. **Kim, S.J., Kwon-Choung, K.J., Milne, G.W.A., Hill, W.B. and Patterson, G.,** Relationship between polyene resistance and sterol compositions in *Cryptococcus neoformans, Antimicrob. Agents Chemother.*, 7, 99, 1975.
22. **Vanden Bossche, H., Marichal, P., Le Jeune, L., Coene, M.C., Gorrens, J. and Cools, W.,** Effects of itraconazole on cytochrome P-450-dependent sterol 14α-demethylation and reduction of 3-ketosteroids in *Cryptococcus neoformans, Antimicrob. Agents Chemother.*, 37, 2101, 1993.
23. **Saag, M.S., Powderly, W.G., Cloud, G.A., Robinson, P., Grieco, M.H., Sharkey, P.K., Thompson, S.E., Sugar, A.M., Tuazon, C.U., Fisher, J.F., Hyslop, N., Jacobson, J.M., Hafner, R., Dismukes, W.E. and The NIAID Mycoses Study Group, and The AIDS Clinical Trials Group,** Comparison of amphotericin B with fluconazole in the treatment of acute AIDS-associated cryptococcal meningitis, *N. Engl. J. Med.*, 326, 83, 1992.

24. **Spitzer, E.D., Spitzer, S.G., Freundlich, L.F. and Casadevall, A.**, Persistence of the initial infection in recurrent cryptococcal meningitis, *Lancet*, 341, 595, 1993.
25. **Casadevall, A., Spitzer, E.D., Webb, D. and Rinaldi, M.G.**, Susceptibilities of serial *Cryptococcus neoformans* isolates from patients with recurrent cryptococcal meningitis to amphotericin B and fluconazole, *Antimicrob. Agents Chemother.*, 37, 1383, 1993.
26. **Currie, B., Sanati, H., Ibrahim, A.S., Edwards, J.E., Jr., Casadevall, A. and Ghannoum, M.A.**, The sterol composition and susceptibility to amphotericin B of environmental *Cryptococcus neoformans* isolates are changed by murine passage, *Antimicrob. Agents Chemother.*, In press,1995.
27. **Nozawa, Y. and Morita, T.**, Molecular mechanisms of antifungal agents associated with membrane ergosterol. Dysfunction of membrane ergosterol and inhibition of ergosterol biosynthesis, in *In vitro and in vivo Evaluation of Antifungal Agents*, Iwata, K. and Vanden Bossche, H., Eds., Elsevier Science Publishers B.V. Amsterdam, 111, 1986.
28. **Fromtling, R.A., Shadomy, H.J. and Jacobson, E.S.**, Decreased virulence in stable, acapsular mutants of *Cryptococcus neoformans*, *Mycopathologia*, 79, 23, 1982.
29. **Kozel, T.R. and Gotschlich, E.C.**, The capsule of *Cryptococcus neoformans* passively inhibits phagocytosis of the yeast by macrophages, *J. Immunol.*, 129, 1675, 1982.
30. **Chang, Y.C. and Kwon-Choung, K.J.**, Complementation of a capsule-deficient mutation of *Cryptococcus neoformans* restores its virulence, *Mol. Cell. Biol.*, 14, 4912, 1994.

Chapter 8

LIPIDS OF DERMATOPHYTES

G. K. Khuller and N. Manchanda

CONTENTS

0-8493-4794-7/96/$0.00+$.50

I. INTRODUCTION

Dermatophytes are a highly specialized group of pathogenic filamentous fungi which have been placed under class fungi imperfecti. According to their imperfect forms, these fungi are grouped into three genera: *Microsporum, Epidermophyton* and *Trichophyton*. Several species of *Microsporum* and *Trichophyton* are known, whereas *Epidermophyton* is a monotype dermatophyte. These are responsible for skin infections called dermatomycoses and mostly infect keratinized surfaces. The prevalence of serious mycoses has dramatically increased in recent years in immunosuppressed patients.[1] Moreover, mycotoxins, the toxic fungal metabolites, have been responsible for major epidemics of poisoning in man and animals.[2] The capacity of dermatophytes to parasitize the host depends on the action of lipases and other ectoenzymes required to degrade keratin.[3,4] Lipids of dermatophytes are involved in pathogenesis[5] and phospholipid components have been shown to elicit allergic reactions.[6] They also have a significant role in the sensitivity of these organisms against the antifungal drugs and help the fungus to adapt to the changed surroundings. This chapter discusses some aspects of dermatophyte lipids.

II. LIPID COMPOSITION

Burdon[7] and Akasaka[8] demonstrated the presence of lipids in dermatophytes with the help of various staining techniques. Lipid composition of ten species of dermatophytes has been reported[9] and a comparison of the gross chemical constitution of some species of *Trichophyton* with several non-pathogenic fungi has also been made.[10] Most dermatophytes contain total lipids in the range of 7-19.8% of dry weight.[5,11] During starvation, lipids represent primary reserve material rather than carbohydrates. The lipid content was observed to decrease from 23% in the spores to 13% in 24 h old mycelia of *M. quinceanum*[12] in starved fungi. A significant variation in total lipids of *E. floccosum* with age has been reported.[11,13] The total lipids of *M. gypseum* and *E. floccosum* were observed to be significantly low as compared to those of *T. mentagrophytes* at all phases of growth.[9] Lipid composition of *M. gypseum* was also reported by Rawat and Das.[14] Lipids act as an important source of energy, which was confirmed by cytochemical studies,[15] and, also, are the prime carbon source and substrate for polysaccharide synthesis in dermatophytes.[16]

The cell wall of dermatophytes has been extensively studied and lipid content of cell wall of all three dermatophyte genera has been found to range between 3.1-6.6% of dry weight.[17,18] The cell wall lipids impart rigidity and thus protect the cells from dying. Neutral lipids and phospholipids constitute the total lipids of dermatophytes. The neutral lipids generally have an energy storage function and polar lipids are important determinants of the structure and function of the cellular membranes. A tentative distribution of various types of lipids in *T. mentagrophytes* was reported by Prince.[19] There are qualitative and quantitative differences between various lipid fractions of the three dermatophyte genera. Each lipid class is further composed of several types which are discussed separately.

A. FATTY ACIDS

Fatty acids of dermatophytes are the most well studied components of these fungi. With the help of various methods, such as paper and gas chromatography and chemical degradation, it was deduced that dermatophyte fatty acids consist of a homologous series of saturated and unsaturated fatty acids ranging from C_6 to C_{22}. The most predominant are $C_{16:0}$, $C_{18:0}$, $C_{18:1}$ and $C_{18:2}$ acids. The percentage content of $C_{16:0}$, $C_{16:1}$, $C_{18:0}$, $C_{18:1}$ and $C_{18:2}$ fatty acids in all the dermatophytes is similar[20] and they constitute more than 80% of the total fatty acids.[13,21-27] The relative proportion of these fatty acids in each genera of the fungi is, nevertheless, variable depending on the species and culture conditions. In various studies, the effects of age or phase of growth on the fatty acid composition of these fungi have been examined.[11,13,23]

Short-chain fatty acids are present in substantial amount throughout the growth cycle of *E. floccosum* whereas in *M. cookei* and *T. mentagrophytes*, they are present in minor or negligible amounts. In Table 1, a comparison of phospholipid fatty acid composition of these fungi has been depicted.[23] Short-chain fatty acids were detected in significant amounts only during the early growth phase. An increase in the unsaturated : saturated fatty acids (U:S) ratio of phospholipids was observed until the early stationary phase of *M. gypseum* and *T. mentagrophytes* which is mainly due to increase in $C_{18:2}$ content with age of the culture. In *E. floccosum*, conflicting results were obtained by different workers. Khuller et al.[24] reported alterations in fatty acid composition with age while no change was seen in fatty acid profile by Yamada et al.[25] Only minor changes were observed in the fatty acid composition in *M. gypseum* and *E. floccosum* when the growth conditions were changed from stationary to shake cultures grown at 27°C,[28-30] while there was no change with varying growth conditions when the growth temperature was maintained at 37°C.[22]

So far, not much work has been done on fatty acid synthesis in dermatophytes except a study by Kostiw et al.,[21] wherein it was postulated that equilibrium between malonyl-CoA (the intermediate in *de novo* synthesis) and acetyl-CoA (the intermediate in chain elongation) might be responsible for regulation of these two processes.

Table 1
Fatty Acid Composition of Phospholipids during Growth Phases of Dermatophytes[a]

Chain length	*E. floccosum* (growth phases)				*M. cookei* (growth phases)				*T. mentagrophytes* (growth phases)				*M. gypseum* (growth phases)			
	EL (10%)	ML (20%)	ES (30%)	S (40%)	EL (5%)	ML (15%)	ES (25%)	S (50%)	EL (15%)	ML (30%)	ES (45%)	S (60%)	EL (7%)	ML (15%)	ES (30%)	S (45%)
10:0	11.95	8.13	18.23	39.28	T	T	3.05	4.44	T	T	T	T	8.60	-	T	T
12:0	4.76	3.12	6.18	22.38	T	T	0.31	-	T	T	T	T	8.20	-	T	T
14:0	3.16	0.42	0.86	3.37	T	1.03	0.59	0.63	T	T	T	T	1.60	1.80	T	2.50
16:0	26.47	22.85	24.86	13.55	27.29	28.85	25.53	32.58	34.73	32.79	31.00	27.53	32.20	38.70	31.00	32.40
16:1	2.03	1.56	0.82	1.56	T	1.38	1.64	0.87	1.89	6.78	3.15	1.44	T	1.00	2.80	1.20
16:2[b]	1.10	1.05	0.39	4.79	2.16	1.49	2.07	1.99	1.72	-	1.01	0.92	1.50	-	-	4.30
18:0	4.99	5.74	3.90	1.43	6.52	7.11	6.27	4.19	21.67	8.40	11.76	2.99	6.80	4.86	5.60	7.20
18:1	2.08	5.18	4.63	3.19	7.89	11.87	6.16	3.95	3.13	3.89	3.75	2.99	4.90	3.90	9.30	7.60
18:2	47.02	46.86	40.01	11.23	55.89	50.02	55.51	53.80	36.89	51.77	51.04	53.73	24.00	46.30	48.90	42.40
S:U	1.03	0.67	1.18	4.00	0.51	0.58	0.56	0.72	1.23	0.96	0.75	0.69	1.99	0.83	0.57	0.71

[a]Values shown are the mean of three independent batches, analyzed in duplicate.
EL, Early log phase; ML, mid log phase; ES, early stationary phase; S, stationary phase;
T, trace amounts; S:U, ratio of saturated to unsaturated fatty acids (Data from Khuller, G.K., Chopra, A., Bansal, V.S. and Masih, R., *Lipids*, 16, 20, 1981a, and Bansal, V.S., Chopra, A., Kasinathan, C. and Khuller, G.K., *Sabouraudia*, 19, 223, 1981. With permission).
[b]Tentative identification

Table 2
Total Phospholipid Content of Various Species of Dermatophytes

Organism	Culture	Total phospholipid content (mg/g dry wt)	Reference
M. gypseum	Stationary	11.15	31
M. gypseum	Shake	16.14	46
E. floccosum	Stationary	8.20	23
E. floccosum	Shake	13.95	26
M. cookei	Stationary	14.30	23
T. mentagrophytes	Stationary	35.90	23

B. PHOSPHOLIPIDS

Phospholipids are important structural components of biological membranes forming the lipid bilayer in which proteins are embedded and are included in the structure and maintenance of membrane functions in the cell. The total phospholipid content of various species of dermatophytes has been presented in Table 2. The major phospholipids are phosphatidylcholine (PC), phosphatidylethanolamine (PE), phosphatidylserine (PS), phosphatidylinositol (PI), phosphatidylglycerol (PG) and some unusual lipids. Das and Banerjee[32] reported the presence of PC, PE, PS, PI, PG and phosphatidic acid (PA) in *T. rubrum*, while PC, PE, PS, minor quantities of PI, lysoPC (LPC) and cardiolipin along with an unknown polar component were detected in *E. floccosum*.[33] The phospholipid composition of *M. gypseum* was essentially similar to that of *E. floccosum*.[11] Phospholipid composition of a wild and mutant strain of *Arthroderma unicatum* showed only minor quantitative differences.[34] The phospholipid content of *M. gypseum* and *E. floccosum* was significantly lower than that of *T. mentagrophytes*.[11] Total and individual phospholipids of *T. mentagrophytes* and *M. cookei*[23] were observed to increase significantly up to the mid-log phase and decreased subsequently. However, in *E. floccosum*, total and individual phospholipids (except LPC) decreased throughout the cell cycle.[35] The relative proportions of individual phospholipids did not change with age of the culture in *T. rubrum*[32] while variations were observed in the proportions of these components in *M. gypseum*.[36] An increase in phospholipid content was seen when the stationary cultures were shifted to shake cultures (Table 2).

C. NEUTRAL LIPIDS

In dermatophytes, triacylglycerol (TG) is the major component of neutral lipids; while monoacylglycerol (MG), diacylglycerol (DG), sterols and sterol esters were present in small amounts. It could be due to the quick esterification of MG and DG to TG. TG has a physiological importance in dermatophytes because it is regarded as an energy reserve during unfavorable conditions and also helps to keep fatty acids under non-toxic limits and stimulates reproduction.[37] Fungal lipases catalyze the formation of acyl glycerols in

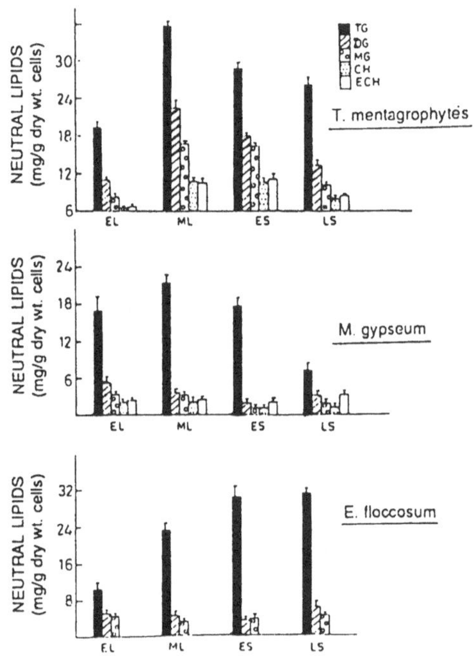

Figure 1. Variations in the individual neutral lipid contents during different phases of growth in different dermatophytes. TG, DG, MG, Tri-, di-, and monoacylglycerols; CH, cholesterol; ECH, esterified cholesterol; EL, early log phase; ML, mid-log phase; ES, early stationary phase; LS, late stationary phase (Bansal, V.S. and Khuller, G.K., unpublished observations).

addition to their lipolytic function.[38] So far, no specific function has been assigned to MG and DG. Acylglycerols were observed to constitute a major fraction of neutral lipids during all phases of growth in different species of dermatophytes, while sterols represented a minor fraction.[11] The relative distribution of these components is given in Figure 1.

Among non-polar fractions, only sterols are known to be essential structural components of fungi.[37] These are responsible for growth, reproduction and maintenance of membrane structure and fluidity.[39] Apparently, only free sterols are involved in controlling membrane stability and permeability.[37] Presence of ergosterol was shown in *T. asteroides,*[40] *T. rubrum* and *T. mentagrophytes.*[41,42] Wirth et al.[43] also reported the presence of brassicasterol in addition to ergosterol in five strains of *T. rubrum*. Blank et al.[44] substantiated the observation of Wirth et al.[43] and they also discovered trace amounts of brassicasterol in *E. floccosum*. Sterol composition in

Table 3
Quantitative Estimation of Free Sterol Components in *T. rubrum*[a]

Sterol component	Amount (mg/g dry wt)	Relative abundance
Ergosterol	5.40±0.30	71.6
Ergocalciferol	1.00±0.09	13.2
Brassicasterol	0.61±0.06	7.9
Campesterol	0.53±0.03	7.0

[a]Data from Das, S.K. and Banerjee, A.B., *Indian J. Med. Res.*, 65, 70, 1977c.

T. rubrum is listed in Table 3. Only 7% of the total sterols of *T. rubrum* were present in esterified form, ergosterol being the primary component to be esterified. Sterol composition of dermatophytes is similar to other fungi except the presence of brassicasterol in the *Trichophyton* group. Fatty acids found in sterol esters of *T. rubrum* were mainly of chain length C_{14}-C_{18}, which included one di- and two mono-unsaturated fatty acids.[45] In shake cultures of *M. gypseum* and *E. floccosum* significant amounts of sterols were present in mid log phase;[26,46] however, esterified sterols were present in minor amounts.

D. UNUSUAL LIPIDS

Besides phospholipids, an amino acid containing lipid was also detected in the polar lipid fraction of *M. gypseum* and *E. floccosum*.[11,47] From physical and chemical analysis, this unusual lipid was found to contain dibasic amino acid, a long-chain alcohol and a fatty acid methyl ester.[11] The unusual lipid of *M. gypseum* was structurally different from *E. floccosum* in which it was 1 (3): 2-diacylglycerol-3 (1)-0-4′(N,N,N-trimethyl) homoserine. Fatty acid composition of unusual lipid in *E. floccosum* showed palmitic (37%) and linoleic (45.9%) as the predominant fatty acids, while palmitoleic (2.9%), stearic (3.4%) and oleic (8.2%) acids represented the minor fractions.[47] In *M. gypseum*, palmitate (58.7%), stearate and oleate (24%) were the major fatty acids.[11] Since lipids are charged molecules, these components probably substitute for some phospholipids that are absent in the organism, as has been suggested for *Desulfovibrio gigas*.[48]

III. BIOSYNTHESIS AND TURNOVER OF LIPIDS

Incorporation of radiolabeled precursors into lipids and chase of the radioactivity incorporated is one of the important tools to study the biosynthesis and turnover of lipids. Various precursors e.g., [^{14}C]-acetate, [^{32}P]-orthophosphoric acid and nitrogenous bases are used for such studies. With [^{14}C]-acetate, synthesis and turnover of phospholipids has been studied in *E. floccosum*[49] and *M. gypseum*.[50] Among phospholipids, PC was observed to exhibit maximum rate of synthesis and degradation, followed by PE and PS. A rapid decrease in total radioactivity was observed in the initial period of chase, followed by a constant loss (Figures 2a and 2b), indicating the existence of two

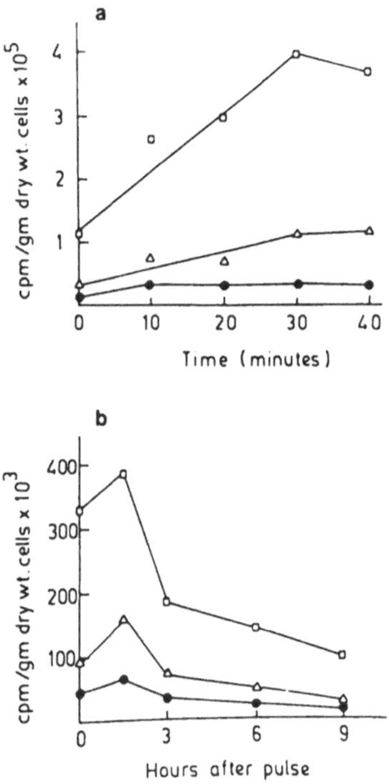

Figure 2. (a) [^{14}C]-acetate incorporation into major phospholipids of *E. floccosum*, (b) loss of radioactivity of *E. floccosum* after pulse labeling (□) PC; (Δ) PE; (●) PS (from Chopra, A. and Khuller, G.K., *Indian J. Med. Res.*, 73, 325, 1981a. With permission).

phospholipid pools with different turnover rates. The pool with fast turnover was proposed to be located in metabolically more active regions of the organism, as compared to the pool with slow turnover.[49] With [^{32}P]-orthophosphoric acid, a significant difference in the rate of synthesis as well as in the pattern of distribution into various phospholipids was observed, which suggested a rapid interconversion of phospholipids in *M. gypseum.*[51] Similar results were obtained with *E. floccosum*[52] and *T. rubrum.*[53] PC was observed to have a maximum turnover rate in *E. floccosum* while in *T. rubrum*, no turnover was observed even after 36 h of chase.[53] However, high turnover rates of PE and PS were observed in both the organisms. With a non-specific precursor, i.e., [^{14}C]-glycerol, about 30% of the total label was detected in *M. gypseum* phospholipids[51] and the order of incorporation into various phospholipids was PS > PE > PC > LPC.

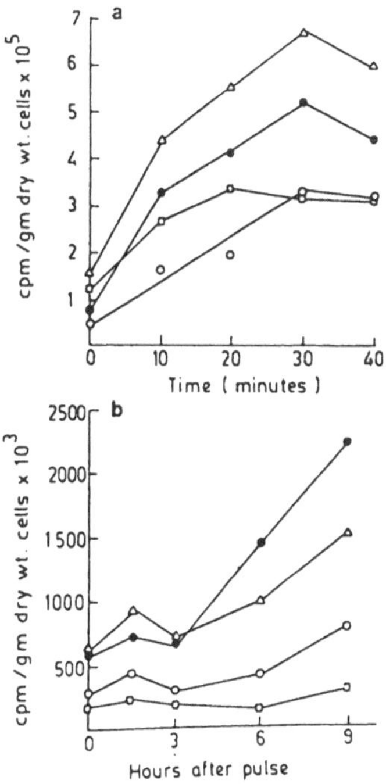

Figure 3. (a) [^{14}C]-acetate incorporation into neutral lipids of *E. floccosum*. (b) Loss of radioactivity from neutral lipids of *E. floccosum* after pulse labeling. (Δ) TG; (□) MG; (●) FFA; (O), DG (from Bansal, V.S., Chopra, A., Kasinathan, C. and Khuller, G. K., *Ind. J. Med. Res.*, 76, 832, 1982b. With permission).

In vivo incorporation studies with nitrogenous bases[51,54] revealed that PE is mainly synthesized by PS decarboxylation as well as by Kennedy and methylation pathways. The results of [^{14}C]-methionine and [^{14}C]-choline incorporation also indicated the presence of two pools of lipids containing choline in *M. gypseum*, similar to *E. floccosum*.[49,52]

Although phospholipid catabolism and turnover are not essential for growth as they influence specific membrane function not related to cell division, they are essential for membrane adaptation to environmental changes.[54]

Neutral lipids constitute a major portion of total lipid pool (discussed in section IIC). Incorporation of [^{14}C]-acetate, a common precursor for polar and nonpolar lipids, revealed that maximum uptake was observed in TG, followed by FFA, DG and MG in *E. floccosum* (Figures 3a and 3b). However, on chase, a continuous synthesis of all the neutral lipid fractions, and, at the same time, degradation of phospholipids fraction were observed, suggesting a recycling of the labeled carbon. These experiments suggested the existence of a common pathway for neutral as well as for phospholipid synthesis in this fungus.[55] An increased degradation observed in polar lipids could be due to highly active phospholipases in the dermatophytes. Consequently, the released FFA and PA may be utilized for acylglycerol synthesis. Moreover, the presence of a large FFA pool in this fungus could easily result in acylation of MG and DG, with simultaneous replenishment by *de novo* synthesis or by phospholipase action on phospholipids.

Bansal et al.,[56] studied the biosynthesis of neutral lipids in *M. gypseum* with [^{14}C]-acetate, [^{14}C]-glycerol and [^{14}C]-glucose as precursors. The incorporation of [^{14}C]-acetate was maximum in neutral lipids (92%), followed by glucose (70%) and glycerol (52%). With [^{14}C]-acetate the order of incorporation into various neutral lipid fractions was cholesterol ester, followed by FFA, TG, MG, cholesterol and DG, which is different from the incorporation pattern of this precursor in *E. floccosum,*[49] in which maximum incorporation was found in TG fraction. With [^{14}C]-glucose and [^{14}C]-glycerol, maximum incorporation was found in TG suggesting high activity of acyltransferase. A continuous synthesis of TG may be required to keep the level of FFA under non-toxic levels. The high rates of incorporation into neutral lipids of *M. gypseum* suggest that neutral lipid synthesizing enzymes are more active than phospholipid-biosynthetic enzymes. [^{14}C]-glycerol was incorporated exclusively into the lipid backbone, while [^{14}C]-glucose was incorporated into the backbone as well as into fatty acyl moieties of phospholipids.[56] In dermatophytes, glycolysis has been shown to be the principal pathway for formation of acetate from glucose, which acts as a precursor for lipid synthesis in eukaryotic systems.[57] From the above studies, it was proposed that glycerol and acetate were directly incorporated into phospholipids, glucose had to pass several metabolic steps before it could enter into lipid moiety.

IV. BIOSYNTHETIC ENZYMES OF PHOSPHOLIPIDS

Studies of the enzymes of certain dermatophytes were performed as early as 1929 by Tate.[58] Among dermatophytes, these enzymes were identified mainly in *M. gypseum* and their intracellular localization and kinetic properties have been investigated.[49,56,59] These are discussed in the following sections.

A. GLYCEROL KINASE

It is the key enzyme in the pathway of phospholipid synthesis and catalyzes the formation of glycerol-3-phosphate from glycerol. Among dermatophytes, this enzyme has been identified and characterized in *M. gypseum* and *E. floccosum.*[60,61] It was found in cytosolic fraction and was activated and stabilized by ammonium sulfate. Kinetic studies of the enzyme showed that it catalyzed the reaction by ping-pong mechanism,[61] as compared to ordered bi-bi sequential mechanism for *C. mycoderma*. Two pH optima 8.0 and 10.5 were observed for this enzyme in both *M. gypseum* and *E. floccosum*. Glucose-6-phosphate and fructose 1,6-bisphosphate were found to inhibit the enzyme competitively while glucose had no effect on the activity of this enzyme. The molecular weight of this enzyme in *M. gypseum* was found to be 450 kDa and ATP was the most effective phosphate donor.[61]

B. ACYLTRANSFERASE

It catalyzes the formation of phosphatidic acid by successive acylation of glycerol-3-phosphate which is a key intermediate in phospholipid synthesis. The first enzyme which causes the acylation of glycerol-3-phosphate is glycerol-3-phosphate acyl transferase and it is localized in the microsomal fraction of *M. gypseum.*[62] LPC acyltransferase from *T. rubrum* was reported to be cytosolic.[63] The activity of this enzyme in *M. gypseum* was observed between pH 6.5 and 7.5 with no requirement for metal ions.

C. ENZYMES OF PC BIOSYNTHESIS

PC is the major phospholipid and is mainly synthesized by two pathways. One is the *de novo* or CDP-base pathway while the other is the methylation pathway. The enzymes involved in these pathways are:

1. CDP Pathway

a) Choline kinase

Choline is brought into the cells by facilitated transport and is then phosphorylated by choline kinase.[64] The formation of phosphorylcholine initiates the subsequent reactions of cytidine pathway. This enzyme was found to be inducible in *M. gypseum* and *E. floccosum.*[60,65] The presence of cytidine pathway in *M. gypseum* was first reported by Bansal et al.[51] On subcellular fractionation, this enzyme was found to be localized in cytosolic fraction of both *M. gypseum* and *E. floccosum,*[59] and ATP is a better substrate for phosphorylation of choline in both these dermatophytes. Concentration of choline is an important factor in controlling the choline phosphorylation as enzyme activity is enhanced with increasing concentrations of choline (Figures 4a and 4b). However, no change in the activity of this enzyme was observed on growing the *M. gypseum* cells in the presence of nitrogenous bases.[54]

b) Phosphorylcholine cytidyltransferase

The synthesis of the activated intermediate i.e., CDP-choline is catalyzed by this enzyme. The membrane bound form of cytidyltransferase is active while the cytosolic form is relatively inactive. This enzyme regulates the PC biosynthesis by its reversible translocation from cytosol to endoplasmic reticulum.[66] Only recently, the activity of this enzyme has been measured in *M. gypseum* by Bindra[67] and was found to be influenced by modulators (aminophylline and atropine) of cAMP levels.

c) Diacylglycerol choline phosphotransferase

This enzyme is responsible for transferring phosphorylcholine from CDP-choline to DG. It has been detected in mitochondrial and microsomal membranes of *S. cerevisiae*, but its presence in dermatophytes is yet to be shown.

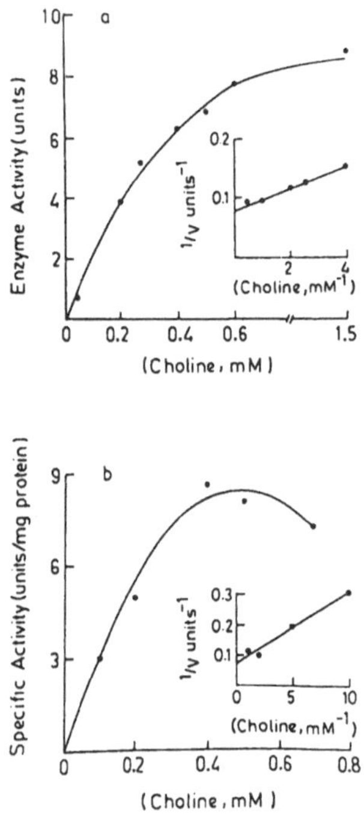

Figure 4. Effects of choline concentration on kinase of (a) *M. gypseum* and (b) *E. floccosum* (from Kasinathan, C., Govindarajan, U., Chopra, A. and Khuller, G.K., *Lipids*, 18, 727, 1983. With permission).

Table 4
Incorporation of [^{14}C]-Methionine and [^{3}H]-Ethanolamine into Different Fractions of *M. gypseum* Lipids

Radiolabeled precursor	Lipid Constituents (cpm/g dry wt)		
	Total lipids	Phospholipids	PC + LPC
[^{14}C]-Methionine	44862±4052	32153±111	14418±182
[^{3}H]-Ethanolamine	110072±11147	19383±637	9151±161

Values are mean ± S.D. of three batches in duplicate (Data from Bindra, A. and Khuller, G.K., *Indian J. Biochem. Biophys.*, 30, 311, 1993. With permission).

2. Methylation Pathway

In addition to CDP-base pathway, methylation pathway also plays a role in the synthesis of PC. In this pathway, there is stepwise transfer of three methyl groups from S-adenosylmethionine to PE. The two distinct methyltransferases are involved in the conversion of PE to PC, one that converts PE to N-methyl phosphatidylethanolamine (PME) with high affinity for S-adenosylmethionine and a second enzyme that converts PME to N,N-dimethylphosphatidylethanolamine and has a low affinity for S-adenosylmethionine.[68-70]

This enzyme has not been studied in dermatophytes, but its presence has been shown indirectly by the incorporation of labeled ethanolamine and methionine into choline containing lipids (Table 4).[71] Both ethanolamine and methionine can get incorporated into choline containing lipids only through methyltransferases, thus establishing the presence of these enzymes in *M. gypseum*.

D. PS SYNTHASE AND PS DECARBOXYLASE

The presence of this enzyme in dermatophytes was reported by Kasinathan et al.,[59] who observed that PS in *M. gypseum* was synthesized by PS synthase, which was localized in the microsomal fraction. Mg^{2+} was essential for the activity of this enzyme, while -SH group reagents had stimulating effect.

The presence of PS decarboxylase has been demonstrated indirectly from *in vivo* studies, wherein the label from serine was maximally incorporated into PE, suggesting the presence of an active decarboxylase enzyme and these observations were further substantiated by the fact that conversion of PS to PE is inhibited by an inhibitor of PS decarboxylase, hydroxylamine.[59]

E. BASE-EXCHANGE ENZYME

Base exchange enzyme for PE synthesis has been investigated in *M. gypseum*. It is localized in the microsomal fraction and requires Ca^{2+} for its activity. The enzyme exhibited saturation kinetics and a very low K_m indicating an active ethanolamine reaction for the formation of PE in *M. gypseum*.[59]

F. ETHANOLAMINE KINASE

It is the first enzyme of CDP-base pathway for *de novo* synthesis of PE. Earlier PS-decarboxylase and base exchange pathways were found to be responsible for the synthesis of PE in *M. gypseum*,[59] but later ethanolamine kinase was detected in crude fractions of *M. gypseum* and CDP-base pathway for the synthesis of PE was established in this fungus.[72] From the various *in vivo/in vitro* incorporation studies and the identification of various enzymes, the biosynthetic pathways which have been elucidated are depicted in Figure 5.

V. CATABOLIC ENZYMES

Catabolism and turnover of phospholipids are essential for membrane adaptation to environmental changes.[55] The different moieties in a given phospholipid class have different turnover rates (Figure 6).[35]

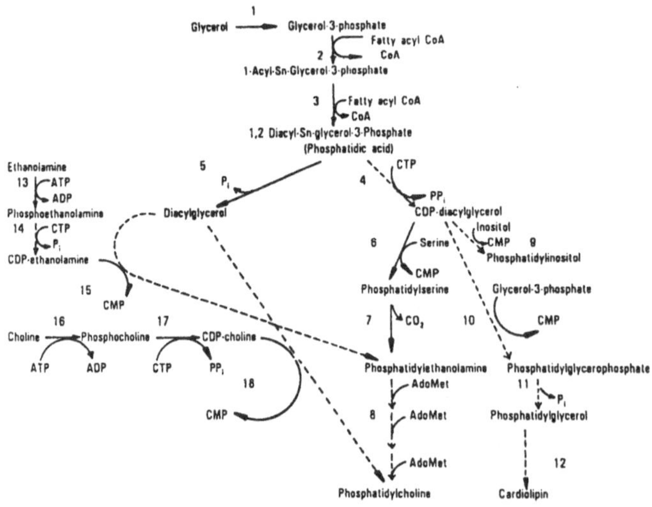

Figure 5. Phospholipid biosynthetic pathways in microorganisms. 1. Glycerol kinase; 2. Glycerophosphate acyltransferase; 3. 1-acyl-sn-glycerolphosphate acyltransferase; 4. CDP-diacylglycerol synthase; 5. Phosphatidic acid phosphatase; 6. Phosphatidylserine synthase; 7. Phosphatidylserine decarboxylase; 8. Phosphatidylethanolamine-N-methyltransferase; 9. Phosphatidylinositol synthase; 10. Phosphatidylglycerophosphate synthase; 11. Phosphatidylglycerophosphate phosphatase; 12. Diphosphatidylglycerol synthase; 13. Ethanolamine kinase; 14. Phosphoethanolamine cytidyltransferase; 15. Diacylglycerolethanolamine phosphotransferase; 16. Choline kinase; 17. Phosphorylcholine cytidyltransferase; 18. Phosphorylcholine diacylglycerol transferase; ---- enzymes identified in dermatophytes; - - - - enzymes not identified in dermatophytes.

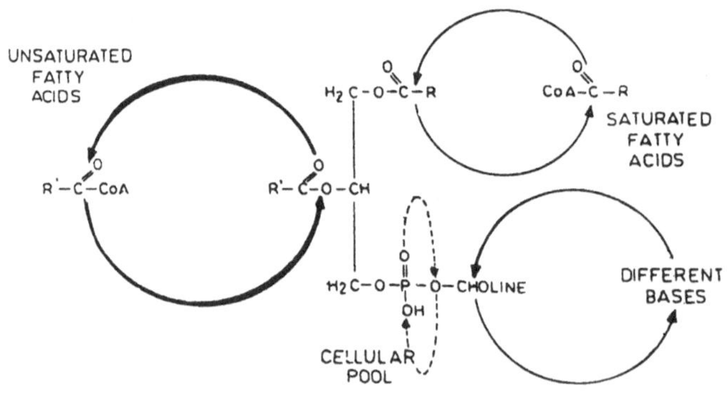

Figure 6. Turnover of phospholipid moieties.

A great deal of interest has been generated in the degradative enzymes of phospholipids because of their possible involvement in initial stages of infection.[3,4,73] Enzymes that catabolize lipids are essential not only for turnover of these molecules but also for diverting their breakdown products towards the synthesis of other cell constituents or for provision of energy.

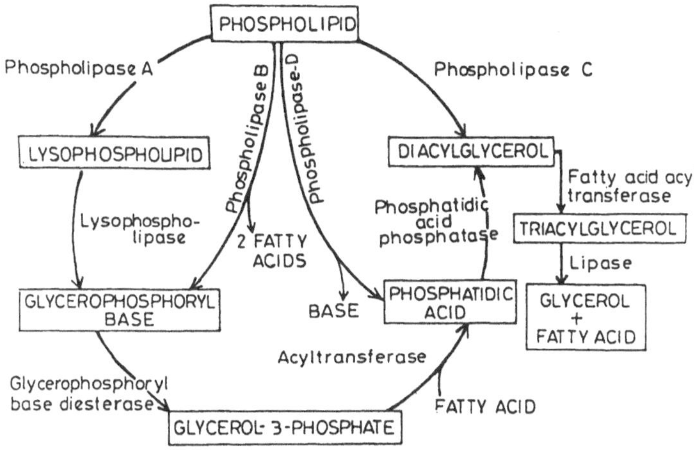

Figure 7. Tentative pathways of phospholipid catabolism in dermatophytes.

A. PHOSPHATIDATE PHOSPHATASE

The content of acylglycerol in *M. gypseum* and *E. floccosum*[49,74] has been observed to be much higher than phospholipids. The fact that incorporation of [^{14}C]-acetate into TG was more than that in other neutral lipids also suggests an active phosphatidate phosphatase in this dermatophyte.[49] This enzyme also provides substrate, DG, for phospholipid synthesis by hydrolyzing PA. The presence of phosphatidate phosphatase has been reported in mitochondrial and microsomal fractions with a pH optimum of 6.0; however, *E. floccosum* enzyme was more active than that of *M. gypseum*. The enzyme from both the dermatophytes required Mg^{2+} for its activity and was sensitive to Mn^{2+}, Fe^{2+} and B^{2+}.[75]

There are three main pathways for the catabolism of phospholipids (Figure 7). The first pathway could be initiated by phospholipase A, resulting in the formation of lysophospholipid which is then degraded by lysophospholipase to give glycerophosphoryl base. It is subsequently hydrolyzed by a diesterase to give glycerol-3-PO_4 and a free base. Second pathway operates through phospholipase C resulting in the formation of DG. Third pathway of phospholipid degradation is through phospholipase D, which forms phosphatidic acid and a free base.

B. PHOSPHOLIPASE A

Phospholipase A activity has been detected in *T. rubrum*,[63] *E. floccosum*, *M. cookei* and *T. mentagrophytes*.[76] In *T. rubrum*, the extracellular phospholipase A was stimulated by undecanoic acid in an undecanoic acid-sensitive strain of *T. rubrum*. Phospholipase A is either A_1 or A_2 type, depending upon the site of ester bond cleaved. The activity in *E. floccosum* was mainly of the A_1 type and A_2 in *M. cookei* while *T. mentagrophytes* had both A_1 and A_2 activities.[77] In stationary phase cultures, the activity of phospholipase increased in *M. cookei* and *E. floccosum* while its activity decreased in *T. mentagrophytes* during this phase of growth.[77] Its activity has been found to be mainly associated with the microsomal fractions of *E. floccosum*[78,79] and was significantly inhibited only by Cu^{2+} and EDTA. The most potent metal ion activator of this enzyme was Hg^{2+}.[35] Ca^{2+} was not required for enzyme activity in *E. floccosum*. This observation is in contrast to the absolute Ca^{2+} requirement known for phospholipase A activity in snake venom and insects,[80] *E. coli* [81] and *Streptomyces griseus.*[82] All sulfhydryl reagents acted as activators of the enzyme in *E. floccosum*.

C. LYSOPHOSPHOLIPASE

Removal of the cytotoxic lysophospholipids produced by the action of intracellular or extracellular phospholipase A could be one of the major functions of lysophospholipase. Lysophospholipase activity in *E. floccosum* was found to be present in both mitochondrial as well as the microsomal fraction.[79] It was stimulated by Ca^{2+}, ethanol and non-ionic detergents but was inhibited by EDTA, Fe^{2+} and Hg^{2+}. Among various -SH group reagents, only sodium fluoride was inhibitory to *E. floccosum* enzyme while others were mild activators.[35]

Another enzyme that combines the function of phospholipase A and lysophospholipase is phospholipase B. It has been detected in many organisms and studied in great detail in *Penicillium notatum*. It first attacks on position-2 and then on position-1.[83,84] This enzyme has not yet been detected in dermatophytes.

D. PHOSPHOLIPASE C

It has been detected in microsomal fraction of *E. floccosum.*[79] The activity of this enzyme was not affected in the presence or absence of divalent metal ions such as Mg^{2+} and Ca^{2+} but was activated by Ba^{2+} and inhibited by Ca^{2+}, Fe^{2+} and EDTA. Differences in the enzyme structure could be one of the reasons for variation in metal ion requirements. Maximum increase in the activity was observed with deoxycholate. In *E. floccosum*, the enzyme was activated to some extent by β-mercaptoethanol, whereas all other-SH reagents were inhibitory indicating involvement of -SH groups in enzyme activity.[35]

E. GLYCEROPHOSPHOBASE DIESTERASE

Glycerophosphoryl base formed after the action of phospholipase A and lysophospholipase is degraded by glycerophosphobase diesterase to yield glycerol-3-PO_4 and base. The base could be choline or ethanolamine or any other as it is not specific for a particular base. In *E. floccosum* glycerophosphorylcholine diesterase, which degrades glycerophosphorylcholine, was detected by Chopra and Khuller.[78] Glycerol-3-PO_4 formed by the action of this enzyme can be utilized for the synthesis of PA, which is again used for lipid synthesis.

F. LIPASE

Lipolytic enzymes are important for the development of dermatomycosis and have been investigated extensively in many dermatophytes.[3,4,73,85] High lipase activity was found in *E. floccosum* and *Trichophyton* species while comparatively low activities were found in *Microsporum* species.[73] To ascertain the contribution of fatty acids from acylglycerols for cellular utilization in *E. floccosum*, the presence of an intracellular lipase was investigated.[85] Lipases are essential for the turnover of neutral lipids, a function similar to phospholipases for phospholipid turnover. Das and Banerjee[63,86] reported the presence of intracellular and extracellular lipases. They also showed the effect of undecanoic acid on the extracellular lipases of undecanoic acid sensitive and resistant strains of *T. rubrum*. However, extracellular lipase was detected in *M. canis*, *E. floccosum* and *T. mentagrophytes* but not in *T. rubrum*.[3] Hellgren and Vincent[4] characterized the extracellular lipase of *E. floccosum* and observed that its activity was not significantly altered by skin lipids or body temperature. Detection of an intracellular lipase in *E. floccosum* showed that this enzyme participates in cellular lipid metabolism of the fungus, besides assisting it in the possible invasion of host tissues, as suggested earlier.[3,4] Variations in the lipolytic activity of various laboratory strains of dermatophytes with time were found which could be partially attributed to pleomorphous mutation, leading to loss or acquisition of different enzymatic equipment.[73] Phospholipase D has not been identified in dermatophytes though its presence has been shown in other fungi e.g., *Neurospora crassa*.[87,88]

VI. LIPIDS AND ANTIFUNGAL DRUGS

Dermatophytes are responsible for the superficial mycoses called dermatomycosis in animals and human beings. When a eukaryote infects a eukaryotic organism, the range of physiological differences between host and parasite are much smaller than observed with bacterial infections. So, it becomes necessary to have antifungal drugs especially effective against these organisms and also it is important to know the mode of action of these drugs. Various classes of antifungal drugs are available which exert their toxic effects at different levels. Some of these drugs act at the membrane level, leading to permeability changes and altered membrane function, or inhibit a particular enzyme of a biosynthetic pathway, which can be used to bring about selective alterations in the lipid composition to a certain extent (for detail on antifungals see chapters 11 and 12 of this volume).

Cerulenin, an antilipogenic antibiotic, was used to modulate the lipid composition in *E. floccosum*.[26,27] This antibiotic had a strong inhibitory effect on *in vivo* synthesis of fatty acid, sterol and phospholipids.[89] The inhibitory effect of cerulenin on growth and lipid synthesis could be reversed by

Table 5
Effects of Nystatin on L-Lysine Uptake in *E. floccosum* Cells[a]

	Uptake (nmol/mg dry wt/7 min)		
	Sterol content (mg/g dry wt)	Minus nystatin	Plus nystatin (25 μg/ml)
Control	8.04±0.91	5.25±0.46	3.92±0.23[b]
Palmitic acid	7.81±0.39[c]	1.85±0.38	1.38±0.17[c]
Palmitic acid + Oleic acid	9.03±0.28[c]	4.95±0.29	3.69±0.21[d]
Cerulenin + Palmitic acid	13.74±0.10[d]	2.52±0.52	1.28±0.29[b]
Cerulenin + Palmitic acid + Oleic acid	13.32±0.94[d]	3.31±0.31	3.00±0.15[c]
Cholesterol	10.26±0.57[b]	4.80±0.34	1.78±0.26[e]
Ergosterol	10.13±0.57[b]	3.81±0.35	1.96±0.17[d]
Cerulenin + Cholesterol	14.25±1.59[d]	8.67±0.93	2.87±0.24[e]
Cerulenin + Ergosterol	10.88±1.88[c]	5.99±0.42	4.20±0.29[d]

[a]Data are compiled from Sanadi, S., Pandey, R. and Khuller, G.K., *Lipids*, 23, 435, 1988a; Sanadi, S., Pandey, R. and Khuller, G.K., *Indian J. Biochem. Biophys.*, 25, 442, 1988b; Sanadi, S., Pandey, R. and Khuller, G.K., *Biochim. Biophys. Acta*, 921, 341, 1987.
[b]$p \leq 0.05$; [c]Not significant; [d]$p \leq 0.01$; [e]$p \leq 0.001$.

supplementation with fatty acids/sterols in the growth medium[26,27] showing that phospholipids, fatty acids and sterols are necessary for growth of the organism. Cerulenin, however, does not have any clinical application, but provides a convenient method for manipulating the lipid composition and determining the effect of altered lipid composition on membrane structure and function.

The most effective antibiotics widely used in the treatment of fungal infections are polyenes. The target of these drugs is the plasma membrane and their selectivity is based upon the difference in sterol composition of the fungus and mammalian cells. It has been postulated that the strong affinity of the hydrophilic portion of the polyene for ergosterol via hydrogen bonds allows the drug to penetrate the membrane with the hydrophobic portion sitting beside the sterol ring.[90,91] These antibiotics are highly effective in the treatment of systemic mycoses but are ineffective in superficial and cutaneous mycoses caused by dermatophytes. Capek and Simek[92] observed that the loss of sensitivity towards polyenes in *M. gypseum* and *T. mentagrophytes* was associated with decreased sterol content. In *E. floccosum*, in the presence of nystatin, irrespective of the sterol content, uptake of amino acid was decreased (Table 5), which indicates that lipids other than sterols influence the overall sensitivity of the *E. floccosum* membrane towards the polyenes (Sanadi and Khuller, unpublished observations).

Another class of antifungals, the azoles, interfere with sterol synthesis and are classified as ergosterol biosynthesis inhibitors. The specific point of inhibition of azole drugs is the microsomal cytochrome P450 dependent lanosterol 14α-demethylase system.[93,94] Azoles disturb the permeability characteristics of the membranes and allow leakage of essential precursors, metabolites, ions and other intracellular compounds with consequent inhibition of macromolecular synthesis.[95,96] Azoles have been widely used to treat superficial infections caused by yeasts and dermatophytes and the minimum inhibitory concentration range has also been reported by Iwata and Yamaguchi.[97] Effect of imidazoles on the germination of arthrospores and microconidia of *T. mentagrophytes* was studied by Scott et al.[98] Reduction in ergosterol content and changes in the ultrastructure of the plasma membrane of *T. mentagrophytes* on treatment with clotrimazole have been reported.[99] Efflux of K^+ was observed in the presence of high concentrations of miconazole and clotrimazole, though miconazole was more effective.

Griseofulvin is an orally effective antimicrobial agent for superficial fungal infections of the skin, with highest efficacy against dermatophytes.[100] It is only fungistatic and not fungicidal even at high concentrations.[101] *In vitro*, griseofulvin is active against all the dermatophytes at a concentration of 1 μg/ml. With electron microscopic studies, changes in cytoplasm and destruction of certain intra-cytoplasmic structures[102] were observed without affecting the cell wall chitin.[103] Griseofulvin inhibited nucleic acid synthesis[104] and it did bind to lipids within the cell.[105] The effect of griseofulvin on lipid composition and membrane permeability has been investigated[106] in *M. gypseum*. Mycelia grown in medium containing this drug (IC_{50} concentration) possessed a lower content of total lipids, phospholipids, sterols and an increased U/S fatty acid ratio. This inhibitory effect was further supported by decreased incorporation of [^{14}C]-acetate in total lipids, phospholiplids and sterols. The effect of griseofulvin on membrane permeability has also been investigated and the results indicated increased K^+ efflux and also greater leakage of intracellular [^{32}P]-labeled components from prelabeled cells.[106]

Recently, a new class of antimycotic agents has been discovered which are allylamine derivatives. This group includes naftifine, terbinafine, tolnaftate and the more potent orally active derivative SF86-327. These are particularly active against dermatophytes *in vitro* and *in vivo.*[107,108] These antifungals inhibit ergosterol biosynthesis with concomitant accumulation of squalene[109,110] due to inhibition of squalene epoxidase.[108-111] The antifungal spectrum and maximum inhibitory concentrations of the antidermatophyte activity of SF86-327 have also been reported.[108]

When lipids were extracted from *T. mentagrophytes* cells grown in the presence of naftifine[109] and SF86-327,[112,113] it was found that there is accumulation of nonpolar compounds that co-migrated with squalene on thin layer chromatography.[114] This increased content of squalene was due to inhibition of squalene epoxidase. When the incorporation of [^{14}C]-acetate into ergosterol and squalene was studied in *T. mentagrophytes*, it was found that incorporation increased in squalene while it decreased in ergosterol fraction[113] (Figure 8). Ryder[111] developed a system to measure ergosterol biosynthesis towards the end of the pathway in fungal cells by exploiting the C-24 methylation of the side-chain. As the methyl group at C-24 is derived by transmethylation from S-adenosylmethionine,[113] this approach was used for measurement of sterol synthesis in cells treated with SF86-327 (1mg/litre).

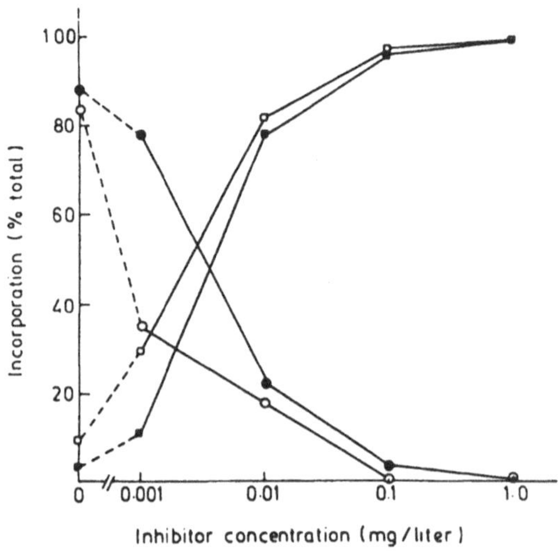

Figure 8. Effects of SF 86-327 and naftifine on incorporation of [^{14}C]-acetate into ergosterol and squalene in *T. mentagrophytes* cells. Shown are ergosterol (O) and squalene (□) in presence of SF 86-327 and ergosterol (●) and squalene (■) in the presence of naftifine (from Ryder, N.S., *Antimicrob. Agents Chemother.*, 27, 252, 1985. With permission).

Squalene epoxidation was observed to be completely inhibited in *T. mentagrophytes* in the presence of SF86-327.[111] Ryder et al.[115] also reported that *T. mentagrophytes* was killed at allylamine concentrations well below those required for 95% inhibition of ergosterol biosynthesis. Exposure of the cells to very high concentrations of allylamine indicated that cell death resulted from high intracellular squalene levels[112] which increased membrane permeability by insertion between the membrane lipids.[116] In *T. mentagrophytes*, the time of cell death coincided with squalene accumulation (after 12 h of incubation), whereas ergosterol depletion was maximum within the first few hours of incubation.[111]

At a concentration that inhibits growth of *T. mentagrophytes* naftifine decreased phospholipid content[114] and caused swelling of the hyphal tips.[117] Studies on the topically administered antibiotics, tolnaftate, revealed a high level of *in vitro* activity against dermatophytes.[118] In *T. quinckeanum*, it inhibited at the level of squalene epoxidation and squalene was observed to accumulate in dermatophytes,[119] similar to the effects observed with naftifine[109] and terbinafine.[113] It also caused a dose-dependent inhibition of [^{14}C]-acetate incorporation into sterols and there was accumulation of radioactivity in squalene fraction in *T. mentagrophytes.*[115] In addition, experiments using radiolabelled squalene-2,3-epoxide as substrate (thus by-passing the epoxidase) have failed to detect any significant effect on later enzymes of the sterol pathway, thereby ruling out the possibility of inhibition of steps subsequent to squalene epoxidation.[112]

Studies were also performed to see the effect of allylamine derivative, tolnaftate, on lipids and macromolecular synthesis in *M. gypseum.*[120] Cells grown in the presence of tolnaftate (at the IC_{50}) showed a decrease in the content of phospholipids and sterols, showing at the same time an increase in RNA content. Synthesis of different macromolecules (lipids, proteins, RNA and DNA) was inhibited in the presence of tolnaftate except for RNA synthesis which increased.[120] Activity of membrane bound enzymes, phosphodiesterase and 5'-nucleotidase, did not change on treatment with tolnaftate whereas there was greater leakage of intracellular [^{32}P]. Macromolecular composition of sensitive and resistant strains revealed higher content of phospholipids in resistant strains while the content of other macromolecules (sterols, proteins, DNA and RNA) was comparable to that of a susceptible strain. Phosphodiesterase enzyme was less active in resistant strain which could be responsible for higher content of phospholipid.[121] It was suggested that fungistatic action of tolnaftate was due to its effect on various target sites in different ways.

VII. CONCLUSION

From the information available on dermatophytes, it seems that extensive work has been done on lipid composition and its biosynthesis using different precursors but some of the enzymes of biosynthetic and catabolic pathways need further attention. From the enzymes identified in dermatophytes, it is clear that the main pathways of lipid metabolism are common with yeast and other fungi. The lipid catabolizing enzymes, lipases and phospholipases, were suggested to be involved in pathogenesis of this fungus, around three decades back, but to date not much work has been done in this direction and the exact role is still not known. Preliminary studies on the role of second messengers cAMP, Ca^{2+} and calmodulin[122-138] in lipid synthesis have recently been carried out; however, more work is needed in this area to elucidate the exact mechanism of action. Another aspect which demands further research is regulation of phospholipid synthesis and its correlation with the synthesis of other macromolecules.[122-138] Further, the role of acyl carrier protein(s) in fatty acid and phospholipid synthesis should also be examined.

VIII. REFERENCES

1. **Kappe, R., Le vetz, M., Cassone, A. and Washburn, R.G.,** Mechanisms of host defence against fungal infection, *J. Med. Vet. Mycol.*, 30, 167, 1992.
2. **Samson, R.A.,** Mycotoxins, a mycologist's perspective, *J. Med. Vet. Mycol.*, 30, 9, 1992.
3. **Hellgren, L. and Vincent, J.,** Lipolytic activity of some dermatophytes, *J. Med. Microbiol.*, 13, 155, 1980.
4. **Hellgren, L. and Vincent, J.,** Lipolytic activity of some dermatophytes. II. Isolation and characterisation of the lipase of *Epidermophyton floccosum*, *J. Med. Microbiol.*, 14, 347, 1981.
5. **Vincent, J.,** Dermatophyte lipids, *Prog. Chem. Fats and Other Lipids*, 16, 171, 1978.
6. **Bansal, V.S. Talwar, P. and Khuller, G.K.,** Allergenic delayed skin reactions from phospholipid fractions of dermatophytes, I.R.C.S, *Med. Sci. Libr. Compend.*, 10, 335, 1982a.
7. **Burdon, K.L.,** Fatty material in bacteria and fungi, *J. Bacteriol.*, 52, 665, 1946.
8. **Akasaka, T.,** Microscopic studies on the fat staining preparations of fungi, *Jap. J. Dermatol.*, 63, 477, 1953.
9. **Al-Doory, Y. and Larsh, H.W.,** Quantitative studies of total lipids of pathogenic fungi, *Appl. Microbiol.*, 10, 492, 1962.
10. **Goetz, H. and Pascher, G.,** Ueber die chemische zusammensetzung des *Trichophyton mentagrophytes var granulosum*, *T. schoenleini* and *T. rubrum* in vergleich zu schimmelpilzen (*Pencillium, Cladosporium Herbarum and Fusarium*), *Dermatologica*, 124, 31, 1962.
11. **Bansal, V.S.,** Studies on the lipids of dermatophytes, Ph.D. thesis, PGIMER, Chandigarh, India, 1981.
12. **Swanson, R. and Stock, J.J.,** Biochemical alterations of dermatophytes during growth, *Appl. Microbiol.*, 14, 438, 1966.
13. **Vincent, J.,** Contribution to the study on fatty acids of *Epidermophyton floccosum*, *Zentralbl. Bacteriol. Parasitenkd: Infektionskr. Hyg. Abt.* I.: Orig. Reihe A, 233, 410, 1975.
14. **Rawat, D.S. and Das, S.K.,** Lipid composition of *M. gypseum*, *Microbiologica (Bologna)*, 5, 361, 1982.
15. **Zalokar, M.,** Integration of cellular metabolism, in *The Fungi*,Vol. 1. Anisworth, G.C. and Sussman, A.F., Eds., Academic Press, New York, London, 1965.
16. **Dill, B.C., Leighton, T.J. and Stock, J.C.,** Physiological and biochemical changes associated with macroconidial germination in *Microsporum gypseum*, *J. Appl. Microbiol.*, 24, 977, 1972.
17. **Shah, V.K. and Knight, S.C.,** Chemical composition of hyphal walls of dermatophytes, *Arch. Biochem. Biophys.*, 127, 229, 1968.

18. **Nozawa, Y., Kitayima, Y. and Ito, Y.**, Chemical and ultrastructural studies of isolated cell walls of *Epidermophyton floccosum*. Presence of chitin inferred from x-ray diffraction analysis and electron microscopy, *Biochim. Biophys. Acta*, 307, 92, 1973.
19. **Prince, H.N.**, Total lipid composition of *Trichophyton mentagrophytes*, *J. Bacteriol.*, 79, 154, 1960.
20. **Jones, M.G. and Noble, W.C.**, A study of fatty acids as a taxonomic tool for dermatophyte fungi, *J. Appl. Bacteriol.*, 50, 577, 1981.
21. **Kostiw, L.L., Vicher, E.E. and Lyon, I.**, Fatty acid synthesis in *Trichophyton rubrum*, *Mycopathol. Mycol. Appl.*, 49, 67, 1973.
22. **Swenson, F.J. and Ulrich, J.A.**, Fatty acids of dermatophytes, *Sabouraudia*, 18, 1, 1980.
23. **Khuller, G.K., Chopra, A., Bansal, V.S. and Masih, R.**, Lipids of dermatophytes, *Lipids*, 16, 20, 1981a.
24. **Larroya, S. and Khuller, G.K.**, Lipids of dermatophytes. II. Effect of growth conditions on the lipid composition and membrane transport of *Microsporum gypseum*, *Lipids*, 20, 11, 1985.
25. **Yamada, T., Watanbe, R., Nozawa, Y. and Ito, Y.**, Lipid composition and its alteration during the growth stage in pathogenic fungus *Epidermophyton floccosum*, *Shinkinto shinkinsho*, 19, 229, 1978.
26. **Sanadi, S., Pandey, R. and Khuller, G.K.**, Lipids of Dermatophytes. III. Sterol induced changes in the lipid composition and functional properties of *Epidermophyton floccosum*, *Lipids*, 23, 435, 1988a.
27. **Sanadi, S., Pandey, R. and Khuller, G.K.**, Role of fatty acids in regulation of membrane properties in *Epidermophyton floccosum*, *Ind. J. Biochem. Biophys.*, 25, 442, 1988b.
28. **Bharti, G.**, Effect of sterol supplementation on lipid composition of *M. gypseum*: Structural and functional aspects, M.Sc. Thesis, PGIMER, Chandigarh, India, 1987.
29. **Chugh, I.B.**, Effect of antifungal drugs on macromolecular composition in *Microsporum gypseum*, M.Sc. Thesis, PGIMER, Chandigarh, India, 1988.
30. **Sanadi, S.**, Involvement of lipids in structure and functions of membranes of *Epidermophyton floccosum*, Ph.D. Thesis, PGIMER, Chandigarh, India, 1987.
31. **Pandey, R., Verma, R.S. and Khuller, G.K.**, Effect of choline, ethanolamine and serine supplementation on the membrane properties of *Microsporum gypseum*, *Lipids*, 22, 530, 1987b.
32. **Das, S.K. and Banerjee, A.B.**, Phospholipids of *Trichophyton rubrum*, *Sabouraudia*, 12, 281, 1974.
33. **Khuller, G.K.**, Phospholipids of *Epidermophyton floccosum*, *Ind. J. Med. Res.*, 69, 60, 1979.
34. **Kish, Z. and Jack, R.C.**, Phospholipids of two strains of dermatophyte, *Arthroderma uncinatum*, *Lipids*, 9, 264, 1974.

35. **Chopra, A.**, Metabolism of phospholipids in *Epidermophyton floccosum*, Ph.D. thesis, Punjab University, Chandigarh, India, 1982.
36. **Khuller, G.K., Verma, J.N., Bansal, V.S. and Talwar, P.**, Changes in the phospholipid composition of *Microsporum gypseum* during growth, *Ind. J. Med. Res.*, 68, 23, 1978.
37. **Brennan, P.J. and Lösel, D.M.**, Physiology of fungal lipids, *Adv. Lipid Res.*, 17, 47, 1978.
38. **Tsujisaka, Y., Okumura, S. and Iwai, M.**, Glyceride synthesis by four kinds of microbial lipase, *Biochim. Biophys. Acta*, 489, 415, 1977.
39. **Rodriguez, R.J., Low, C., Bollema, C.D.K. and Parks, L.W.**, Multiple functions for sterols in *Saccharomyces cerevisiae*, *Biochim. Biophys. Acta*, 837, 336, 1985.
40. **Okazaki, K. and Tamemasa, O.**, Studies on the chemical composition of dermatophytes, *J. Pharm. Soc. Jap.*, 75, 1087, 1955.
41. **Wirth, J.C., O'Brien, P.J., Schmitt, F.L. and Sohler, A.**, The isolation in crystalline form of some of the pigments of *Trichophyton rubrum*, *J. Invest. Dermatol.*, 29, 47, 1957.
42. **Audette, R.C.S., Baxter, R.M. and Walker, G.C.**, A study of the lipid content of *Trichophyton mentagrophytes*, *Can. J. Microbiol.*, 7, 282, 1961.
43. **Wirth, J.C., Beesley, T. and Miller, W.**, The isolation of a unique sterol from the mycelium of a strain of *Trichophyton rubrum*, *J. Invest. Dermatol.*, 37, 153, 1961.
44. **Blank, F., Shortland, F.E. and Just, G.**, The free sterols of dermatophytes, *J. Invest. Dermatol.*, 39, 91, 1962.
45. **Das, S.K. and Banerjee, A.B.**, Sterol and sterol esters of *Trichophyton rubrum*, *Ind. J. Med. Res.*, 65, 70, 1977c.
46. **Vaidya, S., Bharti, G., Pandey, R. and Khuller, G.K.**, Effect of altered sterol levels on the transport of amino acids on membrane structure of *Microsporum gypseum*, *J. Biol. Sci.*, 13, 235, 1988.
47. **Yamada, T. and Nozawa, Y.**, An unusual lipid in the human pathogenic fungus *Epidermophyton floccosum*, *Biochim. Biophys. Acta*, 574, 433, 1979.
48. **Makula, R.A. and Finnerty, W.R.**, Isolation and characterisation of an ornithine containing lipid from *Desulfovibrio gigas*, *J. Bacteriol.*, 123, 523, 1975.
49. **Chopra, A. and Khuller, G.K.**, Metabolism of lipids in *Epidermophyton floccosum*, *Indian J. Med. Res.*, 73, 325, 1981a.
50. **Bindra, A. and Khuller, G.K.**, Correlation between intracellular cAMP levels and phospholipids of *Microsporum gypseum*, *Biochim. Biophys. Acta*, 1124, 185, 1992.
51. **Bansal, V.S., Chopra, A., Kasinathan, C. and Khuller, G.K.**, *In vivo* studies on phospholipids biosynthesis in *Microsporum gypseum*, *Ind. J. Med. Res.*, 76, 832, 1982b.

52. **Chopra, A. and Khuller, G.K.**, Phosphatidate metabolism in *Epidermophyton floccosum*, *FEMS Microbiol. Lett.*, 10, 189, 1981b.
53. **Das, S.K. and Banerjee, A.B.**, Phospholipid turnover in *Trichophyton rubrum*, *Sabouraudia*, 15, 99, 1977b.
54. **Pandey, R.**, Effect of supplementation on nitrogenous bases on the phospholipids of *Microsporum gypseum*: structural and functional aspects, Ph.D. Thesis, PGIMER, Chandigarh, India, 1986.
55. **Raetz, C.R.H.**, Enzymology, genetics and regulation of membrane phospholipid synthesis in *Escherichia coli*, *Bacteriol. Rev.*, 42, 614, 1978.
56. **Bansal, V.S., Chopra, A., Kasinathan, C. and Khuller, G.K.**, Biosynthesis of neutral lipids in *Microsporum gypseum*, *Sabouraudia*, 19, 223, 1981.
57. **Hori, T., Itakasa, O., Sugita, M. and Arakawa, J.**, Isolation of sphingoethanolamine from pupae of the green bottle fly, *Lucillia caesar*, *J. Biochem. (Tokyo)*, 64, 124, 1968.
58. **Tate, P.**, On the enzymes of certain dermatophytes or ringworm fungi, *Parasitology*, 21, 31, 1929.
59. **Kasinathan, C., Govindarajan, U., Chopra, A. and Khuller, G.K.**, Phospholipid synthesizing enzymes of dermatophytes. II. Characterization of choline kinase, *Lipids*, 18, 727, 1983.
60. **Govindrajan, U.**, Studies on enzymes of lipid metabolism in *Epidermophyton floccosum*, M.Sc. Thesis, PGIMER, Chandigarh, India, 1983.
61. **Kasinathan, C. and Khuller, G.K.**, Phospholipid synthesizing enzymes of dermatophytes. III. Glycerol kinase of dermatophytes, *Lipids*, 19, 289, 1984.
62. **Kasinathan, C.**, Some aspects of major phospholipid biosynthesis in *Microsporum gypseum*, Ph.D. Thesis, PGIMER, Chandigarh, India, 1983.
63. **Das, S.K. and Banerjee, A.B.**, Lipolytic enzymes of *Trichphyton rubrum*, *Sabouraudia*, 15, 313, 1977a.
64. **Hosaka, K. and Yamashita, S.**, Choline transport in *Saccharomyces cerevisiae*, *J. Bacteriol.*, 143, 176, 1980.
65. **Khuller, G.K., Kasinathan, C., Bansal, V.S. and Chopra, A.**, Effect of glycerol substitution and choline/ethanolamine supplementation on phospholipid, *Ind. J. Exp. Biol.*, 19, 1054, 1981b.
66. **Vance, D.E. and Pelech, S.L.**, Enzyme translocation in the regulation of phosphatidylcholine biosynthesis, *TIBS*, 9, 17, 1984.
67. **Bindra, A.**, Influence of cyclic AMP levels on phospholipid synthesis in *Microsporum gypseum*, Ph.D. Thesis, PGIMER, Chandigarh, India, 1993.
68. **Hirata, F., Viveros, O.H., Diliberto, E.J. and Axelrod, J.**, Identification and properties of two methyl transferases in conversion of phosphatidylethanolamine to phosphatidylcholine, *Proc. Natl. Acad. Sci. U.S.A.*, 75, 1718, 1978.

69. **Hirata, F. and Axelrod, J.,** Enzymatic methylation of phosphoethanolamine increases erythrocyte membrane fluidity, *Nature*, 275, 219, 1978.
70. **Sastry, B.V.R., Statham, C.N., Axelrod, J. and Hirata, F.,** Evidence for two methyltransferases involved in the conversion of phosphatidylethanolamine to phosphatidylcholine in the rat liver, *Arch. Biochem. Biophys.*, 211, 762, 1981.
71. **Bindra, A. and Khuller, G.K.,** Influence of cyclic AMP on the biosynthesis of phosphatidylcholine in *Microsporum gypseum*, *Ind. J. Biochem. Biophys.*, 30, 311, 1993.
72. **Vaidya, S. and Khuller, G.K.,** Effect of dibutyryl cyclic AMP on lipid synthesis in *Microsporum gypseum*, *Biochim. Biophys. Acta*, 960, 435, 1988.
73. **Nobre, G. and Viegas, M.P.,** Lipolytic activity of dermatophytes, *Mycopathologia et Mycologia Applicata*, 46, 319, 1972.
74. **Khuller, G.K.,** Lipid composition of *Microsporum gypseum*, *Experentia*, 34, 432, 1978.
75. **Kasinathan, C., Chopra, A. and Khuller, G.K.,** Phosphatidate phosphatase of dermatophytes, *Lipids*, 17, 859, 1982.
76. **Kasinathan, C. and Khuller, G.K.,** Biosynthesis of major phospholipids of *Microsporum gypseum*, *Biochim. Biophys. Acta*, 752, 187, 1983.
77. **Chopra, A., Larroya, S. and Khuller, G.K.,** Studies on the phospholipase A of dermatophytes, *Curr. Microbiol.*, 41, 171, 1981.
78. **Chopra, A. and Khuller, G.K.,** Lipids of pathogenic fungi, *Prog. Lipid Res.*, 22, 189, 1983.
79. **Chopra, A. and Khuller, G.K.,** Lipid metabolism in fungi, *CRC Crit. Rev. Microbiol.*, 11, 209, 1984.
80. **Rao, R.H. and Subrahmanyam, D.,** Distribution and properties of phospholipase A of *Culex pipiens* fatigens, *Arch. Biochem. Biophys.*, 140, 443, 1970.
81. **Bernard, M.C., Brisou, J., Denis, F. and Rosenberg, A.J.,** Metabolisme phospholipasique chez les bacteries a gram negatif purification et etude cinetique de la phospholipase A_2 Soluble d' *E coli* 0118, *Biochimie*, 55, 377, 1973.
82. **Verma, J.N., Bansal, V.S., Khuller, G.K. and Subrahmanyam, D.,** Characterization of phospholipase A in *Streptomyces griseus*, *Ind. J. Med. Res.*, 72, 487, 1980.
83. **Saito, Y. and Kates, M.,** Substrate specificity of a highly purified phospholipase B from *Penicillium notatum*, *Biochim. Biophys. Acta*, 369, 245, 1974.
84. **Kawasaki, N., Sugatini, J. and Saito, K.,** Studies on a phospholipase B from *Pencillium notatum*. Purification, properties and mode of action, *J. Biochem.*, 77, 1233, 1975.

85. **Chopra, A., Asotra, S. and Khuller, G.K.**, Intracellular lipase of *Epidermatophyton floccosum*, *IRCS Med. Sci. Libr. Compend.*, 10, 803, 1982.
86. **Das, S.K. and Banerjee, A.B.**, Effect of undecanoic acid on the production of exocellular lipolytic and keratinolytic enzymes by undecanoic acid sensitive and resistant strains of *Trichophyton rubrum*, *Sabouraudia*, 20, 179, 1982b.
87. **Chakravarti, D.N., Chakravarti, B. and Chakravarti, P.**, Studies on phospholipase activities in *Neurospora crassa* mycelia, *Lipids*, 15, 830, 1980.
88. **Chakravarti, D.N., Chakravarti, B. and Chakravarti, P.**, Studies on phospholipase activities in *Neurospora crassa* conidia, *Arch. Biochem. Biophys.*, 206, 392, 1981.
89. **Sanadi, S., Pandey, R. and Khuller, G.K.**, Reversal of cerulenin induced inhibition of phospholipids and sterol synthesis by exogenous fatty acid/sterols in *Epidermophyton floccosum*, *Biochim. Biophys. Acta*, 921, 341, 1987.
90. **Norman, A.W., Spielvogel, A.M. and Wong, R.G.**, Polyene antibiotics-sterol interaction, *Adv. Lipid Res.*, 14, 127, 1976.
91. **Gale, E.F.**, The release of potassium ions from *Candida albicans* in the presence of polyene antibiotics, *J. Gen. Microbiol.*, 80, 451, 1974.
92. **Capek, A. and Simek, A.**, Antimicrobial agents. XIV. Content of ergosterol and the problem of transient resistance of dermatophytes, *Folia Microbiol. (Prague)*, 17, 239, 1972.
93. **Wilkinson, C.F., Hetnarki, K. and Yellin, T.O.**, Imidazole derivative – A new class of microsomal enzyme inhibitors, *Biochem. Pharmacol.*, 21, 3187, 1972.
94. **Vanden Bossche, H., Lauwers, W., Willemsons, G., Marichal, P., Cornelissen, C. and Cols, W.**, Molecular basis for the antimycotic and antibacterial activity of N-substituted imidazoles and triazoles, inhibition of isoprenoid biosynthesis, *Pestic. Sci.*, 15, 188, 1983.
95. **Iwata, K., Yamaguchi, H. and Hiratani, T.**, Mode of action of clotrimazole, *Sabouraudia*, 11, 158, 1973.
96. **Kerridge, D.**, Mode of action of clinically important antifungal drugs, *Adv. Microbiol. Physiol.*, 27, 1, 1986.
97. **Iwata, K. and Yamaguchi, H.**, Mechanism of action of clotrimazole. Survey of the principal action on *Candida albicans* cells, *Jap. J. Bacteriol.*, 28, 513, 1973.
98. **Scott, E.M., Gorman, S.P. and Wright, C.R.**, The effect of imidazoles on germination of arthrospores and microconidia of *Trichophyton mentagrophytes*, *J. Antimicrob. Chemother.*, 13, 101, 1984.

99. **Scott, E.M., Gorman, S.P., Millership, J.S. and Wright, L.R.,** Effect of miconazole and clotrimazole on K^+ release and inhibition of ergosterol biosynthesis in *Trichophyton mentagrophytes* and related ultrastructural observations, *J. Antimicrob. Chemother.*, 17, 423, 1986.
100. **Shah, V.P.,** Griseofulvin absorption per/os and percutaneous Preusser, eds, *Medical Mycology, Zentralbr. Bakteriol. Parasitenkd. Infectionskr*, Suppl. 8, Gustave Fishcer Verlag Stuttgart, New York, 1980.
101. **Rippon, J.W.,** In *Medical Microbiology, The Pathogenic fungi and the Pathogenic Actinomycetes*, 2nd ed., Saunders, Philadelphia, 1982.
102. **Blank, H., Taplin, D. and Roth, F.J., Jr.,** Electron microscopic observations of the effects of griseofulvin on dermatophytes, *Arch. Dermatol.*, 81, 669, 1960.
103. **Brian, P.W.,** Studies on the biological activity of griseofulvin, *Ann. Biol.* (London), 13, 59, 1949.
104. **Glynn, E., Evans, V. and Gentle, J.C.,** *Essentials of Medical Mycology*, Churchill Livingstone, Edinburgh, Scotland, 30,1985.
105. **Weinstein, G.D. and Blank, H.,** Quantitative determination of griseofulvin by a spectrophotometric assay, *Arch. Dermatol.*, 81, 746, 1960.
106. **Chugh, I.B., Gupta, M.P. and Khuller, G.K.,** Effect of griseofulvin on lipid composition and membrane integrity in *Microsporum gypseum*, *J. Bio. Sci.*, 16, 243, 1991.
107. **Georgopoulos, A., Petranyi, G., Mieth, H. and Drews, J.,** *In vitro* activity of naftifine, a new antifungal agent, *Antimicrob. Agents Chemother.*, 19, 386, 1981.
108. **Petranyi, G., Ryder, N.S. and Stiiz, A.,** Allylamine derivatives: New class of synthetic antifungal agents inhibiting fungal squalene epoxidase, *Science*, 224, 1239, 1984.
109. **Ryder, N.S., Seidl, G. and Troke, P.F.,** Effect of the antimycotic drug naftifine on growth and sterol biosynthesis in *Candida albicans*, *Antimicrob. Agents Chemother.*, 25, 483, 1984.
110. **Ryder, N.S. and Dupoint, M.C.,** Inhibition of squalene epoxidase by allylamine antimycotic compounds: a comparative study of the fungal and mammalian enzymes, *Biochem. J.*, 130, 765, 1985.
111. **Ryder, N.S.,** Biochemical mode of action of the allylamine antimycotic agents naftifine and SF86-327, in *In vitro and in vivo Evaluation of Antifungal Agents*, Iwata, K. and Vanden Bossche, H., Eds., Elsevier, Amsterdam, 89, 1986.
112. **Ryder, N.S.,** *Selective Inhibition of Squalene Epoxidation by Allylamine Antimycotic Agents,* Nombela, C., Ed., Elsevier, Amsterdam, 313,1984.
113. **Ryder, N.S.,** Specific inhibition of fungal sterol biosynthesis by SF86-327, a new allylamine agent, *Antimicrob. Agents Chemother.*, 27, 252, 1985.

114. **Paltauf, G., Daum, G., Zuder, G., Hogenauer, G., Schulz, G. and Seidl, G.**, Squalene and ergosterol biosynthesis in fungi treated with naftifine, a new antimycotic agent, *Biochim. Biophys. Acta*, 712, 268, 1982.
115. **Ryder, N.S., Frank, I. and Dupoint, M.C.**, Ergosterol biosynthesis inhibition by the thiocarbamate antifungal agents tolnaftate and tolciclate, *Antimicrob. Agents Chemother.*, 29, 858, 1986.
116. **Lanyi, J.K., Placky, W.Z. and Kates, M.**, Lipid interactions in membranes of extremely halophilic bacteria. II. Modification of the bilayer structure by squalene, *Biochemistry*, 13, 4914, 1974.
117. **Meingassner, J.G., Sleytr, U.B. and Petranyi, G.**, Morphological changes induced by Naftifine, a new antifungal agent in *Trichophyton mentagrophytes*, *J. Invest. Dermatol.*, 77, 444, 1981.
118. **Decarneri, I., Monti, G., Bianchi, A., Castellino, S., Mcinardi, G. and Mandelli, U.**, *Tolciclate against dermatophytes*, Arzneim-Forsch, 26, 769, 1976.
119. **Barret-Bee, K.J., Lane, A.C. and Turner, R.W.**, The mode of antifungal action of tolnaftate, *J. Med. Vet. Mycol.*, 24, 155, 1986.
120. **Gupta, M.P., Kapur, N., Bala, I. and Khuller, G.K.**, Studies on the mode of action of tolnaftate in *Microsporum gypseum*, *J. Med. Vet. Mycol.*, 29, 45, 1991.
121. **Bindra, A. and Khuller, G.K.**, Influence of aminophylline on the lipids of *Microsporum gypseum*, *Biochim. Biophys. Acta*, 1081, 61, 1991.
122. **Niles, R.M. and Makarski, J.S.**, Regulation of phosphatidylcholine metabolism by cyclic AMP in a model alveolar type 2 cell line, *J. Biol. Chem.*, 254, 4324, 1979.
123. **Berry, L.A. and Skett, P.**, The role of cyclic AMP in the regulation of steroid metabolism in isolated rat hepatocytes, *Biochem. Pharmacol.*, 37, 2411, 1988.
124. **Argilaga, C.S., Russell, R.L. and Heimberg, M.**, Enzymatic aspects of the reduction of microsomal glycerolipid biosynthesis after perfusion of the liver with dibutyryl adenosine-3′,5′-monophosphate, *Arch. Biochem. Biophys.*, 190, 367, 1978.
125. **Mooney, R.A., Swicegood, C.L. and Marx, R.B.**, Coupling of adenylate cyclase to lipolysis in permeabilized adipocytes; Direct evidence that an antipolytic effect of insulin is independent of adenylate cyclase, *Endocrinology*, 119, 2240, 1986.
126. **Maziere, C., Maziere, J.C., Salman, S., Auclair, M., Mora, L., Moreau, M. and Polonovski, J.**, Cyclic AMP decreases LDL catabolism and cholesterol synthesis in human hepatoma cell line Hep G2, *Biochem. Biophys. Res. Commun.*, 156, 424, 1988.
127. **Wee, S.F. and Grogan, W.M.**, Temperature lability and cAMP-dependent protein kinase activation of cholesterylester hydrolase as a function of age in developing rat testis, *Lipids*, 24, 824, 1989.

128. **Rickenberg, H.V.**, Cyclic AMP in prokaryotes, *Ann. Rev. Microbiol.*, 28, 357, 1974.
129. **Pall, M.L.**, Adenosine 3′,5′-phosphate in fungi, *Microbiol. Rev.*, 45, 462, 1981.
130. **Gupta, S., Srivastava, D.K., Saxena, J.K., Tandon, J.S. and Shukla, O.P.**, Differentiation of *Acanthamoeba cullertsoni*: cyclic nucleotides and encystation, *Ind. J. Exp. Biol.*, 24, 235, 1986.
131. **Foury, F. and Goffeau, A.**, Stimulation of active uptake of nucleosides and amino acids by cyclic adenosine 3',5'-monophosphate in the yeast, *Schizosaccharomyces pombe*, *J. Biol. Chem.*, 250, 2354, 1975.
132. **Kinney, A.J., Bae Lee, M., Singh, Panghall S., Kelly, M.J., Gaynor, P.M. and Carman, G.M.**, Regulation of phospholipid biosynthesis in *Saccharomyces cerevisiae* by cAMP-dependent protein kinase, *J. Bacteriol.*, 172, 1133, 1990.
133. **Gabrielides, C., Zrike, J. and Scott, W.A.**, Cyclic AMP levels in relation to membrane phospholipid variations in *Neurospora crassa*, *Arch. Microbiol.*, 134, 108, 1983.
134. **Gancedo, J.M., Mazon, M.J. and Eraso, P.**, Biological roles of cAMP: similarities and differences between organisms, *TIBS*, 10, 210, 1985.
135. **Engelhard, V.H., Esko, J.D., Storm, D.R. and Glaser, M.**, Modification of adenylate cyclase activity in LM cells by manipulation of membrane phospholipid composition *in vivo*, *Proc. Natl. Acad. Sci. U.S.A.*, 73, 4482, 1976.
136. **Giri, S., Mago, N., Bindra, A. and Khuller, G.K.**, Possible role of calcium in phospholipid synthesis of *Microsporum gypseum*, *Biochim. Biophys. Acta*, 1215, 337, 1995.
137. **Giri, S., Bindra, A. and Khuller, G.K.**, Calcium induced alterations in structural and functional role of phospholipids in *Microsporum gypseum*, *Ind. J. Biochem. Biophys.*, (in press).
138. **Bindra, A., Giri, S. and Khuller, G.K.**, Identification, localization and possible role of calmodulin-like protein in phospholipid synthesis of *Microsporum gypseum*, *Biochim. Biophys. Acta*, (in press).

Chapter 9

LIPIDS AND LIPID-LIKE COMPOUNDS OF *FUSARIUM*

A. H. Merrill, Jr., A. M. Grant, E. Wang and C. W. Bacon

CONTENTS

0-8493-4794-7/96/$0.00+$.50

I. INTRODUCTION

The genus *Fusarium* Link ex Fr. was defined early in the nineteenth century and validated shortly after to include those species of fungi with nonseptate fusiform spores borne on a stroma or sporodochium. However, currently the major criterion for inclusion into this genus is the presence of fusoid macroconidium with a foot cell bearing some type of heel. The number of species ranges from as few as 31 to as many as 65. This variation in the range of taxa is reflective and dependent on one of several currently available conflicting classification systems for the genus. Regardless of the absolute number of taxa, the *Fusarium* species represent a diverse group of fungi that are widely distributed in nature where they are found in biotic and abiotic environments and where as many as one million colony forming units per gram of substrate may be encountered. Thus, the fusaria are associated with all the major economically important plant groups and are found in cultivated and non-cultivated soils, from the sands of the Sahara to the fertile soils in both temperate and tropical farming regions.

These fungi produce serious diseases of several human foodstuffs and they are also involved in several diseases of animals and humans. It is the threat to animal and human food production that serves as the major focus of interest to this genus. The fusaria have the reputation of undergoing immediate morphological and physiological changes in response to new environments as a survival strategy and associated with this process of change is the production of a variety of diverse secondary metabolites, several of which are biologically active but quite characteristic of specific taxa. The goal of this chapter is to provide a brief summary of the types of lipids that have been found in the genus *Fusarium* and describe a class of lipid-like compounds (termed fumonisins) that are generated by some *Fusarium* species. Fumonisins are interesting because they are potent inhibitors of sphingolipid metabolism and appear to be responsible for diseases of both plants and animals that are exposed to these fungi. This chapter will also describe a number of compounds that are produced by other microorganisms and bear structural similarity to fumonisins and affect sphingolipid metabolism. From the large number and diversity of these very recently discovered compounds, one suspects that many sphingolipid-like agents are produced by other fungi.

II. LIPIDS OF *FUSARIUM*

Several excellent reviews have described the lipids in fungi and although these have not been focused specifically on *Fusarium*, they summarize the general lipid content and metabolism of these organisms.[1-4]

Table 1
Examples of Fatty Acids Found in Lipids in *Fusarium* Molds

Species	Fatty acids (% recovered)				References
	Palmitic (16:0)	Stearic (18:0)	Oleic (18:1)	Linoleic (18:2)	
F. aqueductum	11-12	4-6	25-29	47-58	13
F. aquenaceum [a]	23	4	39	25	14
F. culmorum	24	11	31	33	15
F. graminearum	25	6	33	15	14
F. moniliforme	20-22	6-11	36-48	21-26	6
F. oxysporum [a]	14-16	3-5	15-58	17-54	8
F. oxysporum [b]					12
Phospholipids	15	3	10	54	
Triacylglycerols	16	7	32	29	
F. sambucinum [c]	10	7	33	39	13
F. sulphureum					7
Macroconidium [d]					
Phospholipids	13	2	3	56	
Triacylglycerols	23	9	33	24	
Total	21	9	29	27	
Chlamydospore [e]					
Phospholipids	14	0.1	27	48	
Triacylglycerols	17	4	43	25	
Total	17	3	44	26	
F. solani f. phaseoli					16
Neutral lipids[f]	45	1	15	31	
Polar lipids	17	4	10	45	
Total [f]	28	10	28	22	

[a]Also contains 8% 18:3; this reference also contains data on the fatty acid composition of 14 other species of *Fusarium*
[b]Also contains 1-19% 18:3[8,12]
[c]Also contains 5% 17:1
[d]Also contains 18:3 in phospholipids (26%), triacylglycerols (12%) and total lipids (14%)
[e]Also contains 18:3 in phospholipids (11%), triacylglycerols (11%) and total lipids (9%)
[f]Also contains 6 to 7% 16:1

A. GLYCEROLIPIDS

It has been known for a long time[5] that the *Fusarium* spp. are capable of forming large amounts of fat. Triacylglycerols constitute a large percentage of the total lipid; for example, 73-91% of the total lipid and 21-43% of the dry weight of *F. moniliforme.*[6] The chlamydospores contain more triacylglycerol than the macroconidium (e.g., 19 mg/10^9 cells versus 7.7mg/10^9 cells) and it has been suggested that they are present for use during germination and outgrowth.[7] Monoacylglycerol and diacylglycerol, with free fatty acids, constitute 8 to 12 % of the total lipid.[7]

Shown in Table 1 are the major fatty acids in *Fusarium*. About half of the fatty acids are unsaturated (Table 1), and the fatty acid composition varies with temperature.[1]

In *F. oxysporum*, it has been shown that the percent of oleic acid increases almost 4-fold when the fungi are grown at 37°C versus 15°C and this can probably be explained by an increase in stearoyl-CoA desaturase activity.[8] The further desaturation (e.g.,18:1 to 18:2) appears to occur after the oleic acid has been incorporated into phospholipid.[9] It is not certain whether the fusaria obtain a significant portion of their fatty acids from their environment or primarily make them *de novo*. *F. vasinefectum* (*F. oxysporum*) has been shown to possess the requisite lipase activity but addition of olive oil or coconut oil to the medium did not increase growth.[10] *F. oxysporum f.sp.lini* also has a lipase activity that is specific for saturated fatty acids.[11] When provided with cyclopropenoid fatty acids in the culture medium, *F. oxysporum* incorporated small amounts (ca 1% of the total lipid) into both phospholipids and triacylglycerols, without any detectable change in growth.[12-16]

An essential (volatile) oil from the fresh leaves of *Schinus molle* appears to be toxic for *F. culmorum* and *Alternaria alternata*.[17] Propionic acid treatment of corn inhibited the growth of seven *Fusarium* species and reduced the production of zearalenone by two strains of *F. culmorum* and three strains of *F. graminearum* and the production of T-2 toxin by two strains of *F. tricinctum* and one strain of *F. sporotrichioides*.[18]

Phospholipids account for 4 to 21% of the total lipid depending on the *Fusarium* species[7,14] and generally have a higher percentage of unsaturated fatty acids than the triacylglycerols. The major phospholipids include phosphatidylcholine, phosphatidylethanolamine, phosphatidylglycerol, polyglycerophosphatides, diphosphatidylglycerol and phosphatidic acid,[19] all of which are found in most fungi. The phosphoinositides of *F. graminearum* and a number of other filamentous fungi include two isomers of phosphatidylinositol monophosphate and isomers of inositol *bis*- and *tris*-phosphates that have not been observed in animal or plant cells.[20] Pulse labeling studies have established that the synthesis and turnover of the individual phospholipid classes are rapid under both normal culture conditions and when the cells were starved for nutrients.[19]

Addition of betaine, choline, ethanolamine, monomethylethanolamine or dimethylethanolamine (but not of serine, glycine, dimethylglycine, methylamine, hydroxylamine or β-hydroxyethylhydrazine) to the culture medium of *F. graminearum* increased the colony radial growth rate but no significant difference was observed in the phospholipid composition with addition of 100 μM choline, which increased the mean hyphal extension rate and colony radial growth rate.[21]

B. STEROLS

Ergosterol is the major sterol of *F. aqueductum* and *F. sambucinum*,[13] and sterols can be found as the esters and glycosides.[19] *F. oxysporum* has also been shown to contain several sterol ester hydrolase activity.[22] The sterol contents of *Fusarium* isolates from eighteen patients with

Table 2
Major Long-Chain Bases and Fatty Acids of Sphingolipids in *F. lini*[a]

Long-chain bases[b]	% of Total	Fatty acid	% of Total
Ceramides			
*i*17:0 (16-methyl-4-hydroxy-)	15.5	*n*16:0	2.0
*n*18:0 (4-hydroxy-)	63.4	HO-*n*16:0	12.4
*i*18:0 (17-methyl-4-hydroxy-)	1.6	*n*18:0	6.4
*i*19:0 (17-methyl-4-hydroxy-)	9.1	*n*18:1	1.2
*i*20:0 (19-methyl-4-hydroxy-)	8.8	*n*18:2	1.1
*i*21:0 (20-methyl-4-hydroxy-)	1.6	HO-*n*18:0	71.1
		*n*20:0	2.4
Cerebrosides			
*i*17:0 (16-methyl-4-hydroxy-)	10.5	*n*16:0	0.2
*n*18:0 (4-hydroxy-)	20.9	HO-*n*16:0	2.3
*i*18:0 (17-methyl-4-hydroxy-)	16.3	*n*18:0	0.4
*i*19:0 (17-methyl-4-hydroxy-)	2.8	*n*18:1	0.4
*i*20:0 (19-methyl-4-hydroxy-)	38.4	*n*18:2	0.5
*i*21:0 (20-methyl-4-hydroxy-)	11.1	HO-*n*18:0	94.2
		*n*20:0	0.1

[a]Compiled from Weiss, B. and Stiller, R.L., *Lipids*, 8, 25, 1978.
[b]Some sphinganines were also found.

keratomycosis neither differed nor appeared to be related to the severity of infection.[23]

C. SPHINGOLIPIDS

Ceramides, which are comprised of a long-chain (sphingoid) base and amide-linked fatty acid, and cerebrosides (monoglucosylceramides) constitute approximately 0.2% of the dry weight mass of *F. oxysporum*,[24] with the free ceramides comprising the majority (70%) of the total sphingolipids. The long-chain bases that were characterized were all phytosphingosines (i.e., 1,3,4-trihydroxy-species). However, there were other species present (including the ones with double bonds along the alkyl chain) that were not identified.

A large portion of the phytosphingosines had a branching methyl group along the alkyl chain (designated "i" for "iso" homologs in Table 2). The cerebrosides contained a larger percentage of the phytosphingosines with 20 to 21 carbon atoms. The major fatty acid of both the ceramides and cerebrosides was hydroxy-stearic acid and the ceramides also contained a significant amount of hydroxy-palmitic acid (Table 2). Because saponified lipid extract was used for analyses, it is therefore possible that alkali labile compounds might have been lost.

Studies with *F. oxysporum* have shown that a cytochrome P450 system is capable of the (ω-1), (ω-2), (ω-3)-hydroxylation of fatty acids; therefore, it is probably responsible for formation of these hydroxy-fatty acids.[25]

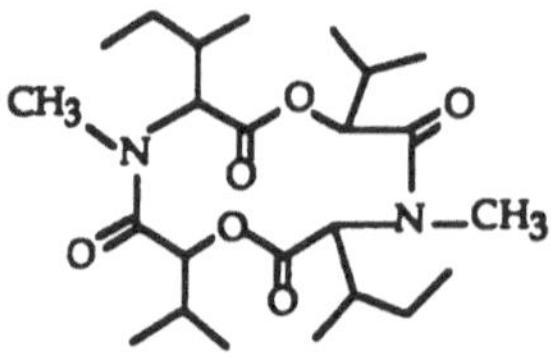

Enniatin A
(*Fusarium oxysporum*)

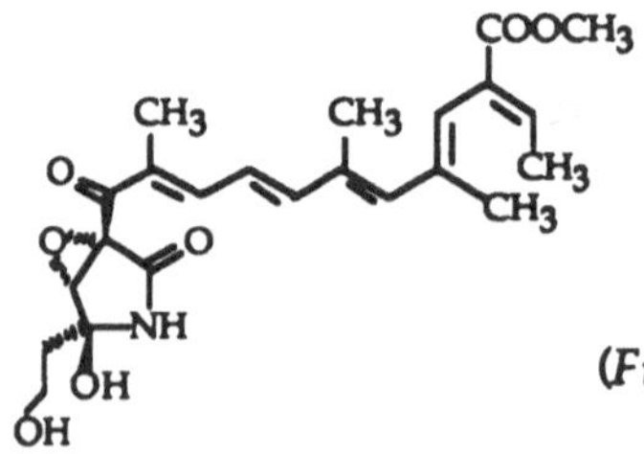

Fusarin C
(*Fusarium moniliforme*)

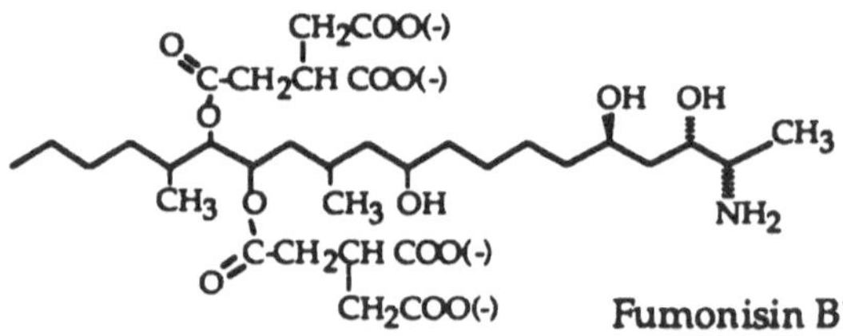

Fumonisin B1
(*Fusarium moniliforme*)

Figure 1. Secondary metabolites produced by some *Fusarium* species.

D. OTHER COMPOUNDS

Fusarium spp. produce a large number of other lipids such as γ-carotene, lycopene and rhodoxanthine.[13] They also produce many secondary metabolites that are hydrophobic, such as the cyclic depsipeptides of alternating α-hydroxy acid residues termed enniatins, which are made by *F. oxysporum* (Figure 1).[26] The high proportion of valine and leucine amino acids in enniatins allows them to partition into membranes, where they can serve as ionophores. Many such depsipeptides have antibiotic activity against bacteria, fungi and insects.[3] Another example is fusarin C (Figure 1), a polyketide produced by *F. moniliforme.*[27] The synthesis of fusarin C is reduced when zinc is added to media limiting in glucose because, in part, this funnels the available carbon into ethanol.[6]

Fusarin C is toxic and mutagenic and represents just one of the many secondary metabolites of *Fusarium* (T-2 toxin, fusaric acid, moniliformin, nivalenol, deoxynivalenol, beauvericin and diacetoxy-scirpenol) and will be discussed in depth because they are important contributors to some of the diseases caused by *F. monoliforme.*

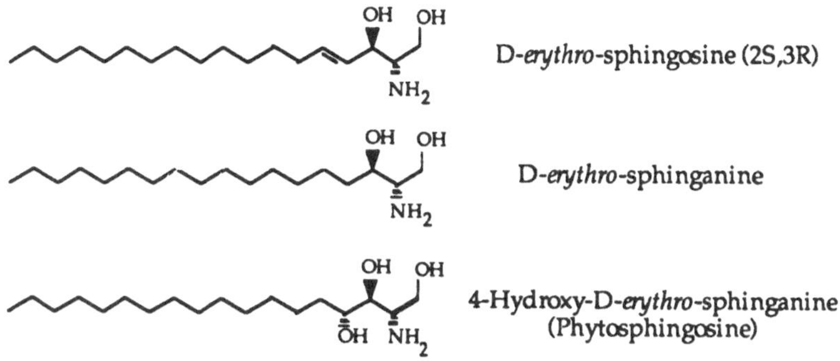

Figure 2. Structures of long-chain (sphingoid) bases.

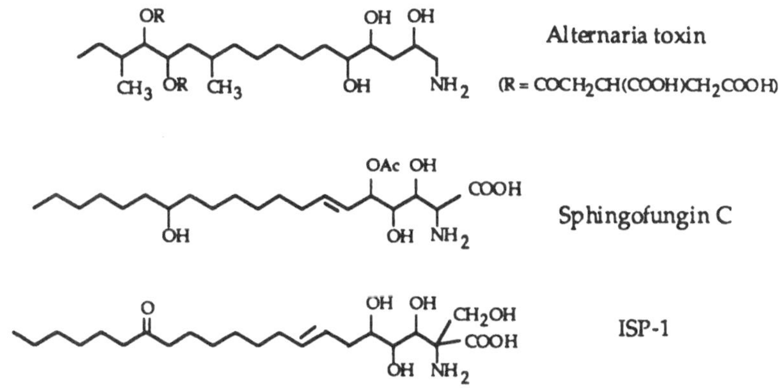

Figure 3. Additional inhibitors of sphingolipid metabolism with structural similarity to long-chain bases.

III. FUMONISINS: SPHINGOLIPID-LIKE COMPOUNDS PRODUCED BY SOME SPECIES OF *FUSARIUM*

Fumonisins are mycotoxins produced by *F. moniliforme*[28] and related fungi[29] on corn, sorghum, millet and other agricultural products.[28] There are several classes of fumonisins, the most prevalent being fumonisin B_1 (Figure 1) followed by B_2, B_3 and B_4. These compounds can also be found acetylated on the nitrogen and are named fumonisin A_1, A_2, etc. Purified fumonisin B_1 has been shown to cause equine leukoencephalomalacia (ELEM)[30] and porcine pulmonary edema,[31] which are diseases that have also been associated with the consumption of contaminated grain. Fumonisin consumption also results in hepatotoxicity and liver tumors in rats[32-34] and it also affects the kidney.[35] The human impact of these mycotoxins is not yet clear; however, the consumption of contaminated maize has been correlated with esophageal cancer in areas of southern Africa and China (for review, see Riley et al).[36]

A. STRUCTURAL SIMILARITIES BETWEEN FUMONISINS AND SPHINGOLIPIDS

The fumonisins bear a remarkable structural similarity to sphingosine, sphinganine and phytosphingosine, as shown in Figure 2. This led us to hypothesize that these mycotoxins may inhibit one of the enzymes of sphingolipid metabolism, and we found that fumonisins are potent inhibitors of the enzyme that adds a long-chain fatty acid to the amino group of long-chain bases (ceramide synthase)[37] (as discussed in greater detail below). Recent reports of stereochemistry of fumonisins B_1[38-40] and B_2[41] have revealed that the configuration at carbons 2 and 3 is *threo*, whereas long-chain bases are *erythro* (cf. Figures 1 and 2). Other microorganisms also produce sphingolipid-like compounds (Figure 3) that inhibit ceramide synthase (alternaria toxin)[42,43] or serine palmitoyltransferase (sphingofungins and ISP-1).[44,45]

B. INHIBITION OF SPHINGOLIPID METABOLISM BY FUMONISINS

Sphingolipid biosynthesis *de novo* proceeds via the reactions shown in Figure 4.[46] In the first and committed step, serine is condensed with palmitoyl-CoA by serine palmitoyl-transferase, a pyridoxal 5'-phosphate-dependent enzyme and the resulting 3-keto-sphinganine is reduced to sphinganine using NADPH. Sphinganine is acylated to dihydroceramide (also called N-acyl-sphinganine) by ceramide synthase using various long-chain fatty acyl-CoAs. Head groups (e.g., phosphorylcholine, glucose, galactose, and hundreds of more complex polysaccharides) are subsequently added to the 1-hydroxyl group in various intracellular compartments. It is not clear when the 4,5-*trans*-double bond of the sphingosine backbone is added, except that it is after acylation of the amino group, and that free sphingosine is not an intermediate of *de novo* sphingolipid biosynthesis (Figure 4). Sphingolipid turnover is thought to involve the hydrolysis of complex sphingolipids to ceramides and then to sphingosine. Sphingosine is either reacylated or phosphorylated and cleaved to a fatty aldehyde and ethanolamine phosphate, which is incorporated into phosphatidyl ethanolamine. For reviews of sphingolipid metabolism, see references.[47-50]

1. Inhibition of Ceramide Synthase *in vitro*

In rat hepatocytes, the IC_{50} for inhibition of ceramide synthase is approximately 0.1 μM for fumonisin B_1 and B_2.[37,43] Removal of the tricarballylic acid side-chains reduces the potency by approximately 10 fold,[43] and we have recently tested fumonisin A_1 and the only inhibition could be attributed to the small amount of fumonisin B_1 that was present in the preparation (Vales, T., manuscript in preparation). A more detailed

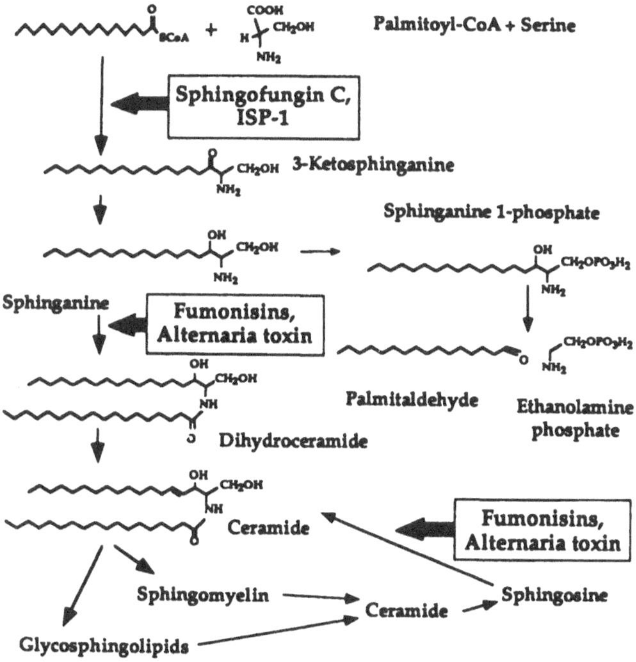

Figure 4. Overview of sphingolipid metabolism and sites of action of naturally occurring inhibitors.

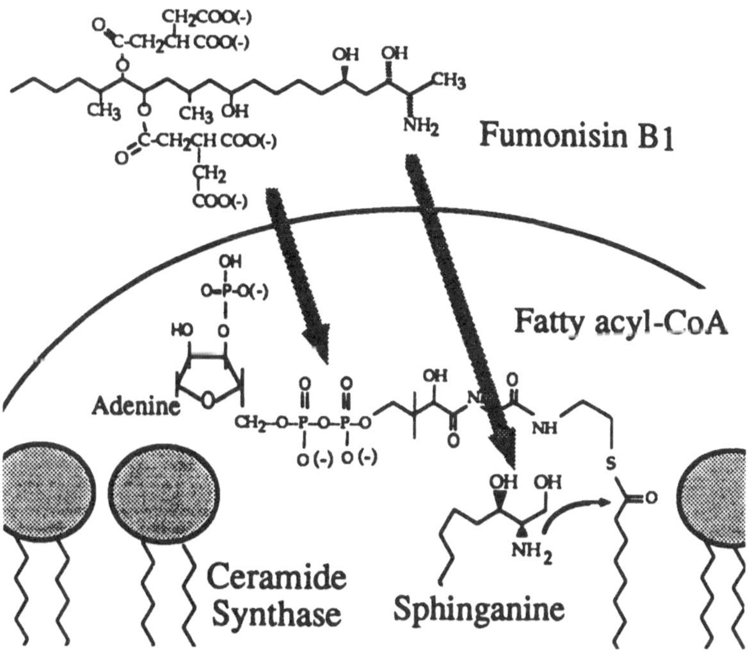

Figure 5. A hypothetical model to explain why inhibition of ceramide synthase by fumonisin B_1 is competitive with respect to sphinganine and fatty acyl-CoA.

kinetic analysis with microsomes from mouse brain[51] detected inhibition at nanomolar fumonisin B_1 concentrations and found that the inhibition was competitive with both the long-chain (sphingoid) base and fatty-acyl-CoA. This might be due to interactions between different domains of the fumonisin molecule and the binding sites for these two substrates, as indicated in Figure 5.

A more practical implication of this kinetic behavior is that the potency of inhibition of ceramide synthesis depends on the levels of the substrates. When the substrates are present in high amounts, higher levels of fumonisins are required; when they are low, significant inhibition can be seen with nanomolar concentrations of fumonisin B_1.[51] As the inhibition of ceramide synthase causes sphinganine to accumulate, higher levels of fumonisins will be needed to block the enzyme activity.[37,43,51]

2. Studies of Inhibition of Ceramide Synthase Using Cells in Culture

Fumonisins have been shown to inhibit sphingolipid biosynthesis in a variety of cells in culture, including primary cultures of rat hepatocytes,[37] a mouse renal cell line (LLC-PK_1 cells),[52] cultured mouse cerebellar neurons,[51] rat hippocampal neurons,[53] Swiss 3T3 cells,[54] and plant cells.[55] Fumonisin B_1 characteristically inhibits [^{14}C]-serine or [^{3}H]-palmitic acid incorporation into complex sphingolipids, although the IC_{50} ranges from approximately 0.1 μM for hepatocytes to two orders of magnitude higher for many other cells. It is not known if this variability is due to differences in fumonisin B_1 uptake, or other features of the cells (such as the amounts of endogenous substrates, as has already been mentioned).

In addition to the inhibition of sphingolipid labeling, a reduction in sphingolipid mass has been found in some studies,[37,52,53] especially after long periods of inhibition. The loss of ganglioside synthesis has been related to neurite extension for the hippocampal neurons;[53] however, a considerable reduction in mass in hepatocytes did not result in a loss of cell viability.[37]

When [^{3}H]-sphingosine was added to hepatocytes treated with fumonisin B_1, a small amount of radiolabel appeared at the solvent front, where the cleavage products of [^{3}H]-sphingosine catabolism (*trans*-4-hexadecenal and fatty acids derived from subsequent metabolism) migrate. This indicates that inhibition of the acylation of [^{3}H]-sphingosine (and presumably sphinganine) increases the amount that is degraded in the reaction shown in Figure 4. We have recently shown this to be the case (Smith, E.R. and Merrill, A. H., Jr., manuscript submitted for publication), and found that not only is there a significant increase in the amounts of sphinganine-1-phosphate and sphingosine-1-phosphate, but, also, the formation of ethanolamine via sphingolipid degradation can make a major contribution to phosphatidylethanolamine synthesis.

Despite the degradation of long-chain bases in the presence of fumonisin B_1, the mass of sphinganine rises significantly in the cells.[37,51,52,54] In treated hepatocytes,[37] the increase in sphinganine was evident within hours, and after several days, the mass of sphinganine increased by 110-fold. In contrast, sphingosine level does not often increase in cell culture since it is not an intermediate in sphingolipid biosynthesis but only appears during turnover of complex sphingolipids. In plant cell cultures, sphinganine and phytosphingosine are elevated.[55]

The effects of fumonisin on the cells are relatively selective, because there is no reduction in the radiolabeling of fatty acids, phosphatidylserine, phosphatidylethanolamine or phosphatidylcholine from [^{14}C]-serine, nor in the mass of these phospholipids.[37] Nonetheless, given the complex interactions between lipid metabolic pathways, it is likely that disruption of sphingolipid metabolism will have secondary effects on the other pathways. One example of this is the increased incorporation of ethanolamine (from sphinganine degradation) into phosphatidylethanolamine (as discussed above). Another example might be the accumulation of hydroxyhexadecanoic acid and trihydroxydocosanoic acid that has been seen in yeast treated with fumonisin B_1.[56]

3. Studies of the Disruption of Sphingolipid Metabolism *in vivo*

Studies of ponies,[57] pigs,[58] rats,[35] chicken,[59] and a few other animals have established that fumonisins also disrupt sphingolipid metabolism *in vivo* and that this can be detected by an elevation of sphinganine level in urine, serum, and tissues. In the study of ponies given fumonisin-contaminated feed,[57] a typical response for a pony consuming a diet containing 44 ppm fumonisin B_1 was: a 2.7 fold increase in sphinganine after 2 days, a 4.5-fold increase by 7 days, and 13-fold elevation in sphinganine and 2-fold in sphingosine by 10 days. There was also a 7-fold increase in serum transaminase activities, indicating that the pony had significant cellular injury. At this stage, at least, the elevation of serum sphinganine is reversible because the sphinganine level began to return to normal when the ponies stopped eating fumonisins.

There was also a reduction in the amount of complex sphingolipids in serum of animals that consume fumonisins;[57] the concentration of total (complex) sphingolipid was reduced by 50 to 95% during the early times after feeding fumonisins to ponies, but increased again when there was also an elevation in hepatic enzyme activities. One explanation for these changes might be that fumonisin B_1 blocks the secretion of sphingolipids with lipoproteins.

In pigs fed with corn screenings containing greater than 23 ppm fumonisin B_1 plus B_2, free sphinganine were elevated in liver, lung, and kidney, and complex sphingolipids were reduced.[58] These changes were accompanied with signs of respiratory distress due to severe pulmonary interstitial edema with pleural effusion for three of five pigs fed 175 ppm fumonisin. Hepatic injury was observed in most of the pigs fed > 39 ppm fumonisins, as characterized by hepatocyte disorganization, single cell necrosis, and mild inflammation.

There have been several recent studies of the pharmacokinetics of fumonisin uptake, transport, and elimination in rats,[60,61] pigs,[62] chicken,[63] and non-human primates.[64]

4. Disruption of Sphingolipid Metabolism in Plants Exposed to Fumonisins

In addition to these studies with animals, the treatment of various plants with fumonisins of *F. moniliforme* has been shown to cause elevation in sphinganine and phytosphinganine levels.[55] Therefore, disruption of sphingolipid metabolism in plants may play a role in the pathology of *F. moniliforme* and related fungi for their usual hosts.

5. Elevation of Sphinganine as a Biomarker for Exposure to Fumonisins

In every animal (and plant) tested to date, it has been possible to detect elevation of sphinganine in tissues, blood and urine upon administration of fumonisins. These changes can often be seen before other signs of injury;[57] therefore, an increase in sphinganine (sometimes expressed as the sphinganine-to-sphingosine ratio) has been proposed as a biomarker for exposure of animals to these compounds.[65] Use of this biomarker might allow fumonisin toxicity to be detected early enough to save animals by removal of the contaminated food or other measures.

The availability of this biomarker may also aid in the evaluation of fumonisins as human toxins and carcinogens. Fumonisins have been found in corn meal and grits in the U.S. at readily detectable levels [55,66,67] that are usually lower than those which cause diseases in veterinary animals. However, the levels of fumonisins can be underestimated if care is not taken in the extraction and analyses.[68] The relative sensitivity of humans to fumonisins is not known.[69] Several studies have considered the possible role of consumption of *F. monoliforme* in esophageal cancer in certain regions of Africa[69,70] and China[71-74] and it has been speculated that some of the inconsistencies encountered in epidemiological and experimental studies on the role of various fats in the etiology of cardiovascular disorders and certain tumors may be related to the presence of contaminants such as *Fusarium* mycotoxins.[75] There is clearly a need for more extensive epidemiological studies on this topic.

IV. ADDITIONAL *FUSARIUM* PRODUCTS THAT AFFECT LIPID METABOLISM IN OTHER ORGANISMS

A potent specific inhibitor of 3-hydroxy-3-methylglutaryl Coenzyme A (HMG-CoA) synthase has been isolated from *Fusarium* spp.[76] The structure of this β-lactone, termed L-659,699, is (E, E)-11-[3-(hydroxy-methyl)-4-oxo-2-oxytanyl]-3,5,7-trimethyl-2,-4-undecadienenoic acid. Continuous fermentation of *F. graminearum* (Schwabe) produces a "mycoprotein" that has been shown to decrease total and low-density-lipoprotein (LDL) cholesterol and increased high-density-lipoproteins (HDL) cholesterol.[77]

Some isolates of *F. solani* produce an extracellular cutinase which can be assayed by the hydrolysis of cutin containing [^{14}C]-palmitic acid. Mutants lacking the cutinase have a reduced virulence measured with a pea stem bioassay, which suggests that cutinase has an important role in infection of plants.[78]

V. CONCLUSION

The *Fusarium* spp. are fascinating organisms from the perspective of "lipidologists" both because of their own lipid composition and the many compounds that they produce to affect lipid metabolism by other organisms. These findings raise many interesting questions: What is the role of the neutral lipid produced by these organisms, and could its production have industrial applications? Are the lipids of *Fusarium* used just for structural purposes or are they part of signal transduction pathways as for many other eukaryotic organisms? How many lipid-altering compounds are made by these fungi? How do they both make sphingolipids and produce compounds that potently disrupt this pathway? What role do these fungi play in the etiology of disease? And, could some of the toxins produced by these fungi be of pharmaceutical use?

ACKNOWLEDGMENTS

The authors are grateful for many valuable collaborators in various aspects of this work, namely Drs. C. Alexander, V. Beasley, W. Haschek, B. Hennig, D.R. Ledoux, D.C. Liotta, F. Meredith, W.P. Norred, R.D. Plattner, S. Ramasamy, L. Rice, R.T. Riley, P.F. Ross, K. Sandhoff, J.J. Schroeder, E.R. Smith, P. Stancel, G. Van Echten, K.A. Voss, T. Wilson and H. Yoo. We also thank Mrs.Winnie Scherer for help in preparing this chapter. The work by Dr. Merrill's lab was supported by funds from the USDA (91-37204-6684), the NIH (grants GM33369and GM46368) and the American Institute for Cancer Research.

VI. REFERENCES

1. **Weete, J.D.**, Fungal lipid biochemistry, in *Monographs in Lipid Research*, Vol. 1, Kritchevsky, D., Ed., Plenum Press, New York, 1974, chap 1.
2. **Weete, J.D.**, *Lipid Biochemistry of Fungi and Other Organisms*, Plenum Press, New York, 1980.
3. **Brennan, P.J., Griffin, P.F.S, Lösel, D.M. and Tyrrell, D.**, The lipids of fungi, *Prog. Chem. Fats and Other Lipids*, 14, 49, 1974.
4. **Brennan, P.J. and Lösel, D.M.**, Physiology of fungal lipids: Selected topics, *Adv. Microbial Physiol.*, 17, 47, 1978.
5. **Foster, J.W.**, *Chemical Activities of Fungi*, Academic Press, New York, 1949.
6. **Jackson, M.A. and Lanser, A.C.**, Glucose and zinc concentration influence fusarin C synthesis, ethanol synthesis and lipid composition in *Fusarium moniliforme* submerged cultures, *FEBS Microbiol. Lett.*, 108, 69, 1993.
7. **Barran, L.R. and Schneider, E.G.**, Effect of thiols on macroconidia of *Fusarium sulphureum*, *Can. J. Microbiol.*, 25, 618, 1979.
8. **Wilson, A.C. and Miller, R.W.**, Growth temperature-dependent stearoyl coenzyme A desaturase activity of *Fusarium oxysporum* microsomes, *Can. J. Biochem.*, 56, 1109, 1978.
9. **Wilson, A.C., Adams, W.C. and Miller, R.W.**, Lipid involvement in oleoyl CoA desaturase activity of *Fusarium oxysporum* microsomes, *Can. J. Biochem.*, 58, 97, 1980.
10. **Ahamed, N.M.M., Meenakshisundaram, S. and Shanmugasundaram, E.R.B.**, Lipid and lipase activity in strains of *Fusarium vasinfectum*, *Ind. J. Exp. Biol.*, 11, 37, 1973.
11. **Joshi, S. and Dhar, D.N.**, Specificity of fungal lipase in hydrolytic cleavage of oil, *Acta Microbiol. Hung.*, 34, 111, 1987.
12. **Schmid, K.M. and Patterson, G.W.**, Effects of cyclopropenoid fatty acids on fungal growth and lipid composition, *Lipids*, 23, 248, 1988.
13. **Gribanovski-Sassu, O. and Foppen, F.H.**, Lipid constituents of some *Fusarium* species, *Archiv fur Mikrobiologie*, 62, 251, 1968.
14. **Gorbik, L.T., Pidoplichko, G.A. and Loiko, Z.I.**, Fungal lipids of the genus *Fusarium*, *Lk. ex Fr.*, *Mikrobiol. Zh*, 42, 191, 1980.
15. **Marchant, R. and White, M.F.**, The carbon metabolism and swelling of *Fusarium culmorum* conidia, *J. Gen. Microbiol.*, 48, 65, 1967.
16. **Gunasekaran, M. and Weber, D.J.**, Polar lipids and fatty acid composition of phytopathogenic fungi, *Phytochemistry*, 11, 3367, 1972.
17. **Gundidza, M.**, Antimicrobial activity of essential oil from *Schinus molle Linn*, *Cent. Afr. J. Med.*, 39, 231, 1993.

18. **Muller, H.M. and Thaler, M.**, Propionic acid preservation of corn following inoculation with molds and yeasts, *Arch. Tierernahr.*, 31, 789, 1981.
19. **Bhatia, I.S. and Arneja, J.S.**, Lipid metabolism in *Fusarium oxysporum*, *J. Sci. Fd. Agric.*, 29, 619, 1978.
20. **Prior, S.L., Cunliffe, B.W., Robson, G.D. and Trinci, A.P.**, Multiple isomers of phosphatidylinositol monophosphate and inositol bis- and tris-phosphates from filamentous fungi, *FEMS Microbiol. Lett.*, 110, 147, 1993.
21. **Wiebe, M.G., Robson, G.D. and Trinci, A.P.**, Effects of choline on the morphology, growth and phospholipid composition of *Fusarium graminearum*, *J. Gen. Microbiol.*, 135, 2155, 1989.
22. **Madho Singh, C. and Orr, W.**, Sterol ester hydrolase in *Fusarium oxysporum*, *Lipids*, 16, 125, 1981.
23. **Raza, S.K., Mallet, A.I., Howell, S.A. and Thomas, P.A.**, An *in vitro* study of the sterol content and toxin production of *Fusarium* isolates from mycotic keratitis, *J. Med. Microbiol.*, 41, 204, 1994.
24. **Weiss, B. and Stiller, R.L.**, Sphingolipids of the fungi *Phycomycetes blakesleeanus* and *Fusarium lini*, *Lipids*, 8, 25, 1978.
25. **Shoun, H., Sudo, Y. and Beppu, T.**, Subterminal hydroxylation of fatty acids by a cytochrome P-450-dependent enzyme system from a fungus, *Fusarium oxysporium*, *J. Biochem. (Tokyo)*, 97, 755, 1985.
26. **Taylor, A.**, The occurrence, chemistry, and toxicology of the microbial peptide-lactones, *Adv. Appl. Microbiol.*, 12, 189, 1970.
27. **Wiebe, L.A., Bjeldancs, L.F. and Fusarin, C.**, A mutagen from *Fusarium moniliforme* grown on corn, *J. Food Sci.*, 46, 1424, 1981.
28. **Bezuidenhout, C.S., Gelderblom, W.C.A., Gorstallman, C.P., Horak, R.M., Marasas, W.F.O., Spiteller, G. and Vleggaar, R.**, Structure elucidation of the fumonisins, mycotoxins from *Fusarium moniliforme*, *J. Chem. Soc. Commun.*, 743, 1988.
29. **Nelson, P.E., Plattner, R.D., Shackelford, D.D. and Desjardins, A.E.**, Fumonisin B_1 production by *Fusarium* species other than *F. moniliforme* in *Sectin liseola* and by some related species, *Appl. Environ. Microbiol.*, 58, 984, 1992.
30. **Marasas, W.F.O., Kellerman, T.S., Gelderblom, W.C.A., Coetzer, J.A.W., Thiel, P.G. and van der Lugt, J.J.**, Leukoencephalomalacia in a horse induced by fumonisin B_1 isolated from *Fusarium moniliforme*, *Onderstepoort J. Vet. Res.*, 55, 197, 1988.
31. **Harrison, L.R., Colvin, B., Greene, J.T., Newman, L.E. and Cole, J.R.**, Pulmonary edema and hydrothorax in swine, produced by fumonisin B_1, a toxic metabolite of *Fusarium moniliforme*, *J. Vet. Diagn. Invest.*, 2, 217, 1990.

32. **Gelderblom, W.C.A., Jaskiewicz, K., Marasas, W.F.O., Thiel, P.G., Horak, R.M., Vleggaar, R. and Kriek, N.P.J.,** Cancer promoting potential of different strains of *Fusarium moniliforme* in a short-term cancer initiation/promotion assay, *Carcinogenesis*, 9, 1405, 1988.
33. **Gelderblom, W.C.A., Kriek, N.P.J., Marasas, W.F.O. and Thiel, P.G.,** Toxicity and carcinogenicity of the *Fusarium moniliforme* metabolite, fumonisin B_1, in rats, *Carcinogenesis*, 12, 1247, 1991.
34. **Marasas, W.F.O., Kriek, N.P.J., Fincham, J.E. and van Rensburg, S.J.,** Primary liver cancer and esophageal basal cell hyperplasia in rats caused by *Fusarium moniliforme*, *Int. J. Cancer*, 34, 383, 1984.
35. **Riley, R.T., Hinton, D.M., Chamberlain, W.J., Bacon, C.W., Wang, E., Merrill, A.H., Jr. and Voss, K.A.,** Dietary fumonisin B_1 induces disruption of sphingolipid metabolism in Sprague-Dawley rats: A new mechanism of nephrotoxicity, *J. Nutr.*, 124, 594, 1994.
36. **Riley, R.T., Norred, W.P. and Bacon, C.W.,** Fungal toxins in foods: Recent concerns, *Annu. Rev. Nutr.*, 13, 167, 1993.
37. **Wang, E., Norred, W.P., Bacon, C.W., Riley, R.T. and Merrill, A.H., Jr.,** Inhibition of sphingolipid biosynthesis by fumonisins: Implications for diseases associated with *Fusarium moniliforme*, *J. Biol. Chem.*, 266, 14486, 1993.
38. **Hoye, T.R., Jimenez, J.I. and Shier, W.T.,** Relative and absolute configuration of the fumonisin B_1 backbone. *J. Am. Chem. Soc.*, 116, 9409, 1994.
39. **ApSimon, J.W., Blackwell, B.A., Edwards, O.E. and Fruchier, A.,** Relative configuration of the C-1 to C-5 fragment of fumonisin B_1, *Tetrahedron Lett.*, 35, 7703, 1994.
40. **Poch, G.K., Powell, R.G., Plattner, R.D. and Weisleder, D.,** Relative stereochemistry of fumonisin B_1 at C-2 and C-3, *Tetrahedron Lett.*, 35, 7707, 1994.
41. **Harmange, J.C., Boyle, C.D. and Kishi, Y.,** Relative and absolute stereochemistry of the fumonisin B_2 backbone, *Tetrahedron Lett.*, 35, 6819, 1994.
42. **Bottini, A.T., Bowen, J.R. and Gilchrist, D.G.,** Phytotoxins II. Characterization of a phytotoxic fraction from *Alternaria alternata f. sp. lycopersici*, *Tetrahedron Lett.*, 22, 2723, 1981.
43. **Merrill, A.H., Jr., Wang, E., Gilchrist, D.G. and Riley, R.T.,** Fumonisin and other inhibitors of *de novo* sphingolipid biosynthesis, in *Advances in Lipid Research: Sphingolipids and Their Metabolites*, Vol. 26, Bell, R.M., Hannun, Y.A. and Merrill, A.H., Jr., Eds., Academic Press, San Diego, CA, 215, 1993.
44. **Zweerink, M.M., Edison, A.M., Wells, G.B., Pinto, W. and Lester, R.L.,** Characterization of a novel, potent and specific inhibitor of serine palmitoyltransferase, *J. Biol. Chem.*, 267, 25032, 1992.

45. **Miyake, Y., Kozutsumi, T. and Kawasaki, T.,** Action mechanism of sphingosine-like immunosuppressant. ISP-1, *Igaku no Ayumi (Tokyo)*, 171, 921, 1994.
46. **Sweeley, C.C.,** Sphingolipids, in *Biochemistry of Lipids, Lipoproteins, and Membranes*, Vance, D.E. and Vance, J.E., Eds., Elsevier Science Publ., Amsterdam, 327, 1991.
47. **Merrill, A.H., Jr. and Jones, D.D.,** An update of the enzymology and regulation of sphingomyelin metabolism. *Biochim. Biophys. Acta*, 1044, 1, 1990.
48. **Merrill, A.H., Jr.,** Cell regulation by sphingosine and more complex sphingolipids. *J. Bioenerget. Biomembr.*, 23, 83, 1991.
49. **Bell, R.M., Hannun, Y.A. and Merrill, A.H., Jr.,** Eds., *Advances in Lipid Research: Sphingolipids and Their Metabolites*, Vol. 25, Academic Press, Orlando, FL, 1993.
50. **Bell, R.M., Hannun, Y.A. and Merrill, A.H., Jr.,** Eds., *Advances in Lipid Research: Sphingolipids and Their Metabolites*, Vol. 26, Academic Press, Orlando, FL, 1993.
51. **Merrill, A.H., Jr., van Echten, G., Wang, E. and Sandhoff, K.,** Fumonisin B_1 inhibits sphingosine (sphinganine) N-acetyltransferase and *de novo* sphingolipid biosynthesis in cultured neurons *in situ*, *J. Biol. Chem.*, 268, 27299, 1993.
52. **Yoo, H., Norred, W.P., Wang, E., Merrill, A.H., Jr. and Riley, R.T.,** Sphingosine inhibition of *de novo* sphingolipid biosynthesis and cytotoxicity are correlated in LLC-PK_1 cells, *Toxicol. Appl. Pharmacol.*, 114, 9, 1992.
53. **Harel, R. and Futerman, A.H.,** Inhibition of sphingolipid synthesis affects axonal outgrowth in cultured hippocampal neurons, *J. Biol. Chem.*, 268, 14476, 1993.
54. **Schroeder, J.J, Crane, H.M., Xia, J., Liotta, D.C. and Merrill, A.H., Jr.,** Disruption of sphingolipid metabolism and stimulation of DNA synthesis by fumonisin B_1. A molecular mechanism for carcinogenesis associated with *Fusarium moniliforme*, *J. Biol. Chem.*, 269, 3475, 1994.
55. **Abbas, H.K., Tanaka, T., Duke, S.O., Porter, J.K., Wray, E.M., Hodges, L., Sessions, A.E., Wang, E., Merrill, A.H., Jr. and Riley, R.T.,** Fumonisin- and AAL-toxin-induced disruption of sphingolipid metabolism with accumulation of free sphingoid bases, *Plant Physiol.*, 106, 1085, 1994.
56. **Kaneshiro, T., Vesonder, R.F., Peterson, R.E. and Bagby, M.O.,** 2-hydroxyhexadecanoic and 8,9,13-trihydroxydocosanoic acid accumulation by yeasts treated with fumonisin B_1, *Lipids*, 28, 397, 1993.

57. **Wang, E., Ross, P.F., Wilson, T.M., Riley, R.T. and Merrill, A.H., Jr.,** Alteration of serum sphingolipids upon dietary exposure of ponies to fumonisins, mycotoxins produced by *Fusarium moniliforme*, *J. Nutr.*, 122, 1706, 1992.
58. **Riley, R.T., An, N.H., Showker, J.L., Yoo, H. S., Norred, W.P., Chamberlain, W.J., Wang, E., Merrill, A.H., Jr., Motelin, G., Beasley, V.R. and Haschok, W.M.,** Alteration of tissue and serum sphinganine to sphingosine ratio: an early biomarker of exposure to fumonisin-containing feeds in pigs, *Toxicol. & Appl. Physiol.*, 118, 105, 1993.
59. **Weibking, T.S., Ledoux, D.R., Bermudez, A.J., Turk, J.R., Rottinghaus, G.E., Wang, E. and Merrill, A.H., Jr.,** Effects of feeding *Fusarium moniliforme* culture material, containing known levels of fumonisin B_1, on the young broiler chick, *Poultry Sci.*, 72, 456, 1993.
60. **Norred, W.P., Plattner, R.D. and Chamberlain, W.J.,** Distribution and excretion of [^{14}C]-fumonisin B_1 in male Sprague-Dawley rats, *Natural Toxins*, 1, 341, 1993.
61. **Shephard, G.S., Thiel, P.G., Sydenham, E.W. and Alberts, J.F.,** Biliary excretion of the mycotoxin fumonisin B_1 in rats, *Food. Chem. Toxicol.*, 32, 489, 1994.
62. **Prelusky, D.B., Trenholm, H.L. and Savard, M.E.,** Pharmacokinetic fate of ^{14}C-labelled fumonisin B_1 in swine, *Nat. Toxins*, 2, 73, 1994.
63. **Vudathala, D.K., Prelusky, D.B., Ayroud, M., Trenholm, H.L. and Miller, J.D.,** Pharmacokinetic fate and pathological effects of ^{14}C-fumonisin B_1 in laying hens, *Natural Toxins.*, 2, 81, 1994.
64. **Shephard, G.S., Thiel, P.G., Sydenham, E.W., Vleggaar, R. and Alberts, J.F.,** Determination of the mycotoxin fumonisin B_1 and identification of its partially hydrolysed metabolites in the faeces of non-human primates, *Food. Chem. Toxicol.*, 32, 23, 1994.
65. **Riley, R.T., Wang, E. and Merrill, A.H., Jr.,** Liquid chromatography of sphinganine and sphingosine: Use of sphinganine to sphingosine ratio as a biomarker for consumption of fumonisins, *J. AOAC. Int.*, 77, 533, 1993.
66. **Sydenham, E.W., Shephard, G.S., Thiel, P.G., Marasas, W.F.O. and Stockenstrom, S.,** Fumonisin contamination of commercial corn-based human foodstuffs, *J. Agric. Food Chem.*, 39, 2014, 1991.
67. **Chamberlain, W.J., Bacon, C.W., Norred, W.P. and Voss, K.A.,** Levels of fumonisin B_1 in corn naturally contaminated with aflatoxins, *Food Chem. Toxicol.*, 31, 995, 1993.
68. **Scott, P.M. and Lawrence, G.A.,** Stability and problems in recovery of fumonisins added to corn-based foods, *J. AOAC. Int.*, 77, 541, 1994.
69. **Marasas, W.F.O.,** Occurrence of *Fusarium moniliforme* and fumonisins in maize in relation to human health [editorial], *S. Afr. Med. J.*, 83, 382, 1993.

70. **Marasas, W.F.O.**, Mycotoxicological investigations on corn produced in esophageal cancer areas in Transkei, in *Cancer of the Oesophagus*, Vol. 129, Pfeiffer, C.J., Ed., CRC Press, Boca Raton, FL, 129, 1982.
71. **Lin, M., Lu, S., Ji, C., Wang, Y., Wang, M., Cheng, S. and Tian, G.**, Experimental studies on the carcinogenicity of fungus-contaminated food from Linxian County, in *Genetics and Environment Factors in Experimental and Human Cancer* (Gelboin, H.V., ed.) Japan Science Society Press, Tokyo, 139, 1980.
72. **Yang, C.S.**, Research on esophageal cancer in China: A review, *Cancer Res.*, 40, 2633, 1980.
73. **Chu, F.S. and Li, G.Y.**, Simultaneous occurrence of fumonisin B_1 and other mycotoxins in moldy corn collected from the People's Republic of China in regions with high incidences of esophageal cancer, *Appl. Environ. Microbiol.*, 60, 847, 1994.
74. **Yoshizawa, T., Yamashita, A. and Luo, Y.**, Fumonisin occurrence in corn from high- and low-risk areas for human esophageal cancer in China, *Appl. Environ. Microbiol.*, 60, 1626, 1994.
75. **Schoental, R.**, *Fusarium* mycotoxins and the effects of high-fat diets, *Nutr. Cancer*, 3, 57, 1981.
76. **Greenspan, M.D., Yudkovitz, J.B., Lo, C.Y., Chen, J.S., Alberts, A.W., Hunt, V.M., Chang, M.N., Yang, S.S., Thompson, K.L., Chiang, Y.C., et al.**, Inhibition of hydroxymethylglutaryl-coenzyme A synthase by L-659, 699, *Proc. Natl. Acad. Sci. USA*, 84, 7488, 1987.
77. **Turnbull, W.H., Leeds, A.R. and Edwards, D.G.**, Mycoprotein reduces blood lipids in free-living subjects, *Am. J. Clin. Nutr.*, 55, 415, 1992.
78. **Dantzig, A.H., Zuckerman, S.H. and Andonov-Roland, M.M.**, Isolation of a *Fusarium solani* mutant reduced in cutinase activity and virulence, *J. Bacteriol.*, 168, 911, 1986.

Chapter 10

LIPIDS AND DIMORPHISM OF *CANDIDA ALBICANS* AND *SPOROTHRIX SCHENCKII*

Y. Kitajima and Y. Nozawa

CONTENTS

0-8493-4794-7/96/$0.00+$.50

I. INTRODUCTION

Candida albicans is a saprophyte found on mucous membrane surfaces of the oral cavity, gastrointestinal tract or vagina, and frequently causes opportunistic infections in humans with diminished host defenses. This organism is a dimorphic fungus showing yeast and mycelial morphological forms. It is believed that dimorphism has great relevance to virulence, since the mycelial form predominates in invasive disease state and the yeast phase predominates in saprophytic state, suggesting that the mycelial form is more pathogenic. *Sporothrix schenckii* is a pathogenic fungus, which causes diseases by accidental inoculation in the skin tissues of humans, and has been isolated as a saprophyte from soil, fertilizer and timbers. This fungus also exerts a dimorphic conversion. In contrast to *C. albicans*, however, it grows only in the yeast-like or budding spherical form in tissues. The mycelial form produces conidia in culture at 27°C, while the yeast-like form is produced in culture at 37°C. Therefore, the yeast-phase growth appears to be associated with virulence more than the mycelial phase.

Fungal dimorphism, an important factor in determining pathogenicity, is categorized into three types: i) temperature-dependent, ii) temperature as well as nutrition-dependent and iii) nutrition-dependent.[1] Dimorphism of *C. albicans* is nutrition-dependent and that of *S. schenckii* is temperature- and nutrition-dependent. Despite many studies, mechanisms of the morphological conversion between yeast and mycelial forms are unclear. In this chapter, involvement of lipids in fungal dimorphism is discussed in *C. albicans* in comparison to *S. schenckii*. A major factor involved in determining the morphology of fungus is the chemical composition of cell walls which, in turn, is regulated by enzymes localized in plasma membranes, such as chitin synthetase. For full activity of the membrane-bound enzymes, the optimal physical state and lipid composition of the membrane are required. In this context, it is important to study the involvement of lipids in the fungal dimorphism, which is closely associated with virulence.

II. LIPID CONTENT AND DIMORPHISM IN *C. ALBICANS* AND *S. SCHENCKII*

Although the lipid content of *C. albicans* fluctuates due to variations in culture medium and extraction method, it usually ranges from approximately 2% to 5% of the dry weight as shown in Table 1. By the extraction method of Folch et al.,[2] the lipid content is 3.0% at 37°C and 2.4% at 25°C in cells (morphology is not defined) grown in a shaking medium (4% glucose, 1% peptone, and 1 % yeast extract),[3] 5.2% (morphology is not defined)[4] at 37°C in a shaking salt medium containing 0.001% biotin and 0.5% glucose.

Table 1
Total Lipid Content of *C. albicans* and *S. schenckii*

Organism	Yeast	Mycelia	Temp.	Medium	Method	Ref.
C. albicans 3153	3.95 (pH 4.5)	4.75 (pH 6.8)	37°C	0.001% biotin 0.8% glucose	Folch	5
C. albicans 3153			37°C	0.001% biotin 0.5% glucose	Folch	4
C. albicans 0583			27°C 37°C	1% yeast extract 1% peptone 4% glucose	Folch	3
C. albicans 10273				0.1% glucose	Folch	6
12h-Culture	0.4	1.3	37°C	0.1% starch		
96h-Culture	1.3	6.3	37°C	0.001% biotin		
C. albicans 3153	2.73		37°C	yeast nitrogen base 2% glucose	Bligh & Dyer	8
S. schenckii	1.22 5.70	3.53	27°C 37°C	brain, heart infusion	Bligh & Dyer	10

Expressed as percentage based on the dry weight.

Table 2
Recovery of Total Phospholipids from *C. albicans*

Methods	Content (μg Pi* per 10^{10} cells)
Folch	22 ± 8
Modified Folch	102 ± 25
Bligh-Dyer	64 ± 13
Angus-Lester	55 ± 11

*Phospholipid phosphorus
Compiled from Mirbod, F., Mori, S. and Nozawa, Y., *J. Med. Vet. Mycol.*, 31, 403, 1993.

According to Goyal and Khuller,[5] the lipid content is 4.0% and 4.8% in stationary phase of yeast at pH 4.5 and in mycelial forms at pH 6.8 respectively, cultured in a defined mineral salt medium containing 0.8% glucose and 0.1 μg/l biotin. Growth phase also affects the total lipid content. In exponential phase, the lipid content is as low as 0.4% in the yeast form and 1.3% in the mycelial form cells, while it is as high as 1.3% in the yeast form and 6.3% in the mycelial form cells in the late stationary phase,[6] when *C. albicans* is cultured in a medium containing 0.1% glucose, 0.1% starch and 0.001% biotin. By the extraction with Bligh and Dyer's[7] method, total lipid content of this organism is 2.73% (yeast phase) grown at 37°C in yeast nitrogen base containing 2% glucose.[8] Such variations in lipid content would be, at least in part, due to the usage of different extraction methods. However, as far as phospholipids are concerned, a modified Folch's method, which includes 1-min pre-treatment with methanol, is much more effective for lipid extraction as compared to the methods of Folch, Bligh and Dyer, and Angus-Lester (Table 2).[9]

In *S. schenckii*, the total lipid content measured by the method of Bligh and Dyer is 1.2% in the yeast form at 27°C, 3.5% in the mycelial form at 27°C, and 5.7% in the yeast form at 37°C grown on brain heart infusion agar medium.[10]

From these observations, it appears that the mycelial form contains more lipids as compared to the yeast form, when grown in the same media, and also that at a higher growth temperature level of lipids is higher regardless of the morphology of these two species. However, it should be noted that starvation increases the total lipid content whereas germination decreases it, because membrane biogenesis for germination gives rise to rapid turnover of the lipid pool.[11]

III. STEROLS AND DIMORPHISM IN *C. ALBICANS* AND *S. SCHENCKII*

Sterols are one of the major lipid constituents of fungal plasma membranes also observed in other eukaryotic cells. Ergosterol is a principal component of plasma membranes of fungi, while cholesterol is the main sterol component in mammalian cell membranes. Cells that contain sterols strictly control its level in membranes. The main function of sterols in membranes is to bioregulate membrane fluidity under changing growth conditions.[12] In *C. albicans*, the sterol content is known to decrease when germination is initiated.[11] In *C. albicans* grown in the medium supplemented with ergosterol, which may be incorporated into the plasma membrane, germ-tube formation and membrane-bound chitin synthetase activity are inhibited.[13] Thus, although the precise mechanisms are not known, sterols seem to be closely related to the morphogenesis in *C. albicans.*

A. STEROL COMPOSITION OF *C. ALBICANS*

The major sterol in *C. albicans* is ergosterol both in the yeast and mycelial forms and the minor components include 4-methylenezymosterol, 4,14-dimethylzymosterol and 24-methylene-dihydrolanosterol[14,15] (Table 3). In a study on the plasma membrane sterols,[16] the yeast cell membrane has been shown to contain 50% ergosterol and 12% calciferol of total free sterols, while the mycelial cell membrane contains 42% ergosterol and 16% zymosterol. It should be, however, noted that the major esterified sterol is zymosterol with only small amount of esterified ergosterol. The esterified zymosterol is possibly transported to the cell wall.[16] However, esterified sterols could be contaminants of stored lipids, because they are thought to be a storage form of sterols in the cytosolic inclusions.[11] On the other hand, Ghannoum et al. showed by gas chromatography and mass spectrometry that cholesterol is also present in lipid fraction extracted from mycelial cells by Folch's method, but

Table 3
Sterol Composition in *C. albicans*

	MEM		MM2002		ATCC10231	
Sterols	**Yeast**	**Mycelia**	**Yeast**	**Mycelia**	**Yeast**	**Mycelia**
Cholesterol					ND	38
Ergosterol	86	88	86	87	15	6
24,28-dihydroergosterol					43	19
Zymosterol					17	14
4-methylzymosterol	9	8	9	9		
4,4-dimethylzymosterol					10	0
Lanosterol	3	2	3	3		
24-methylenedihydrolanosterol	2	2	2	1		
3-β-hydroxy-24-methylcholest-5,7-diene					12	22
References	14	14	14	14	6	6

Expressed as percentage of the total sterol content. ND: not determined.

Table 4
Phospholipid/Sterol Ratio in *C. albicans* and *S. schenckii*

	Yeast	Mycelium	References
C. albicans			
NCPF3153			
Exponential	15.4		8
Stationary	8.6		8
B2630	23.0		8
Azole-resistant NCPF3302	7.8		8
Azole-resistant NCPA3303	7.5		8
Azole-resistant Darlington	16.3		8
0269			
Growth		0.80	11
Starvation		0.85	
Germination		0.67	
10231			
12h	5.06		6
98h	2.95	0.40	
S. schenckii			
27°C	2.88	0.89	10
37°C	2.04		

not in the yeast form of *C. albicans* 10231.[6] They suggested that most of cholesterol exists as cholesteryl mannoside, whose biological function(s) is unknown.

B. POSSIBLE INVOLVEMENT IN DIMORPHISM OF *C. ALBICANS* AND *S. SCHENCKII*

Mycelial cells contain less sterol than yeast cells in *C. albicans* and *S. schenckii* (Table 4). This is also true of lipids of the isolated plasma membranes of *C. albicans* 6406.[16] However, large differences are not found in the sterol composition between the wild mycelia-producing cells MEM and the mycelia-deficient mutant MM2002 of *C. albicans*.[14] These observations suggest that sterol composition may not be a critical determinant of morphogenesis of *Candida*, but the level of sterol content may be related to dimorphism. This view is also supported by the fact that inhibition of sterol biosynthesis by imidazoles correlates well with the formation of abnormally short mycelia.[14]

On the other hand, it is of interest to note that polyene-resistant mutants of *C. albicans* accumulating 14-methyl sterols, instead of ergosterol, are defective in hyphal growth, suggesting that 14-methyl sterols alter the physical properties of the membrane.[15] Clotrimazole, which causes accumulation of 14-methyl sterols and inhibits the biosynthesis of ergosterol, also causes the inhibition of mycelial growth.[15] These observations indicate that sterols may be involved in the morphogenesis of *C. albicans*.

Concerning virulence, clotrimazole exerts clinical effects in candidiasis, probably via inhibition of ergosterol synthesis followed by repressed hyphal growth. This supports the notion that ergosterol, necessary for hyphal growth, plays a certain role in virulence of *C. albicans*. However, this may not be true of *S. schenckii* since it, even though it has lower ergosterol in yeast form, causes skin infection in yeast form, but not in mycelial form. Therefore, ergosterol does not appear to be implicated in virulence, even though it has a relation with morphology in these fungi.

Starved *C. albicans* cells show a low proportion of sterols in the total lipid, although the total lipid content is increased by starvation. Germination increases the proportional content of sterols with a simultaneous decrease in the total lipid content.[11] In this regard, it must be noted that exponential phase cells of *C. albicans* NCPF 3153 contain a much lower content of sterols than that of stationary phase cells as reflected in the phospholipid/sterol ratio of 15.4 and 8.6, respectively.[8] This is also the case with another strain 10231, showing the phospholipid/sterol ratio of 5.06 in 12 h and 2.95 in 98 h cultured cells.[6]

IV. PHOSPHOLIPIDS AND DIMORPHISM IN *C. ALBICANS* AND *S. SCHENCKII*

Phospholipids are the major components of cell membranes. Most membranes contain different classes of phospholipids and their compositions are rather constant in a given membrane under the same growth conditions. Some membrane-bound enzymes require specific lipids and proper physical states, i.e., fluidity of membranes for their optimal activities.[12] In addition to these functions, phospholipids play a key role in signal transduction.[17] Morphological interconversion between the yeast and mycelial forms of dimorphic fungus depends on the growth conditions, and their changes are related to intracellular signal transduction. Therefore, analysis of the membrane phospholipid composition of the yeast and mycelial cells should shed light on the mechanism of dimorphism and virulence.

Table 5
Phospholipid Composition of *C. albicans* and *S. schenckii*

	PC	LysoPC	PE	PS	PI	CL	PG	DPG	Others	Ref.
				C. albicans						
0583										
37°C	45		24	13	13	3	2			3
25°C	43		26	11	12	4	4			3
ATCC3153										
Yeast	20	10	14	17	12	15			17	5
Mycelia	19	12	13	13	11	12			13	5
NCPF3153										
Exponential	25		27	18	20		3	7		8
Stationary	32		23	22			3	21		8
3153										
Stationary	16		13	35	11	8			13	4
ATCC10231										
12h Culture										
Yeast	34		15	12	15		12		12	6
Mycelia	36		16	12	15				13*	6
96h Culture										
Yeast	29		18	13	17		14	9		6
Mycelia	40								60*	6
6406 (Plasma membrane)										
Yeast	4		70	11		15				16
Mycelia	50		50							16
				S. schenckii						
Yeast 37°C	52		15	15	10	2			3**	10
Yeast 27°C	48		20	10	12	1			4**	10
Mycelia 27°C	43		18	5	10	7			8**	10

Expressed as percentage of the total phospholipid content.
PC, phosphatidylcholine; PE, phosphatidylethanolamine; PS, phosphatidylserine; PI, phosphatidylinositol; CL, cardiolipin; PG, phosphatidylglycerol; DPG, diphosphatidylglycerol; * digalactosyldiacylglycerols, ** phosphatidic acid

A. PHOSPHOLIPID COMPOSITION OF *C. ALBICANS* AND *S. SCHENCKII*

The major phospholipids of *C. albicans* and *S. shenckii* are phosphatidylcholine (PC), phosphatidylethanolamine (PE), phosphatidylserine (PS) and phosphatidylinositol (PI) in both the yeast and mycelial forms.

In *C. albicans*, PC and PE range from 15% to 45% and from 12% to 26% of total phospholipids respectively as shown in Table 5. Although previous studies show large variations, profound differences are not observed in the phospholipid composition between the yeast and mycelial forms in *C. albicans*, when the results obtained from the same laboratory are compared. However, the phospholipid analysis of plasma membrane fraction isolated from *C. albicans* 6406 revealed a great difference in PC content between the yeast and mycelial forms.[16] While the PC and PE contents of yeast plasma membranes are 4% and 70%, those of mycelial plasma membranes contain equal proportions of both the phospholipids.[18]

In *S. schenckii* no significant differences are observed in PC and PE contents between the yeast and mycelial cells grown at 27°C in the same culture. However, elevation of the growth temperature to 37°C to obtain the yeast cell growth causes an increase in PC and PS with a simultaneous decrease in PE and PI content.[10] These findings suggest that differences in phospholipid contents are due to the elevated growth temperature and not because of the morphological changes.

B. POSSIBLE INVOLVEMENT OF PHOSPHOLIPIDS IN DIMORPHISM IN *C. ALBICANS*

As mentioned in the previous section, no significant differences are observed in phospholipid composition between the yeast and mycelial forms in *C. albicans*. However, the proportion of components of phospholipids is altered during the process of yeast to mycelial conversion induced by changing the medium.[18]

In a strain of *C. albicans* 3125, the morphological transition gets completed within 4 to 5 h after the change of medium, which can induce germination (Figure 1).[18] In association with the morphological transition, the composition of major phospholipids is dramatically altered as shown in Figure 2.[18] The levels of both PC and PE increase from 28% to 40% and 16% to 27%, respectively, whereas those of PI and PS decline from 33% to 7%. Thereafter, the phospholipid composition returns gradually to the initial profile within 12 h.[18]

Similar observations have been reported in strain ATCC 10261 during starvation and germ-tube formation[11] (Figure 3). At 3 h following the germ-tube formation, the content of PI decreases transiently from 9% to 5% in starved cells and increases to 16% at 2 h after the initiation of germination. In contrast, the PC content increases from 36% to 56% at 3 h after germination,[11] which is also observed in a strain of *C. albicans* 3125.[18] These transient fluctuations in the PI content are of great interest in terms of transmembrane signaling.[17]

Actually, PI turnover is enhanced in association with filamentous growth caused by ethanol in *C. tropicalis* PK 233.[19] Although this strain grows in a filamentous form in a defined medium containing ethanol (1.5% or above), addition of *myo*-inositol (1-5 µg/ml) prevents the ethanol-induced morphological changes. The PI content is extremely low in the ethanol-induced filamentous cells at the logarithmic phase of growth, indicating the enhancement of PI turnover during filamentous growth. However, the PI turnover is reduced to the initial level after the filamentous transformation is completed.[19] It is also conceivable that the reduction of PI content in *C. albicans* 3125 and ATCC 10261 (which is mentioned above) is due to enhanced PI turnover.

Myo-inositol-supplementation inhibits the ethanol-induced morphological changes and also PI turnover. These findings suggest that PI turnover may be

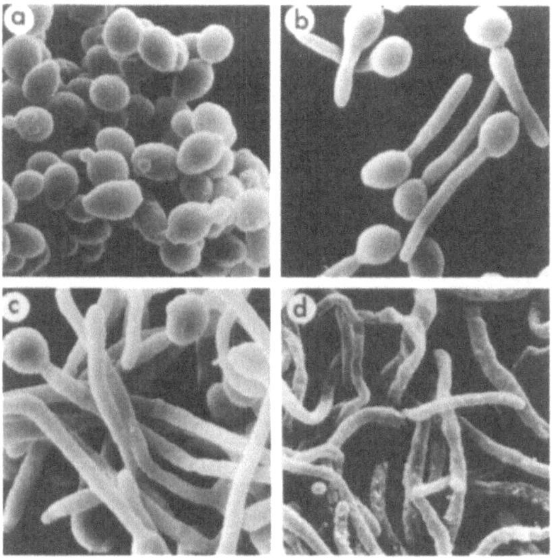

Figure 1. Scanning electron microscopic observations of *C. albicans*. (A) Typical yeast form cells in the early stationary phase. Pictures of *C. albicans* at 1.5 h (B), 4.5 h (C) and 12 h (D) after the early stationary phase cells were transferred into L-methionine-containing medium, which induces germination, illustrate the representative stages of the morphological transition from the yeast cell to hypha. From Yano, K., Yamada, T., Banno, Y., Sekiya, T. and Nozawa, Y., *Jpn. J. Med. Mycol.*, 23, 159, 1982. With permission.

involved in morphogenesis of *Candida* . Thus, it is tempting to speculate that activation of PI turnover-mediated signal transduction plays a role in induction of the mycelial form that may be associated with virulence.

C. PHOSPHOLIPID METABOLISM DURING YEAST TO MYCELIA TRANSFORMATION IN *C. ALBICANS*

To investigate phospholipid biosynthesis during the yeast to mycelial conversion, a procedure has been developed for induction of the transformation in *C. albicans*. Fungal cells grown at 35°C under shaking to the logarithmic phase are kept at 28°C without shaking for 4 h, and then are reshaken at 35°C. Following the reshaking of culture, cells initiated the production of mycelial forms. Cells are labeled with [^{32}P]-orthophosphate for 30 min prior to reshaking and [^{32}P]-incorporation into phospholipids is examined at different intervals after initiation of reshaking. As shown in Figure 4, most of the radioactivity is incorporated into the acidic phospholipids, phosphatidylserine (PS) and phosphatidylinositol (PI) in cells kept in the resting state. However, while reshaking the cultures at 35°C biosynthesis of phosphatidylcholine (PC) is highly activated as reflected in a progressively increased [^{32}P]-incorporation into PC. The profile showing the mirror image of the increase in PC and decrease in PE indicates the precursor- product relationship. The marked decrease in PS indicates that it is first

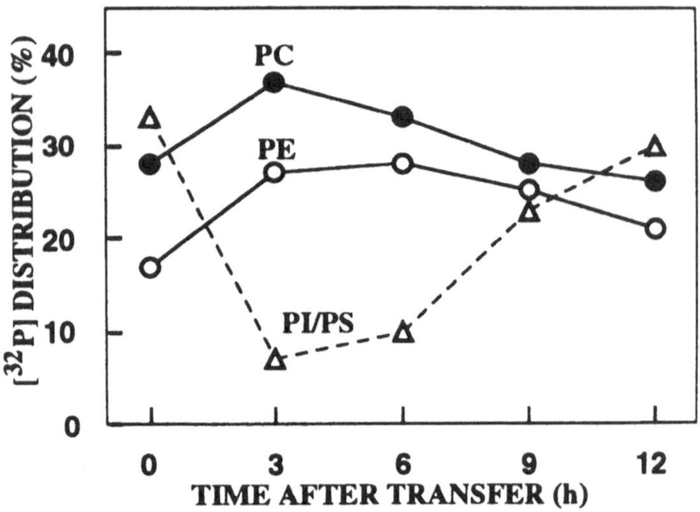

Figure 2. Alterations in phospholipid composition during the yeast cell to hypha conversion in *C. albicans* 3125. *C. albicans* cells grown in the synthetic medium are transferred to the L-methionine-containing medium to induce the morphological transformation. Symbols: (●) PC, phosphatidylcholine; (o) PE, phosphatidylethanolamine; (Δ) PI/PS, phosphatidylinositol and phosphatidylserine. From Yano, K., Yamada, T., Banno, Y., Sekiya, T. and Nozawa, Y., *Jpn. J. Med. Mycol.*, 23, 159, 1982. With permission.

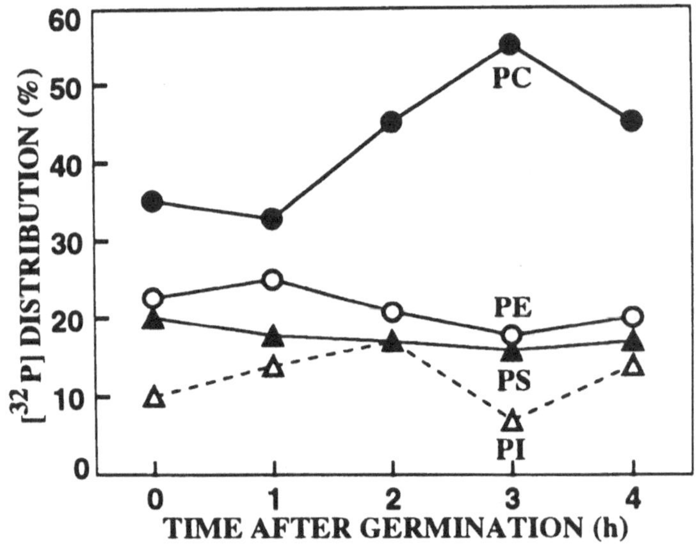

Figure 3. Alterations in phospholipid composition during germination in *C. albicans* ATCC 10261. Starved *C. albicans* cells were transferred to the GlcNAc-containing medium to induce germination. Symbols: (●) PC, phosphatidylcholine; (o) PE, phosphatidylethanolamine; (Δ) PI, phosphatidylinositol; (Δ) PS, phosphatidylserine. Compiled from Sundaram, S., Sullivan, P.A. and Shepherd, M.G., *Exp. Mycol.*, 5, 140, 1981.

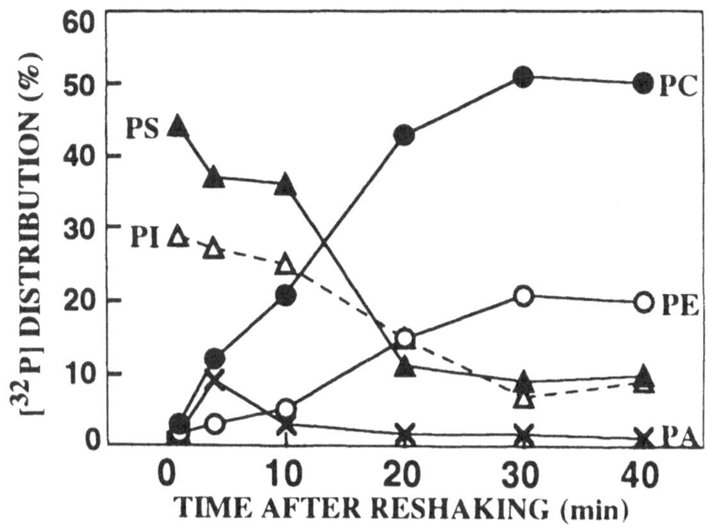

Figure 4. Phospholipid metabolism during the yeast to mycelial transformation induced by reshaking *C. albicans* 1594 IFO. After cells grown at 35°C under shaking for 12 h, cultures were kept without shaking for 4 h and then reshaken at 35°C. Symbols: (●) PC, phosphatidylcholine; (o) PE, phosphatidylethanolamine; (Δ) PI, phosphatidylinositol; (Δ) PS, phosphatidylserine; (×) PA, phosphatidic acid.

converted into PE by decarboxylation in mitochondria which is then utilized to produce PC via stepwise methylation.

V. FATTY ACIDS AND DIMORPHISM IN *C. ALBICANS* AND *S. SCHENCKII*

The principal function of fatty acids is to modulate the physical state of the membrane. When organisms are exposed to altered environmental factors, such as temperature and nutrition, they modify the fatty acid composition to adapt to the new environments to maintain the membrane fluidity. Therefore, fatty acid composition must be regulated in a sophisticated manner. In this context, the study of the relationship between the fatty acid composition and dimorphism in *C. albicans* and *S. schenckii* may provide a clue to better understanding of the role of fatty acids in the morphological changes.

A. EFFECTS OF CULTURE TEMPERATURE

The major fatty acid components of various lipids in *C. albicans* are palmitic (16:0), palmitoleic (16:1), oleic (18:1), linoleic (18:2) and linolenic (18:3) acids, while those in *S. schenckii* are palmitic, oleic and linoleic acids.

Table 6
Fatty Acid Composition of *C. albicans* and *S. schenckii* Cultured at Different Temperatures

	14:0	16:0	16:1	17:1	18:0	18:1	18:2	18:3	U.I.
C. albicans 0583									
37°C	0.8	31.6	12.3	2.6	0.4	20.6	25.9	5.1	104
25°C	0.3	19.4	7.4	2.0	1.1	22.4	26.8	19.8	146
S. schenckii									
37°C			22.3	1.7	0.7	40.7	32.5		108
27°C			18.2	0.6	0.5	33.1	45.9	0.7	128

Expressed as percentage of the total fatty acid content. U.I., unsaturation index.
Compiled from Nishi, K., Ichikawa, H., Tomochika, K., Okabe, A. and Kanemasa, Y., *Acta Med. Okayama*, 27, 73, 1973; Shikano, Y., *Acta School. Med. Uni. in Gifu*, 34, 1007, 1986.

Cells grown at a lower temperature contain a larger proportion of polyunsaturated fatty acids in total lipids (Table 6), as observed in other organisms also. Such profiles are reflected in phospholipid as well as in triacylglycerol fractions. The increase in fatty acid unsaturation is thought to compensate for the decreased membrane fluidity caused by lowering of the environmental temperature so as to maintain the optimal membrane fluidity. Since the increase in fatty acid unsaturation is not associated with morphological conversion in these pathogenic fungi, the degree of unsaturation does not appear to play a key role in morphogenesis.

B. FATTY ACID COMPOSITION AND DIMORPHISM

Since culture conditions, especially growth temperature, are known to affect significantly the fatty acid composition, it is important to obtain the yeast and mycelial cells under the same culture conditions to study the role of fatty acids in morphogenesis. Ghannoum et al.[7] separated yeast and mycelial cells in the same cultures, and showed that the polar lipids of mycelial cells of *C. albicans* ATCC 10261 contain higher levels of polyunsaturated fatty acids (18:2 and 18:3) but lower levels of oleic acid (18:1) as compared to yeast cells in 12 h-cultures (Table 7). This profile is very distinct in 96 h-cultures, where mycelial lipids contain 33.7% linolenic acid (18:3) in the phospholipids. In a strain of *C. albicans* 3153, higher levels of palmitoleic (16:1) and oleic acid (18:1) are observed in mycelial cells, but linoleic (18:2) and linolenic (18:3) acids are less than 1% of total fatty acids in both yeast and mycelial forms[5]. In *S. schenckii*, however, there is no difference in fatty acid composition between yeast and mycelial cells isolated from the same cultures grown at 27°C. Thus, the degree of fatty acid unsaturation does not appear to be an important factor in determining morphology of *S. schenckii*.

Modification of fatty acid unsaturation occurs at an early stage in the mycelial formation in *C. albicans* 3125 (Table 7).[18] At 1.5 h following germination, oleic (18:1) and linolenic acid (18:3) levels increase from 12.2% to 21.5% and from 1.4% to 4.5%, respectively.[18] After 6 h, they reach 29% and 6.4%, respectively. This rapid response suggests that unsaturation may be

Table 7
Fatty Acid Composition and Morphology in *C. albicans* and *S. schenckii*

	14:0	14:1	15:0	16:0	16:1	16:2	18:0	18:1	18:2	18:3	U.I.
C. albicans (3153)											
Yeast; phospholipids	36		10	22	18		1	12	1		45
Mycelia; phospholipids	7		2	36	23		2	30	1		115
C. albicans (ATCC10231)											
12 h culture											
Yeast; polar lipids	4	6		23	10	tr	10	46	11	4	
Mycelia; polar lipids	4	tr		15	13	tr	15	7	14	13	
96 h culture											
Yeast; polar lipids	7	2		25	10	3	7	22	19	4	
Mycelia; polar lipids	7	8		21	tr	10	12	8	tr	34	
C. albicans (3125)						**(17:1)**					
Yeast; phospholipids				24	13	10	2	32	12	1	82
1.5h after Y-M switch				20	12	3	8	23	22	4	94
6h after Y-M switch				22	10	2	4	25	29	6	114
12h after Y-M switch				24	12	1	3	29	24	4	102
S. schenckii; phospholipids											
Yeast at 27°C				18	1		1	33	46	1	127
Mycelia at 27°C				18	1		1	27	46	1	129

Expressed as percentage of the total fatty acids. Y-M switch, yeast to mycelium switch. U.I., unsaturation index.

Compiled from Goyal, S. and Khuller, G.K., *J. Med. Vet. Mycol.*, 30, 355, 1992; Ghannoum, M.A., Janini, G., Khamis, M. and Radwan, S.S, *J. Gen. Microbiol.*, 132, 2367, 1986; Shikano, Y., *Acta School. Med. Uni. In Gifu*, 34, 1007, 1986; Yano, K., Yamada, T., Banno, Y., Sekiya, T. and Nozawa, Y., *Jpn. J. Med. Mycol.*, 23, 159, 1982.

involved in germination of mycelial form in *C. albicans*. Alternatively, unsaturation of phospholipids may reflect an epiphenomenon associated with the transition from a non-growing to growing state, as has been shown in *C. lipolytica*.[20]

VI. CONCLUSION

As described above, it is conceivable that lipids play a role in dimorphism which is related to virulence in *C. albicans* and *S. schenckii*. However, direct evidence is not available to indicate the intrinsic relationship between lipid composition and morphological conversion. Further studies are required to prove the implication of qualitative and/or quantitative modifications in phospholipids, fatty acids and sterols in morphological changes in these pathogenic fungi.

VII. REFERENCES

1. **Gole, G.T. and Kendrick, B.,** Dimorphism, in *Biology of Conidial Forch*, vol. 1., Academic Press Inc., New York, 5, 1981.
2. **Folch, J., Lees, M. and Stanley, G.H.,** A simple method for the isolation and purification of total lipids from animal tissues, *J. Biol. Chem.*, 226, 497, 1957.
3. **Nishi, K., Ichikawa, H., Tomochika, K., Okabe, A. and Kanemasa, Y.,** Lipid composition of *Candida albicans* and effect of growth temperature on it, *Acta Med. Okayama*, 27, 73, 1973.
4. **Nago, N. and Khuller, G.K.,** Lipids of *Candida albicans*: subcellular distribution and biosynthesis, *J. Gen. Microbiol.*, 136, 993, 1990.
5. **Goyal, S. and Khuller, G.K.,** Phospholipid composition and subcellular distributiŏn in yeast and mycelial forms of *Candida albicans*, *J. Med. Vet. Mycol.*, 30, 355, 1992.
6. **Ghannoum, M.A., Janini, G., Khamis, M. and Radwan, S.S.,** Dimorphism associated variations in the lipid composition of *Candida albicans*, *J. Gen. Microbiol.*, 132, 2367, 1986.
7. **Bligh, E. and Dyer, W.J.,** A rapid method of total lipid extraction and purification, *Can. J. Biochem. Physiol.*, 37, 911, 1959.
8. **Hitchcock, C.A., Barrett-Bee, K.J. and Russell, N.J.,** The lipid composition of azole-sensitive and azole-resistant strains of *Candida albicans*, *J. Gen. Microbiol.*, 132, 2421, 1986.
9. **Mirbod, F., Mori, S. and Nozawa, Y.,** Methods for phospholipid extraction in *Candida albicans*: an extraction method with high efficacy, *J. Med. Vet. Mycol.*, 31, 403, 1993.
10. **Shikano, Y.,** Dimorphism and lipid composition of *Sporothrix schenckii*, *Acta School. Med. Uni. in Gifu*, 34, 1007, 1986.
11. **Sundaram, S., Sullivan, P.A. and Shepherd, M.G.,** Changes in lipid composition during starvation and germ-tube formation in *Candida albicans*, *Exp. Mycol.*, 5, 140, 1981.
12. **Thompson, G.A., Jr.,** Sterol metabolism, in *The Regulation of Membrane Metabolism*, CRC Press, Boca Raton, FL, 5, 1992.
13. **Chiew, Y.Y. and Sullivan, P.A.,** The effects of ergosterol and alcohols on germ-tube formation and chitin synthetase in *Candida albicans*, *Can. J. Biochem.*, 60, 15, 1982.
14. **Cannon, R.D. and Kerridge, D.,** Correlation between the sterol composition of membranes and morphology in *Candida albicans*, *J. Med. Vet. Mycol.*, 26, 57, 1988.
15. **Shimokawa, O., Kato, Y. and Nakayama, H.,** Accumulation of 14-methylsterols and defective growth in *Candida albicans*, *J. Med. Vet. Mycol.*, 24, 327, 1986.
16. **Marriott, M.S.,** Isolation and chemical characterization of plasma membranes from the yeast and mycelial forms of *Candida albicans*, *J. Gen. Microbiol.*, 86, 115, 1975.

17. **Nishizuka, Y.**, The role of protein kinase C in cell surface signal transduction and tumor promotion, *Nature*, 308, 693, 1984.
18. **Yano, K., Yamada, T., Banno, Y., Sekiya, T. and Nozawa, Y.**, Modification of lipid composition in a dimorphic fungus, *Candida albicans*, during the yeast cell to hypha transformation, *Jpn. J. Med. Mycol.*, 23, 159, 1982.
19. **Uejima, Y., Koga, T. and Kamihara, T.**, Enhanced metabolism of phosphatidylinositol in *Candida tropicalis* in association with filamentous growth caused by ethanol, *FEBS Letters*, 214, 127, 1987.
20. **Kates, M. and Paradis, M.**, Phospholipid desaturation in *Candida lipolytica* as a function of temperature and growth, *Can. J. Biochem.*, 51, 184, 1973.

Chapter 11

ANTIFUNGALS

A. S. Ibrahim, R. Prasad and M. A. Ghannoum

CONTENTS

0-8493-4794-7/96/$0.00+$.50

I. INTRODUCTION

The most important antifungal agents that are used clinically for the treatment of mycotic infections can be divided into two major groups on the basis of their origin: (1) polyene antibiotics, e.g., amphotericin B and nystatin; (2) synthetic agents, e.g., azole derivatives and flucytosine. Most antibiotics active against fungi act at the level of membrane lipids by inhibiting either their function (polyenes) or their biosynthesis (azoles, allylamines, azasterols, cerulenin).[1-4] Thus, fungal lipids are of obvious interest as therapeutic targets and have been studied extensively in recent years.[5-10]

In this chapter, we discuss the mode of action of the major chemical classes of antifungals. However, a detailed discussion on polyenes is avoided in view of Chapter 12 of this volume. The effect of certain experimental "unorthodox" compounds on the lipids of *Candida* is also described.

II. POLYENES

Polyenes constitute a major class of anti-*Candida* drugs and are known to bind with the plasma membrane leading to altered permeability and subsequent death of the organism.[11-15] There appears to be a direct association between the sensitivity of an organism to a polyene and the presence of sterols in the plasma membrane of the cells. All organisms susceptible to polyenes, e.g., yeasts, algae, protozoa and mammalian cells, contain sterols in their outer membrane, while all the resistant organisms do not contain sterols.[16] Studies with *Acholeplasma laidlawaii*, which is also unable to synthesize sterols, provide further proof of the association between the sensitivity of an organism to polyene and to the presence of sterols.[17,18] When *A. laidlawaii* was grown in a sterol-deficient medium, the cells were resistant to polyenes; when grown in sterol-containing media, sterols were incorporated into the plasma membrane and the organism became sensitive to polyenes.[18]

The importance of membrane sterols vis-a-vis polyenes is also supported by earlier studies,[19-21] wherein it was shown that fungi can be protected from the inhibitory action of certain polyenes by the addition of sterol to the growth medium. This effect is due to a physicochemical interaction between added sterols and the polyenes, which prevents the antibiotics from interacting with the cellular sterols. Fatty acids also protect sensitive organisms against the action of polyene antibiotics, presumably through a similar mechanism.[22]

The interaction between the sterols and polyene antibiotics is further supported by direct evidence based on spectrophotometric data. Lampen et al.[20] reported that when sterols were added to aqueous solutions of the polyene antibiotics, filipin or nystatin, the UV absorbance significantly decreased, suggesting a direct interaction between the added sterol and the polyenes.[23,24] Schroeder et al.[25] provided evidence for binding between sterols and filipin by using a fluorimetric technique involving the measurement of partial quantum efficiency.

Correlation also exists between the degree of membrane damage caused by individual polyene antibiotics and the observed defects seen in the electron micrographs. Freeze-etch electron microscopy demonstrated that filipin, not amphotericin B, induces the formation of aggregates 15 to 25 nm in diameter (or "pits") in cholesterol-containing membranes from *A. laidlawaii*. These aggregates can not be considered as pores because no "through and through" holes were visible on the etched faces.[26,27] Studies carried out on rat erythrocytes also gave similar results,[26,28] although structural alterations (pits, doughnut-shaped craters, and protrusions) in the erythrocyte membrane were apparent. For the larger polyenes (amphotericin B and nystatin), it has been proposed that the interaction of the antibiotic with membrane sterol results in the production of aqueous pores consisting of an annulus of eight amphotericin B molecules linked hydrophobically to the membrane sterols.[27,29] This gives rise to an aqueous pore in which the hydroxyl residues of the polyene face inward to give an effective pore diameter of 0.4 to 1.0 nm. The length of the annulus is such that two half pores are required to span the plasma membrane. The carbohydrate moiety of the polyene, and the hydroxyl group of the membrane sterols, are at one end of the complex located at the lipid-water interface, whereas at the other end of the complex within the lipid bilayer there is a single hydroxyl group from the polyene molecule.[3,30]

Factors other than sterols have also been shown to play a role in polyene sensitivity of yeast cells.[31-33] Hsu Chen and Feingold[34] reported that the presence of cholesterol in liposomes derived from egg lecithin was more sensitive to nystatin or amphotericin B than those derived from dipalmitoyl or distearoyl lecithins. Thus, it appears that polyene antibiotic toxicity may also depend on the fatty acyl composition of the phospholipids. The correlation between polyene sensitivity of different cell lines and the cholesterol:phospholipid molar ratio reflects a relationship between sensitivity and membrane fluidity. Any change in this ratio is expected to affect the internal viscosity and molecular motion of lipid within membranes, which may result in the differences in sensitivity of cells to polyenes. Oxidative damage by polyenes is yet another factor suggested to contribute to the killing of *C. albicans*.[35,36] The reader is referred to chapter 12 for full discussion of the interaction of polyenes with lipids.

III. AZOLES

The first reports of the fungicidal properties of N-substituted imidazoles were made towards the end of the 1960s by a research group at Bayer and Janssen.[37] These original compounds proved to be important drugs for combating human fungal infections. The great success of azole derivatives led to a continuing search for superior azole antifungal. So far, over 40 of the β-substituted 1-phenethylimidazole derivatives have been reported that are active against yeast-like fungi, dermatophytes and also against Gram-positive bacteria.[38] The common imidazoles in current clinical use are clotrimazole, miconazole, econazole, ketoconazole and recently fluconazole and itraconazole (Figure 1).

Iwata and co-workers[39,40] showed that clotrimazole inhibits the synthesis of protein, RNA, DNA, lipids, mannan and glucan of *C. albicans*. Further studies with cell-free extracts prepared from *C. albicans* revealed that these drugs do not directly affect protein synthesis.[39] Clotrimazole at fungicidal (50 μM) but not fungistatic (13 μM) concentration caused a massive release of intracellular phosphorus and potassium, which occurred soon after the addition of the drug to grown cultures of *C. albicans*. This finding suggested that clotrimazole acts primarily on the cell membrane and that depletion of essential ions and metabolic precursors leads to the inhibition of macromolecular synthesis.[41,42]

Vanden Bossche et al.[43] demonstrated that miconazole at low concentrations could block the synthesis of ergosterol in *C. albicans* and that several 14α-methylsterols (24-methylene-dihydrolanosterol, lanosterol, obtusifoliol, 4,14-dimethyl-zymosterol, 14-methyl-fecosterol) accumulated in the cells. In addition, these workers showed that miconazole interfered with triacylglycerol synthesis in *C. albicans*, albeit to a lesser extent.[43] This observation suggested that miconazole acts as a potent inhibitor of the 14α-demethylation step in fungal sterol biosynthesis. Similar inhibition of 14α-demethylation by low concentrations of azoles was subsequently shown to be a common effect of all azole compounds so far tested.[9,44] Other workers showed that the primary target of imidazole derivatives was the heme protein, which co-catalyzes cytochrome P450-dependent 14α-demethylation of lanosterol.[45] The differences in the extent of inhibition of fungal and mammalian cytochromes P450 account for the selective toxicity of azoles for fungi.[46,47] Although mammalian cholesterol synthesis is also blocked by azoles at the stage of 14α-demethylation, the dose required to effect the same degree of inhibition is much higher than that required for fungi.[43]

Further evidence supporting the interactions between the azoles and the plasma membrane was presented by Yamaguchi.[48] He showed that the antifungal activity of both clotrimazole and miconazole is antagonized by several classes of lipids containing unsaturated fatty acids. Similar results were reported by Vanden Bossche and co-workers with ketoconazole.[49] The

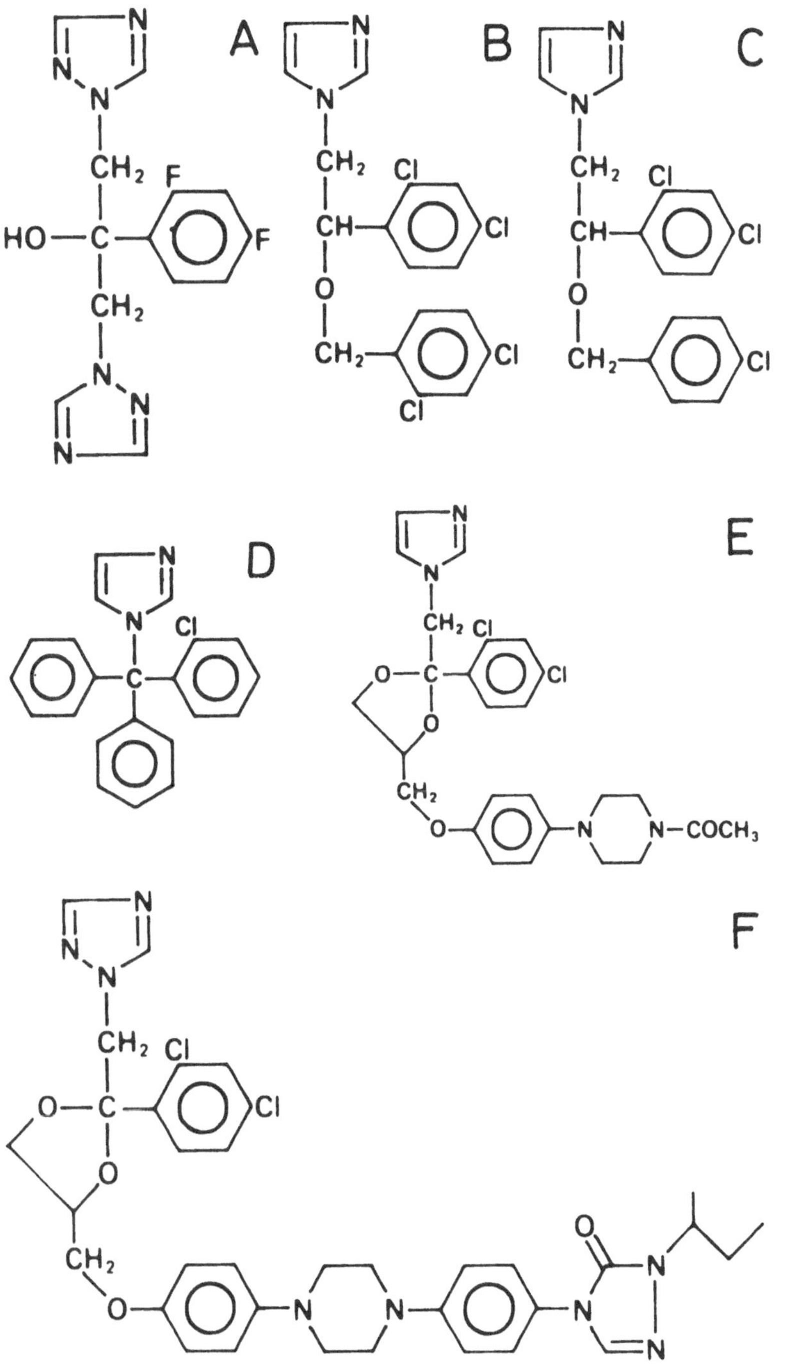

Figure 1. Some common azole antifungals: A. Fluconazole; B. Miconazole; C. Econazole; D. Clotrimazole; E. Ketoconazole, F. Itraconazole.

drugs may alter the organization of the membrane lipids without necessarily binding to them.[49] In addition to its direct action on the plasma membrane, azoles were shown to inhibit plasma membrane ATPase of *C. albicans* and other yeasts.[50] Whether such an effect accounts for the rapid collapse of electrochemical gradients and the fall in intracellular ATP levels has not been ascertained. Surarit and Shepherd[51] reported that miconazole and ketoconazole, at growth inhibitory concentrations, extensively inhibited plasma membrane ATPase, glucan synthase, adenylate cyclase, and 5′-nucleotidase of *C. albicans* when assayed *in situ*. In the same study, they also reported similar inhibition of plasma membrane enzymes by polyenes (nystatin and amphotericin B).

Fluconazole and itraconazole are the most recent addition to the list of triazole. The two drugs are orally active agents with less potency for toxicity as compared to the other azoles. Similar to other azoles, the primary action of both antifungal agents is believed to inhibit the cytochrome P450-dependent 14α-demethylase by selectively interacting with the enzyme, leading to the accumulation of 14-methyl sterols and inhibiting ergosterol synthesis.[52] The mode of action of fluconazole on the sterol composition of *Cryptococcus neoformans* has been discussed in detail in chapter 7. Vanden Bossche and his group studied in detail the mode of action of itraconazole against a variety of fungal species including *Candida albicans, C. glabrata, C. lusitaniae, Aspergillus fumigatus, Trichophyton mentagrophytes, Histoplasma capsulatum* and *Cryptococcus neoformans*. In all these organisms itraconazole was shown to inhibit the synthesis of ergosterol.[53-56] For instance, treating *C. albicans* with 10 nM of itraconazole for 1 h resulted in complete inhibition of ergosterol biosynthesis.[55] In addition, after 24 h, ergosterol synthesis was completely blocked when 30 nM[53] of itraconazole was used and at 300 nM, ergosterol levels decreased below the detection levels (< 5 fg per cell).[54] For the effect of itraconazole on cryptococcal sterols the reader is referred to chapter 7.

It is well recognized that ergosterol serves as a bioregulator of membrane fluidity, asymmetry and consequently of membrane integrity in eukaryotic cells.[57] Integrity of the cell membrane requires that inserted sterols should lack C-4 methyl groups. The inhibition of ergosterol synthesis and the accumulation of 4-methyl sterols in fungi due to treatment with azole antifungals led to the formation of a plasma membrane with altered structure and function.

IV. ALLYLAMINES

Allylamines represent an entirely new class of antifungal agents. Naftifine is one of those allylamines that have been developed to the point of clinical usage with other related compounds reaching various preclinical

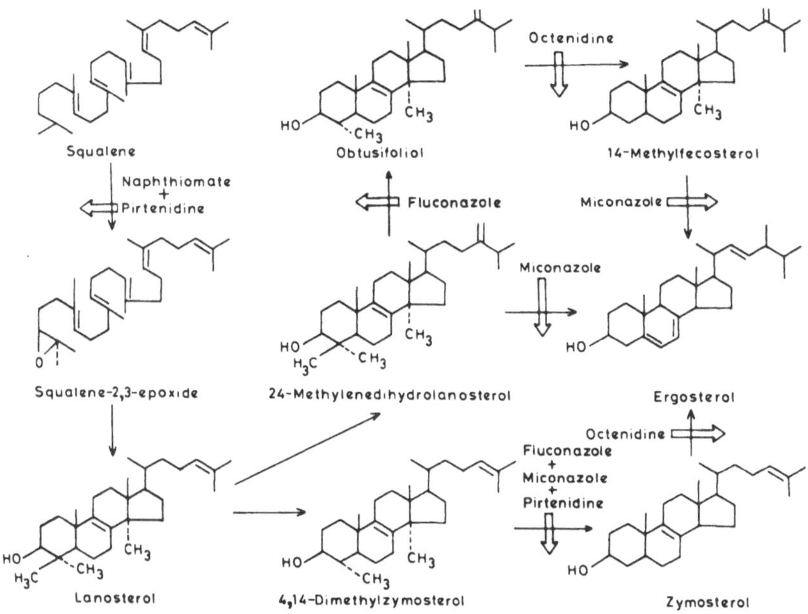

Figure 2. Ergosterol biosynthetic pathways and inhibition sites of octenidine, pirtenidine, miconazole, fluconazole and naphthiomate in *C. albicans*.

stages of development. Allylamines were shown to be highly effective against dermatophytes but less effective against *C. albicans*.[58,59]

Ryder et al.[60] showed that naftifine was a potent inhibitor of ergosterol biosynthesis of *C. albicans*. They showed that naftifine-treated yeast cells had a dose-dependent drop in ergosterol content causing total inhibition of growth. The inhibition coincided with accumulation of the sterol precursor, squalene, and absence of any other sterols. Their findings suggest that the inhibition of sterol synthesis by naftifine occurs solely at the point of squalene epoxidation. Isolated squalene epoxidase from *C. albicans* showed that indeed it was the target enzyme of naftifine.[61] Morita and Nozawa,[62] who investigated the effects of naphthiomate on ergosterol biosynthesis in *C. albicans* and *T. mentagrophytes*, also confirmed the findings of Ryder et al.[60] Allylamines interfere with fungal ergosterol biosynthesis by preventing the conversion of squalene into squalene epoxide, which is mediated by squalene epoxidase. Thus, the action of allylamine compounds is clearly distinct from that of azole antimycotics, which inhibit ergosterol biosynthesis at a later stage in the pathway (lanosterol 14α-demethylation process). Figure 2 shows a simplified pathway of ergosterol biosynthesis in fungi under normal growth conditions and when exposed to antifungal agents.

Most of the studies on the effect of antifungal agents on the lipid of *C. albicans* deal with only a few antifungals at a time due to the fact that these studies are laborious and time consuming. However, Georgopapadakou et al.[63] examined eight antifungals to study their effects on lipid biosynthesis and membrane integrity in *C. albicans*. Such investigations are vital because they carry out comparative work under the same experimental conditions and, therefore, reduce the inter- and intra-laboratory variations. These authors showed that the imidazole antifungal agents miconazole, econazole, clotrimazole and ketoconazole, at concentrations inhibitory to ergosterol biosynthesis, decreased the ratio of unsaturated to saturated fatty acids *in vivo* and *in vitro*. Similarly, naftifine, tolnaftate and azasterol A25822B decreased this ratio *in vivo* only. This suggests that the effect on fatty acids observed with ergosterol biosynthesis inhibitors may be secondary to the effect on ergosterol. With imidazoles, oleic acid antagonized inhibition of cell growth but had no inhibitory effect on ergosterol biosynthesis. This suggests that with the C-14 demethylase inhibitors, decreased unsaturated fatty acids rather than decreased ergosterol are responsible for growth inhibition.

V. CERULENIN

Cerulenin is an antifungal isolated from the culture filtrate of *Cephalosporium caerulens*. It strongly inhibits the growth of yeast-type fungi such as *Saccharomyces*, *Candida* and *Cryptococcus*, and moderately inhibits the growth of filamentous fungi and Gram-positive and -negative bacteria *in vitro*.[64] Cerulenin is a potentially useful and convenient modulator of lipid composition in microorganisms. A number of investigators have studied its mechanism of action.[64-68]

In a comprehensive investigation, Nomura et al.[64] studied the effect of cerulenin on respiration and the biosynthesis of nucleic acid, proteins, cell wall, sterol and fatty acids in yeast. They showed that the antifungal activity of cerulenin was not reversed by amino acids, and purine and pyrimidine derivatives, but it was reversed especially by ergocalciferol, and to a certain extent by retinol, thiamine, pantothenic acid, lauric acid and oleic acid. However, the amount of ergosterol in the cells of *C. stellatoidea* decreased at the growth inhibitory concentration of cerulenin. In addition, cerulenin markedly inhibited the incorporation of [^{14}C]-acetate into sterols and fatty acids in whole cells of the yeast. In contrast, the antibiotic did not inhibit significantly incorporation of [^{32}P]-phosphoric acid into nucleic acid, incorporation of [^{14}C]-glucosamine and [^{14}C]-mannose into the cell wall and had no effect on respiration of the yeast. Based on these findings, it was suggested that cerulenin affected lipid metabolism, especially the biosynthesis of sterols and fatty acids.

Further work showed that cerulenin inhibition of overall sterol synthesis may be accounted for by specific inhibition of HMG CoA, while inhibition of fatty acid synthesis is accounted for by specifically blocking the activity of β-ketoacyl-ACP synthetase.[66] The reaction of cerulenin with the peripheral -SH groups of the synthetase was proposed to be responsible for its inactivation.[65]

In another study,[63] cerulenin, contrary to earlier reports, did not inhibit ergosterol biosynthesis, although it inhibited fatty acids synthesis. It was suggested that the reported inhibition of ergosterol biosynthesis was probably due to the inhibition of sterol ester biosynthesis, secondary to inhibition of fatty acid biosynthesis. This discrepancy was explained on the basis of differences in methodologies used and growth phase of the cells. Therefore, it is important to repeat these experiments using a common methodology and cells of different growth phases to ascertain the exact mode of action of cerulenin.

The effect of cerulenin on lipid biosynthesis led to the suggestion of its potential use as a convenient modulator of lipid composition in microorganisms.[67] In this respect, two spontaneous *Candida albicans* mutants resistant to cerulenin *in vitro* are worth mentioning. These mutants were reported to have markedly reduced ability to adhere to host tissues.[69,70] This finding is quite interesting and may give further credence to the role of lipids on the adherence of *Candida* to epithelial cells. To establish such a correlation, however, further experimentation is necessary (see chapter 5 of this volume for the role of lipids in Candidal adherence).

VI. OCTENIDINE AND PIRTENIDINE

The two drugs octenidine and pirtenidine have recently been developed as potential mouth washes having both antibacterial and antimycotic properties.[71,72] These drugs were shown to cause extensive leakage of cytoplasmic contents from yeast cells, which was correlated with gross morphological and ultrastructural changes in the cell envelope of *C. albicans*.[71] When compared to control cells, the total lipid and sterol contents of *C. albicans* grown in the presence of either octenidine or pirtenidine were reduced (Table 1); however, the levels of phosphatidyl glycerol (PG), phosphatidylcholine (PC) and mono-galactosyldiacylglycerol were increased (Table 1). Drug-grown cells had higher proportions of palmitic and linolenic acids, but lower proportions of oleic acid. The C_{16}:C_{18} ratio was higher for octenidine and pirtenidine-grown cells than that in the control cells. Differences in the fatty acid composition of major phospholipids and neutral lipids between drug-grown and control cells were also observed. Sterol analysis of control cells showed that the major sterol present was ergosterol (65.9%). A significant increase of squalene and 4,14-dimethylzymosterol was also observed in pirtenidine-treated cells, while octenidine-treated cells showed an increase in zymosterol and obtusifoliol contents (Table 1). These results show that octenidine and

Table 1
Lipid Composition of *C. albicans* Grown in the Absence or Presence of Octenidine and Pirtenidine at MIC_{50}

Lipids	Control	Octenidine	Pirtenidine
Total lipid (% dry wt)	1.8 ± 0.06	1.1 ± 0.03	1.6 ± 0.05
Sterol (% dry wt)	0.12 ± 0.002	0.06 ± 0.002	0.1 ± 0.001
Apolar Lipids			
Steryl esters	1.0 ± 0.01	0.5 ± 0.01	1.8 ± 0.1
Alkyl esters	Tr	0.4 ± 0.001	1.0 ± 0.02
Triacylglycerols	14.5 ± 1.20	19.8 ± 2.50	13.2 ± 1.40
Fatty Acids	8.90 ± 0.80	6.7 ± 0.10	7.8 ± 1.10
Diacylglycerols	6.60 ± 0.70	5.5 ± 0.30	5.2 ± 0.20
Sterols	11.70 ± 1.00	10.5 ± 1.20	2.4 ± 0.01
Monoacylglycerols	8.10 ± 0.30	4.8 ± 0.40	6.2 ± 0.07
Polar Lipids			
Ceramide monohexoside	2.1 ± 0.10	1.9 ± 0.01	5.4 ± 0.40
Sterylglycosides	5.5 ± 0.50	1.4 ± 0.02	4.5 ± 0.10
Phosphatidylethanolamine	7.3 ± 0.30	6.7 ± 0.40	9.4 ± 0.70
Phosphatidylglycerol	2.5 ± 0.10	8.2 ± 0.20	8.5 ± 0.60
Phosphatidylcholine	8.8 ± 0.20	9.6 ± 0.40	11.1 ± 1.20
Phosphatidylinositol plus phosphatidylserine	9.0 ± 0.60	13.8 ± 0.40	8.9 ± 0.02
Monogalactosyl diacylglycerol	6.2 ± 0.30	7.4 ± 0.10	10.4 ± 0.80
Phosphatidic acid	5.9 ± 0.10	1.3 ± 0.70	2.8 ± 0.01
Cardiolipin	1.9 ± 0.02	1.5 ± 0.03	1.4 ± 0.01
Sterols			
Squalene	8.6 ± 0.08	6.9 ± 0.07	20.4 ± 0.80
Calciferol	Tr	ND	Tr
Zymosterol	Tr	7.8 ± 0.10	Tr
Ergosterol	65.9 ± 0.90	61.3 ± 0.80	45.6 ± 0.9
4,14-dimethylzymosterol	Tr	Tr	31.0 ± 0.70
Obtusifoliol	15.0 ± 0.30	17.5 ± 0.90	Tr
Lanosterol	10.5 ± 0.09	6.5 ± 0.04	2.7 ± 0.01
24-Methylene dihydrolanosterol	ND	Tr	0.4 ± 0.001
Fatty acids			
16:0	24.5 ± 0.90	35.1 ± 4.00	30.3 ± 2.10
16:1	17.8 ± 1.00	9.8 ± 0.80	17.6 ± 1.90
16:2	1.8 ± 0.01	1.6 ± 0.01	1.7 ± 0.40
18:0	6.6 ± 0.70	9.4 ± 0.80	6.9 ± 0.20
18:1	37.2 ± 2.80	28.8 ± 1.90	29.9 ± 1.40
18:2	12.1 ± 1.00	13.7 ± 0.90	12.5 ± 0.70
18:3	Tr	1.6 ± 0.02	1.1 ± 0.01
% UFA	68.90	55.50	62.80
SFA/UFA	0.45	0.80	0.60
C_{16}/C_{18}	0.79	0.87	0.98

Note: %UFA = percentage of unsaturated fatty acid; Tr = Traces; ND = Not Detected. SFA/UFA = saturated/unsaturated fatty acid ratio. Values are expressed as the percentage (w/w) of the total amount of lipid and are the mean ± SD of three determinations.
Compiled from Ghannoum, M.A., Moussa, N.M., Whittaker, P.A., Swairjo, I. and Abu Elteen, K., *Chemotherapy*, 38, 46-56 (1992). With permission.

Table 2
Lipid Composition of *C. albicans* Grown in Presence or Absence of Aqueous Garlic Extract

Lipid	No added AGE	Plus AGE (0.4 mg/ml)
Apolar Compounds		
Steryl esters	18.0 ± 0.7	7.0 ± 0.2
Alkyl esters	2.0 ± 0.1	2.0 ± 0.05
Triacylglycerols	5.5 ± 0.2	3.0 ± 0.1
Fatty Acids	2.0 ± 0.0	1.5 ± 0.05
Diacylglycerols	3.0 ± 0.1	8.0 ± 0.3
Sterols	5.0 ± 0.1	8.5 ± 0.6
Polar Compounds		
Esterified steryl glycoside	ND	8.0 ± 0.7
Monogalactosyldiacylglycerol	ND	7.0 ± 0.5
Steryl glycoside	1.5 ± 0.0	ND
Ceramide monohexoside	3.0 ± 0.1	ND
Phosphatidylethanolamine	11.5 ± 0.8	13.0 ± 1.1
Phosphatidylglycerol	Tr	1.5 ± 0.1
Phosphatidylchloline	21.5 ± 1.1	9.0 ± 0.8
Phosphatidylserine	13.0 ± 0.9	19.5 ± 1.0
Phosphatidylinositol	8.0 ± 0.3	7.0 ± 0.4
Phosphatidic acid	6.5 ± 0.4	5.0 ± 0.3
Fatty acids		
14:0	3.0 ± 0.2	2.5 ± 0.1
14:1	2.7 ± 0.0	1.5 ± 0.05
16:0	11.6 ± 1.0	26.2 ± 2.3
16:1	Tr	Tr
16:2	11.0 ± 1.8	8.8 ± 1.2
18:0	15.0 ± 1.5	14.8 ± 0.9
18:1	27.2 ± 1.1	37.0 ± 3.4
18:2	13.6 ± 0.8	9.2 ± 0.3
18:3	15.9 ± 2.1	Tr

Note: Values are expressed as percentage (w/w). Tr = Traces; ND = Not detected.
Compiled from Ghannoum, M. A., *J. Gen. Microbiol.*, 134, 2917, 1988. With permission.

pirtenidine affect the lipids and sterols of *C. albicans*. Changes brought about by both drugs on the phospholipids and fatty acids were similar.

The inhibition of ergosterol biosynthesis by pirtenidine may be due to the blockage of squalene cyclization, possibly through interference with epoxidase activity, either by a direct interaction with the enzyme or with its suitable activator.[60] The accumulation of 4,14-dimethylzymosterol in pirtenidine-treated cells indicates that there may be the inhibition of sterol C-14 demethylation in a manner similar to that found for imidazole derivatives.[43] Fryberg et al.[73] suggested that methylation at C-24 can precede nuclear demethylation and that saturation of the 24(28) double bond can also occur very early in the pathway. The presence of 24-methylene-dihydrolanosterol in pirtenidine-treated *C. albicans* cells is in line with the former suggestion. Therefore, the antifungal action of pirtenidine may be attributed to a combined effect of ergosterol deficiency and the accumulation of squalene and 4,14-dimethylzymosterol (Figure 2). In contrast to pirtenidine-treated cells, octenidine-treated cells show an increase in zymosterol and obtusifoliol contents, suggesting that an inhibition of the biosynthesis of ergosterol by this drug occurs after the formation of these two sterols, preventing them from converting to ergosterol (Figure 2). It should be noted that ergosterol biosynthesis in yeast cannot be discussed in terms of a single pathway. Instead, it appears that most yeasts produce ergosterol by several alternative routes[74] and that octenidine and pirtenidine affect the ergosterol biosynthesis by different routes.

VII. *ALLIUM SATIVUM* (GARLIC)

Garlic has been used as a spice, food and folk medicine since ancient times. Medicinal uses of garlic suggested include insecticidal, antimicrobial, antiprotozoal and antitumoricidal properties. Several reports show that interest persists in this activity and its possible extension to therapy *in vivo*. This culminated in the First World Congress on the Health Significance of Garlic and Garlic Constituents in August, 1990 in Washington, D.C., where various aspects of the medicinal application of garlic were discussed.[75] One of these applications was anticandidal activity.

Barone and Tansey[76] proposed that garlic acts by inactivating essential thiols. Adetumbi et al.[77] showed that it can specifically inhibit lipid synthesis of *C. albicans*.[78] Using SEM and cell leakage studies, Ghannoum et al.[78] showed that garlic treatment affected the structure and integrity of the outer surface of *C. albicans* cells. Growth of this yeast in the presence of aqueous garlic extract (AGE) affected its lipid composition. The total lipid content was decreased, while a higher level of phosphatidylserine and a lower level of phosphatidylcholine was observed in garlic-grown cells. In addition to free sterols and sterol esters, *C. albicans* accumulated esterified sterylglycosides in the presence of AGE (Table 2). Analysis of the fatty acid composition of yeast in the presence and absence of AGE showed that cells grown in the presence of AGE had higher proportions of linoleic (18:2) and linolenic (18:3) acids (Table 2). To reveal the effect of garlic on essential thiols, a number of interactions were studied between AGE and thiols including growth antagonism, enzymatic inhibition and interference of two linear zones of inhibitions. All three approaches suggested that garlic exerts its effect by the oxidation of a thiol group present in essential proteins, causing inactivation of enzymes and subsequent growth inhibition.[76]

VIII. REFERENCES

1. **Borgers, M.**, Mechanism of action of antifungal drugs with special reference to the imidazole derivatives, *Rev. Infect. Dis.*, 2, 520, 1980.
2. **Gravestock, M.B. and Riley, J.**, Antifungal chemotherapy, *Ann. Rev. Med. Chem.*, 19, 127, 1984.
3. **Kerridge, D.**, The plasma membrane of *Candida albicans* and its role in the action of antifungal drugs, in *The Eukaryotic Microbial Cell*, Gooday, G.W., Lloyd, D. and Trinci, A.P.J., Eds., Cambridge University Press, Cambridge, 103, 1980.
4. **Medoff, G., Brajtburg, J., Kobayashi, G.S. and Bolard, J.**, Antifungal agents useful in therapy of systemic infections, *Ann. Rev. Pharmacol. Toxicol.*, 23, 303, 1983.
5. **Marriott, M.S.**, Isolation and chemical characterization of plasma membranes from the yeast and mycelial forms of *Candida albicans*, *J. Gen. Microbiol.*, 86, 115, 1975.
6. **Sundaram, S., Sullivan, P.A. and Shepherd, M.G.**, Changes in lipid composition during starvation and germ tube formation in *Candida albicans*, *Exp. Mycol.*, 5, 140, 1981.
7. **Ghannoum, M.A., Janini, G., Khamis, L. and Radwan, S.S.**, Dimorphism-associated variations in the lipid composition of *Candida albicans*, *J. Gen. Microbiol.*, 132, 2367, 1986.
8. **Pierce, A.M., Pierce, H.D., Jr., Unrau, A.M. and Oehlschlager, A.C.**, Lipid composition and polyene antibiotic resistance of *Candida albicans* mutants, *Can. J. Biochem.*, 56, 135, 1978.
9. **Marriott, M.S.**, Inhibition of sterol biosynthesis in *Candida albicans* by imidazole-containing antifungals, *J. Gen. Microbiol.*, 117, 253, 1980.
10. **Taylor, F.R., Rodriguez, R.J. and Parks, L.W.**, Relationship between antifungal activity and inhibition of sterol biosynthesis in miconazole, clotrimazole and 15-azasterol, *Antimicrob. Agents Chemother.*, 23, 515, 1983.
11. **Medoff, G. and Kobayashi, G.A.**, The polyenes, in *Antifungals Chemotherapy*, Speller, D.C.E., Ed., John Wiley & Sons, London, 1980, chap. 3.
12. **Kinsky, S.C.**, The effect of polyene antibiotic on permeability in *Neurospora crassa*, *Biochem. Biophys. Res. Commun.*, 4, 353, 1961.
13. **Kinsky, S.C.**, Alterations in the permeability of *Neurospora crassa* due to polyene antibiotics, *J. Bacteriol.*, 82, 889, 1961.
14. **Kobayashi, G.S. and Medoff, G.**, Antifungal agents: recent developments, *Ann. Rev. Microbiol.*, 31, 291, 1977.
15. **Hurley, R., deLoouvois, J., and Mulhall, A.**, Yeasts as human and animal pathogens, in *The Yeast*, Vol. 1, 2nd ed., Rose, A.H. and Harrison, J.S., Eds., Academic Press, London, 207, 1987.

16. **Norman, A.W., Demel, R.A., de Kruijff, B., Guerts-van Kessel, W.S.M. and Van Deenen, L.L.M.,** Studies on the biological properties of polyene antibiotics: comparison of other polyenes with filipin in their ability to interact specifically with sterol, *Biochim. Biophys. Acta*, 290, 1, 1972.
17. **Feingold, D.S.,** The action of amphotericin B on *Mycoplasma laidlawaii*, *Biochem. Biophys. Res. Commun.*, 19, 261, 1965.
18. **Weber, M.M. and Kinsky, S.C.,** Effect of cholesterol on the sensitivity of *Mycoplasma laidlawaii* to the polyene antibiotic filipin, *J. Bacteriol.*, 89, 306, 1965.
19. **Gottleib, D.H., Carter, H.E., Sloneker, J.G. and Immann, A.,** Protection of fungi against polyene antibiotics by sterols, *Science*, 128, 361, 1958.
20. **Lampen, J.O., Arnow, P.M. and Saferman, R.S.,** Mechanism of protection by sterol against polyene antibiotics, *J. Bacteriol.*, 80, 200, 1960.
21. **Zygmunt, W.A. and Tavorminu, P.A.,** Steroid interference with antifungal activity of polyene antibiotics, *App. Microbiol.*, 14, 865, 1966.
22. **Iannitelli, R.C. and Ikawa, M.,** Effect of fatty acids on action of polyene antibiotics, *Antimicrob. Agents Chemother.*, 17, 861, 1980.
23. **Norman, A.W., Demel, R.A., de Kruijff, B. and Van Deenen, L.L.M.,** Studies on the biological properties of polyene antibiotics. Evidence for the direct interaction of filipin with cholesterol, *J. Biol. Chem.*, 247, 1918, 1972.
24. **Kleinschmidt, M.G., Chough, K.S. and Mudd, J.B.,** Effect of filipin on liposomes prepared with different types of steroids, *Plant Physiol.*, 49, 852, 1972.
25. **Schroeder, F., Holland, J.F. and Bieber, L.L.,** Fluorimetric evidence for the binding of cholesterol to the filipin complex, *J. Antibiot.*, 24, 846, 1971.
26. **Verklelij, A., de Kruijff, B., Gerritsen, W.F., Demel, R.A., Van Deenen, L.L.M. and Ververgart, P.,** Freeze-etch electron microscopy of erythrocytes, *Acholeplasma laidlawaii* cells and liposomal membranes after the action of filipin and amphotericin B, *Biochim. Biophys. Acta*, 291, 577, 1973.
27. **de Kruijff, B. and Demel, R.A.,** Polyene antibiotic-sterol interactions in membranes of *Acholeplasma laidlawaii* cells and lecithin liposomes. III. Molecular structure of the polyene antibiotic-cholesterol, *Biochim. Biophys. Acta*, 339, 57, 1974.
28. **Tillack, T.W. and Kinsky, S.C.,** A freeze-etch study of the effect of filipin on liposomes and human erythrocyte membranes, *Biochim. Biophys. Acta*, 323, 43, 1973.

29. **Holz, R.W.**, The effects of the polyene antibiotics nystatin and amphotericin B on the thin lipid membranes, *Ann. N.Y. Acad. Sci.*, 235, 469, 1974.
30. **Kerridge, D.**, The protoplast membrane and antifungal drugs, in *Fungal Protoplasts, Applications in Biochemistry and Genetics*, Peberdy, J.F. and Ferenczy, L., Eds., Marcel Dekker, New York, 135, 1985.
31. **Vanden Bossche, H., Willemsens, G., Cools, W., Marichal, P. and Lauwers, W.**, Hypothesis on molecular basis of the antifungal activity of β-substituted imidazoles and triazoles, *Biochem. Soc. Trans., U.K.*, 11, 665, 1983.
32. **Rao, T.V.G., Das, S. and Prasad, R.**, Effect of phospholipid enrichment on nystatin action: differences in antibiotic sensitivity between *in vivo* and *in vitro* conditions, *Microbios*, 42, 145, 1985.
33. **Rao, T.V.G., Trivedi, A. and Prasad, R.**, Phospholipid enrichment of *Saccharomyces cerevisiae* and its effect on polyene sensitivity, *Can. J. Microbiol.*, 31, 322, 1985.
34. **Hsu Chen, C.C. and Feingold, D.S.**, Polyene antibiotic action on lecithin liposomes: effect of cholesterol and fatty acyl chains, *Biochem. Biophys. Res. Commun.*, 51, 972, 1973.
35. **Sokol-Anderson, M.L., Brajtburg, J. and Medoff, G.**, Amphotericin B-induced oxidative damage and killing of *Candida albicans*, *J. Infect. Dis.*, 154, 76, 1986.
36. **Sokol-Anderson, M.L., Brajtburg, J. and Medoff, G.**, Sensitivity of *Candida albicans* to amphotericin B administered as single or fractionated doses, *Antimicrob. Agents Chemother.*, 29, 701, 1986.
37. **Holt, R.J.**, The imidazoles, in *Antifungal Chemotherapy*, Speller, D.C.E., Ed., John Wiley & Sons, Chichester, England, 107, 1980.
38. **Van Cutsem, J.M. and Thienpont, D.**, A broad spectrum antimycotic agent with antibacterial activity, *Chemotherapy*, 17, 392, 1972.
39. **Iwata, K., Yamaguchi, H. and Hirantani, T.**, The mode of action of clotrimazole, *Sabouraudia*, 11, 158, 1973.
40. **Iwata, K., Kanda, Y., Yamaguchi, H. and Oumi, M.**, Electron-microscopic studies on the mechanism of action of clotrimazole on *Candida albicans*, *Sabouraudia*, 11, 205, 1973.
41. **Swamy, K.H.S., Sirsi, M. and Rao, G.R.**, Studies on the mechanism of action of miconazole: effect of miconazole on respiration and cell permeability of *Candida albicans*, *Antimicrob. Agents Chemother.*, 5, 420, 1974.
42. **Vanden Bossche, H.**, Biochemical effects of miconazole on fungi. I. Effects on the uptake and/or utilization of purines, pyrimidines, nucleosides, amino acids and glucose by *Candida albicans*, *Biochem. Pharmacol.*, 23, 887, 1974.

43. **Vanden Bossche, H., Willemsens, G., Cools, W., Lauwers, W.F.J. and Lejeune, L.**, Biochemical effect of miconazole on fungi. II. Inhibition of ergosterol biosynthesis in *Candida albicans*, *Chem. Biol. Interact.*, 21, 59, 1978.
44. **Berg, D. and Plempel, M.**, Bifonazole, a biochemist's view, *Dermatologia*, 169, 3, 1984.
45. **Koh, T.Y., Marriott, M.S., Taylor, J. and Gale, E.F.**, Growth characteristics and polyene sensitivity of a fatty acid auxotroph of *Candida albicans*, *Antimicrob. Agents Chemother.*, 102, 105, 1977.
46. **Hitchcock, C.A., Dickinson, K., Brown, S.B., Evands, E.G. and Adams, D.J.**, Interaction of azole antifungal antibiotics with cytochrome P450-dependent 14α-sterol demethylase purified from *Candida albicans*, *J. Biochem.*, 266, 475, 1990.
47. **Vanden Bossche, H. and Willemsens, G.**, Effect of the antimycotics, miconazole and ketoconazole on cytochrome P450 in yeast microsomes and rat liver microsomes, *Arch. Int. Physiol. Biochem.*, 90, B218, 1982.
48. **Yamaguchi, H.**, Antagonistic action of lipid components of membranes from *Candida albicans* and various other lipids on two imidazole antimycotics, clotrimazole and miconazole, *Antimicrob. Agents Chemother.*, 12, 16, 1977.
49. **Vanden Bossche, H., Ruyschaert, J.M., Defrise-Quertain, F., Willemsens, G., Cornelissen, F., Marichal, P., Cools, W. and Van Cutsem, J.**, The interaction of miconazole and ketoconazole with lipids, *Biochem. Pharmacol.*, 31, 2609, 1982.
50. **Portillo, F. and Gancedo, C.**, Model of action of miconazole on yeast: inhibition of mitochondrial ATPase, *Eur. J. Biochem.*, 143, 273, 1984.
51. **Surarit, R. and Shepherd, M.G.**, The effect of azole and polyene antifungals on the plasma membrane enzymes of *Candida albicans*, *J. Med. Vet. Mycol.*, 25, 403, 1987.
52. **Ghannoum, M.A., Spellberg, B.J., Ibrahim, A.S., Ritchie, J.A., Currie, B., Spitzer, E., Edwards, J.E., Jr. and Casadevall, A.**, Sterol composition of *Cryptococcus neoformans* in the presence and absence of fluconazole, *Antimicrob. Agents Chemother.*, 38, 2029, 1994.
53. **Vanden Bossche, H., Heeres, J., Backx, L., Marichal, P. and Willemsens, G.**, Discovery, chemistry, mode of action and selectivity of itraconazole, in *Cutaneous Antifungal Agents*, Rippon, J.W. and Fromtling, R.A., Eds., Marcel Dekker, Inc., New York, 263,1993.
54. **Vanden Bossche, H., Marichal, P., Gorrens, J., Bellens, D., Coene, M.C., Lauwers, W., Le Jeune, L., Moereels, H. and Janssen, P.A.J.**, Mode of action of antifungals of use in immunocompromised patients. Focus on *Candida glabrata* and *Histoplasma capsulatum*, in *Mycoses in AIDS Patients*, Vanden Bossche, H., Mackenzie, D.W.R., Cauwenbergh, G., Van Cutsem, J., Drouhet, E. and Dupont, B., Eds., Plenum Press, New York, 223,1990.

55. **Vanden Bossche, H., Marichal, P., Gorrens, J., Coene, M.C., Willemsens, G., Bellens, D., Roels, I., Moereels, H. and Janssen, P.A.J.,** Biochemical approaches to selective antifungal activity. Focus on azole antifungals, *Mycoses*, 32 (Suppl. 1), 35,1989.
56. **Vanden Bossche, H., Marichal, P., Gorrens, J., Geerts, H. and Janssen, P.A.J.,** Mode of action studies. Basis for the search of new antifungal drugs, *Ann. N. Y. Acad. Sci.*, 544, 191, 1988.
57. **Nozawa, Y. and Morita, T.,** Molecular mechanisms of antifungal agents associated with membrane ergosterol. Dysfunction of membrane ergosterol and inhibition of ergosterol biosynthesis, in *In vitro and in vivo Evaluation of Antifungal Agents*, Iwata, K. and Vanden Bossche, H., Eds., Elsevier Science Publishers, B.V. Amsterdam, 111, 1986.
58. **Ganzinger, U., Stephen, A. and Gumhold, G.,** Treatment of dermatophytosis with naftifine: a new topical antifungal agent, *Clin. Trials J.*, 19, 342, 1982.
59. **Georgopoulos, A., Petranyi, G., Mieth, H. and Drews, J.,** *In vitro* activity of naftifine, a new antifungal agent, *Antimicrob. Agents Chemother.*, 19, 386, 1981.
60. **Ryder, N.S., Seidl, G., and Troke, P.,** Effect of the antimycotic drug naftifine on growth of and sterol biosynthesis in *Candida albicans*, *Antimicrob. Agents Chemother.*, 25, 483, 1984.
61. **Ryder, N.S.,** 15th FEBS Meeting, Brussels, Belgium, Abstr. No. 083, 1983.
62. **Morita, T. and Nozawa, Y.,** Effects of antifungal agents on ergosterol biosynthesis in *Candida albicans* and *Trichophyton mentagrophytes*: differential inhibitory sites of naphthiomate and miconazole, *J. Invest. Dermatol.*, 85, 434, 1985.
63. **Georgopapadakou, N.H., Dix, B.A., Smith, S.A., Freundenberger, J. and Funke, P.T.,** Effects of antifungal agents on lipid biosynthesis and membrane integrity in *Candida albicans*, *Antimicrob. Agents Chemother.*, 31, 46, 1987.
64. **Nomura, S., Horiuchi, T., Omura, S. and Hata, T.,** The action mechanism of cerulenin. I. Effects of cerulenin on sterol and fatty acid biosynthesis in yeast, *J. Biochem.*, 71, 783, 1972.
65. **Kawaguchi, A., Tomoda, H., Nozoe, S., Omura, S. and Okuda, S.,** Mechanism of action of cerulenin on fatty acid synthase: effect of cerulenin on iodoacetamide-induced malonyl CoA decarboxylase activity, *J. Biochem.*, 92, 7, 1982.
66. **Ohno, T., Kesado, T., Awaya, J. and Omura, S.,** Target of inhibition by the anti-lipogenic antibiotic cerulenin on sterol synthesis in yeast, *Biochem. Biophys. Res. Commun.*, 57, 1119, 1974.
67. **Sanadi, S., Pandey, R. and Lullez, G.K.,** Reversal of cerulenin-induced inhibition of phospholipids, and sterol synthesis by exogenous fatty acid/sterols in *Epidermatophyton floccosum*, *Biochim. Biophys. Acta*, 921, 341, 1987.

68. **Mago, N. and Khuller, G.K.,** Influence of lipid composition on the sensitivity of *Candida albicans* to antifungal agents, *Ind. J. Biochem. Biophys.*, 26, 30, 1989.
69. **Lehrer, N., Segal, E., Cihlar, R.L. and Calderone, R.A.,** Pathogenicity of vaginal candidiasis: studies with a mutant which has reduced ability to adhere *in vitro*, *J. Med. Vet. Mycol.*, 24, 127, 1986.
70. **Calderone, R.A., Cihlar, R.L., Lee, D., Hoberg, K. and Scheld, W.M.,** Yeast adhesion in the pathogenesis of *Candida albicans* endocarditis: studies with adherence negative mutants, *J. Infect. Dis.*, 154, 710, 1985.
71. **Ghannoum, M.A., Abu Elteen, K., Ellabib, M. and Whittaker, P.A.,** Antimycotic effects of octenidine and pirtenidine, *J. Antimicrob. Chemother.*, 25, 237, 1990.
72. **Sedlock, D.M. and Bailey, D.M.,** Microbicidal activity of octenidine hydrochloride, a new alkanediylbis (pyridine) germicidal agent, *Antimicrob. Agents Chemother.*, 28, 786, 1985.
73. **Fryberg, M., Oehlschlager, A.C. and Unrau, A.M.,** Sterol biosynthesis in antibiotic resistant yeast strains, *Arch. Biochem. Biophys.*, 160, 83, 1974.
74. **Weete, J.D.,** *Lipid Biochemistry of Fungi and Other Organisms*, Plenum Press, New York, 1980.
75. 1st World Congress on Health Significance of Garlic and Garlic Constituents, Washington, D.C., August, 1990.
76. **Barone, F.E. and Tansey, M.R.,** Inhibition of growth of zoopathogenic fungi by garlic extract, *Mycologia*, 67, 882, 1975.
77. **Adetumbi, M., Javor, G.T. and Lau, B.H.S.,** *Allium sativum* (garlic) inhibits lipid biosynthesis in *Candida albicans*, *Antimicrob. Agents Chemother.*, 30, 499, 1986.
78. **Ghannoum, M.A.,** Studies on the anticandidal mode of action of *Allium sativum* (garlic), *J. Gen. Microbiol.*, 134, 2917, 1988.

Chapter 12

INTERACTION OF THE ANTI-*CANDIDA* AMPHOTERICIN B (AND OTHER POLYENE ANTIBIOTICS) WITH LIPIDS

J. Bolard and J. Milhaud

CONTENTS

0-8493-4794-7/96/$0.00+$.50

I. INTRODUCTION

The polyene antibiotic amphotericin B (AmB) (Figure 1) remains, forty years after its discovery, the "golden standard" for the treatment of candidosis, (this situation may be changing slowly with the introduction of triazoles). However, the clinically used formulation of AmB, fungizone, has many serious side effects, including severe nephrotoxicity. It is necessary to have good knowledge of the mechanism of action of the drug to design new formulations with decreased host toxicity. Indeed, new derivatives of AmB are currently in clinical trials, as well as new delivery systems, particularly liposomes. Nystatin (Figure 1), on the other hand, is still used for topical purposes. Aromatic polyene antibiotics (Figure 2), such as vacidin A, candicidin D and hamycin, are more active but are considered too toxic for clinical use. Their methyl esters, such as mepartricin, do not present this drawback and are potentially good candidates for antifungal chemotherapy. Filipin (Figure 3A) is not clinically used but its study should bring supplementary information on the interaction of polyene antibiotics with membranes.

It is generally assumed that AmB acts by interacting with sterol-containing membranes (ergosterol in fungal cells and cholesterol in mammalian cells) and by forming transmembrane channels through which essential components leak.[1] The selectivity of AmB for fungal cells is due to the higher affinity of AmB for ergosterol containing membranes. However, recent observations have shown that other mechanisms may also be operative such as lipid peroxidation and enzyme inhibition. Furthermore, it has been suggested that, at least in mammalian cells, endocytosis of the drug may occur and result in blockade of cell activity. On the other hand, recent studies have shown that the nature of the permeability pathways induced by AmB in ergosterol and cholesterol-containing membranes are different from each other. In the first case, monomers of AmB interact directly with ergosterol. In the second case, only oligomers of AmB are responsible for inducing leakage of K^+, possibly without direct interaction with cholesterol. It could be assumed that the second type of channels become operative upon by direct interaction with phospholipids.

All these observations lead to a reappraisal of the interaction of AmB with lipids, the origin of AmB activity being certainly not limited to the association of the drug with membrane ergosterol. Due to limited studies, the mechanism of action of the other polyene antibiotics is still not well understood. The aromatic polyene antibiotics present distinct molecular characteristics (kink in the polyene chain, larger length) which may be at the origin of the differences observed in the permeability pathways.

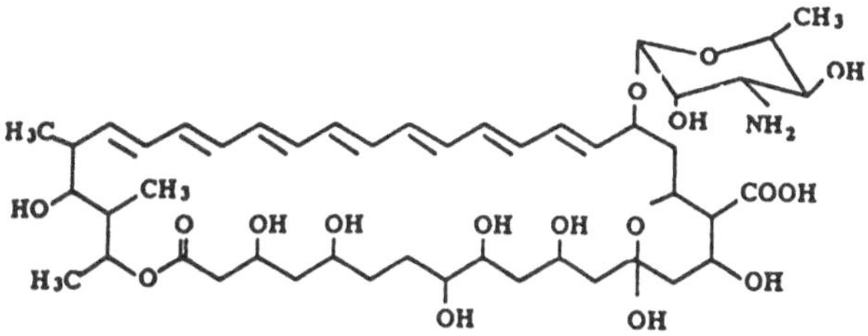

Amphotericin B

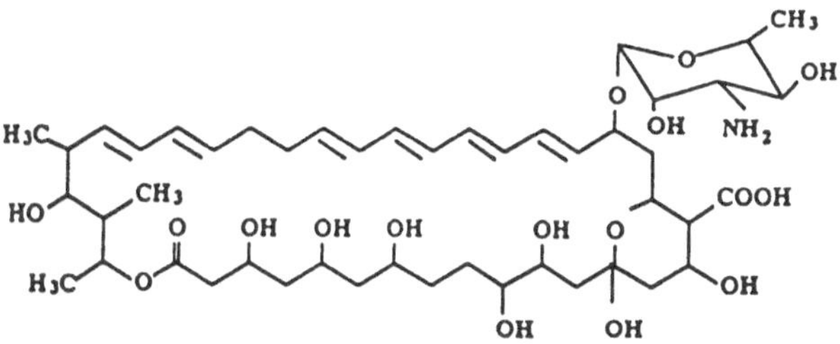

Nystatin A_1

Figure 1. Chemical structures of amphotericin B and nystatin A_1.

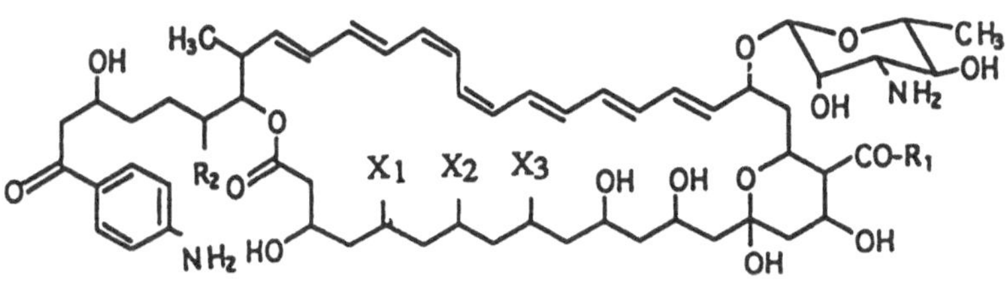

	R_1	R_2	X_1	X_2	X_3
Hamycin	OH	H	H,OH	H,H	H,OH
Candicidin D (Levorin A2)	OH	CH_3	=O	H,H	=O
Partricin B (Vacidin A)	OH	H	H,OH	=O	H,OH
Mepartricin B	OCH_3	H	H,OH	=O	H,OH

Figure 2. Chemical structures of aromatic polyene antibiotics.

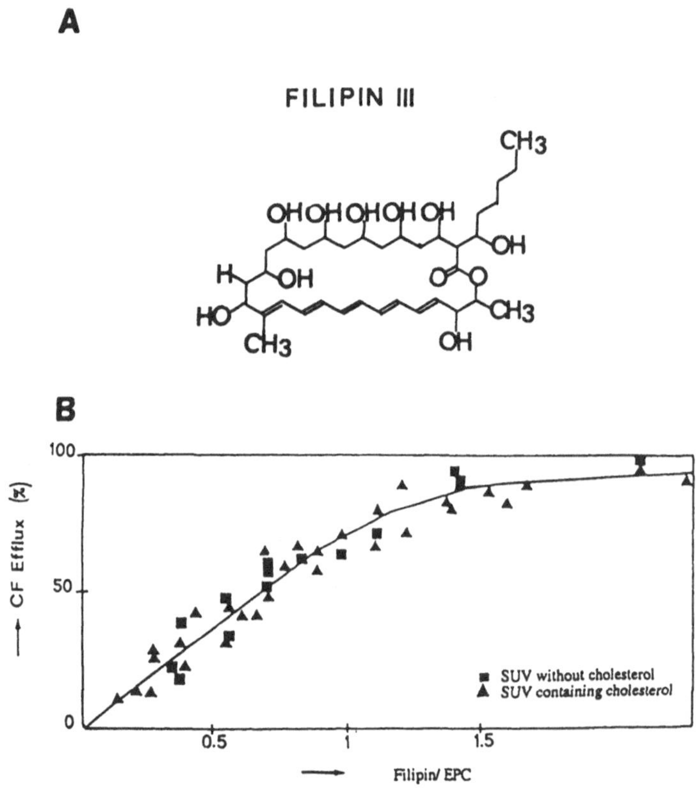

Figure 3. (A) Chemical structure of filipin; (B) Release of carboxyfluorescein (CF) from loaded EPC vesicles as a function of the filipin III to EPC molar ratios. From Whyte, B.S., Peterson, R.P. and Hartsel, S.C., *Biochem. Biophys. Res. Comm.*, 164, 609, 1989. With permission.

II. METHODS TO STUDY POLYENE ANTIBIOTIC-MEMBRANE INTERACTIONS

The polyene antibiotic-membrane interactions can be studied from two angles: focusing on the drug conformational changes or on the perturbations in membrane properties.

A. PROBING AT THE LEVEL OF THE MEMBRANE

The methods of choice to probe any impairment of the overall organization of a membrane upon drug incorporation are differential scanning calorimetry (DSC) and nuclear magnetic resonance (NMR). They, respectively, give insights into the overall thermodynamic state of the membrane and the dynamics (the degree of order) of the fatty acyl chains. In DSC, the phospholipid bilayers are essentially characterized by a sharp endothermic peak, at the temperature Tm, which signifies the gel→fluid transition which corresponds to a change in fatty acyl chains, from a fully extended all-*trans* conformation to more or less disordered conformations due to *trans-gauche* isomerisms. As a consequence of the bulkiness of chains, this process is highly cooperative. In the presence of any additive, this peak is generally broadened. Such a broadening is often interpreted as a diminution of cooperativity. However, the existence of a more or less extended endothermic region, wherein two phases coexist, directly arises from the Gibbs phase rule. Indeed, if the coexistence of two phases is theoretically impossible for a pure compound, for a two-component mixture, the main transition necessarily occurs via a two-phase coexistence region wherein the temperature and the composition are correlated. The plots, which describe the correlation of the onset and the completion temperatures of the endothermic region with composition, respectively are the solidus and the liquidus which delineate the two-phase zone in a temperature = f (composition) phase diagram.

^{2}H-NMR provides maximum information on the dynamic state of acyl chains as compared to other NMR spectroscopic techniques. The deuterium nucleus has an electric quadrupolar movement and, therefore, the interaction with a magnetic field is dominated by the quadrupolar coupling, leading to splitting of the resonance lines. For a C-D bond in an unoriented sample, the quadrupolar splitting, $\Delta\nu_Q$, is directly correlated to its local order parameter, as follows:

$$\Delta\nu_Q = \frac{3}{4} A_Q S_{CD}$$

where A_Q is the static quadrupolar constant. S_{CD} is a measure of the motional anisotropy of the C-D bond. If θ denotes the instantaneous angle between this bond and the normal to the bilayer, S_{CD} is:

$$S_{CD} = \frac{3\overline{\cos^2\theta} - 1}{2}$$

where the bar corresponds to time-averaging. The maximum quadrupolar splitting of a C-D bond, in the absence of any anisotropic movement, is about 250 KHz. The establishment of a rotation movement around an axis and a "wobbling" of this axis, like that in the fluid phase, decreases this quadrupolar splitting.

By using Raman spectroscopy it was demonstrated that in dimyristoyl phosphatidylcholine (DMPC) bilayers the ratio of the intensities at 2880 and 2850 cm^{-1} is increased in the presence of amphotericin B (AmB), which suggests an ordering of the bilayer surface.[2]

All these methods present some limitation as they are insensitive to the presence of the drug. Biological activity of polyene antibiotics develops at low antibiotic/lipid ratios (10^{-4} for instance) and, under these conditions, the perturbations brought about by the drug are local and cannot be detected at the global level which can be seen by former methods. To what extent are the effects detected at higher ratios, relevant to polyene antibiotic antifungal activity, needs to be determined. In contrast, these methods are perfectly suited to study the AmB-lipid interaction in liposomal formulations.[1]

The methods used to probe any alteration in the membrane permeability also depend on the nature of the membranes. Concerning the membranes of whole cells (animal or fungal), a direct method for studying their permeabilization to K^+ is required. After incubation in the presence of the drug, the cells are harvested by centrifugation, lysed and the internal K^+ concentration is directly measured by flame photometry.[3] With vesicular model membranes, more sophisticated methods can be used. In one of them, the changes in the volume of vesicles after their mixing with an hyperosmotic solute solution containing the drug were followed by a stop-flow measurement of the 90° light scattering. At first, shrinkage of vesicles was observed due to the very fast water efflux, followed by swelling, due to the drug-induced influx of the solute solution.[4]

Another elegant method exploits use of the fluorescence changes in the pH-sensitive fluorescent probe pyranine, encapsulated in unilamellar vesicles, to determine the permeability to saline ions.[3,5,6] If, for instance, the salt is composed of K^+ and an impermeant anion, an electroneutral exchange with H^+ through the membrane is necessary to maintain the K^+ flux. When the H^+ current is facilitated by a protonophore, the drug-induced K^+ flux becomes the rate-limiting step. The exchange is triggered either by a transmembrane pH gradient[3] or a salt gradient.[6] Another, more popular, spectrofluorimetric method involves measurement of an increase in fluorescence due to the drug-induced efflux of self-quenched encapsulated carboxyfluorescein or calcein.[7] Such dyes release freely through membranes in their anionic forms and it is the efflux of their alkaline counter-ion which controls their release.[8]

Finally, the AmB-induced changes in calcium permeability were studied using different methods depending upon the nature of the membrane. Using intact cells, the entry of extracellular calcium was monitored as the fluorescence change in quin-2, a specific probe, incorporated during incubation. In model vesicles, the modification of the absorbance of encapsulated Arsenazo III was followed.[9]

A method to study the mechanism of the leakage of alkaline ions A^+ is based on the electroneutral A^+/H^+ exchange through LUV membranes, which uses the chemical shift of the phosphate ion in the internal leaflet of the membrane, in ^{31}P-NMR.[10] In vesicles, the mechanism of the A^+ release depends on the comparative rates of the following processes: the "channel" formation, by accumulation of the drug on the vesicles, and exchange of the drug between vesicles. Either the characteristic signal of pH progressively gets shifted (synchronous permeabilization of the entire population) evoking a mobile-carrier mechanism or this signal progressively splits into two sub-populations, and the mechanism can be called a "channel-type". The latter case occurs when the binding of the drug to vesicle is so strong that its exchange rate becomes the limiting factor.[11]

B. PROBING AT THE LEVEL OF THE DRUG

As a heptaene AmB is wonderfully suited to optical spectroscopy. It has a high extinction coefficient between 300 and 450 nm, allowing its spectrum to be followed without interference of the membrane components (with the exception of erythrocyte ghosts which often contain traces of hemoglobin absorbing around 400 nm. However, this limitation does not present a problem for studying AmB-lipid interaction with model membranes). UV-visible absorption is therefore widely applied to study the AmB-lipid interactions. The absorption spectrum of AmB in aqueous media is concentration dependent. In the presence of lipids this spectrum is modified. The exact characteristics of this modification depend on the nature of lipids, the AmB-lipid ratio, temperature and time of incubation. They will not be detailed here. Grosso modo, a new band, is observed around 415 nm. The band around 340 nm observed in free self-associated AmB decreases in intensity or is shifted to blue.[11]

As far as circular dichroism (CD) is concerned, it represents the most interesting spectroscopic method. The fact that CD spectra present positive and negative bands increases the sensitivity to conformational changes in comparison to electronic absorption. Furthermore, it is the instrument of choice to detect the self-association of chiral molecules: self-association gives rise to intense biphasic dichroic signals (or "excitonic doublet"), easily observed. The higher the extinction coefficient of the molecule, the more intense this doublet. As a matter of fact, AmB with an ε reaching 120000 cm^{-1} mol^{-1} presents one of the highest excitonic doublet observed ($\Delta\varepsilon$ amplitude as high as 1000 cm^{-1} mol^{-1}, in water).

Finally, AmB is not fluorescent. Therefore, fluorescence could not be used directly. In contrast, it should be noted that filipin and lucensomysin, two other polyenes with shorter chains, are fluorescent. This characteristic has enabled studies of their interactions with membranes. Nystatin is very faintly fluorescent but no information is gained about its interaction with lipids. Aromatic heptaenes, such as vacidin A or candicidin D, are expected to fluoresce due to the presence of the aromatic ring, but it has not been observed. The membrane fluorescent probe trimethylammonium-diphenylhexatriene (TMA-DPH) has been used to study the interaction of AmB with membranes. The absorption spectra of AmB and TMA-DPH partly overlap. Under these conditions, if both the molecules are in close proximity of the membrane, upon excitation of TMA-DPH, an energy transfer to AmB will occur. However, since AmB is not fluorescent, this transfer is not radiative: AmB-TMA-DPH proximity will result in a decrease of TMA-DPH fluorescence from which the affinity of AmB for membrane can be determined.[12]

III. THE POLYENE ANTIBIOTIC BOUND TO THE MEMBRANE

A. AFFINITY FOR STEROLS AND PHOSPHOLIPIDS

Several studies were undertaken on model membranes to analyze the role of physical state of phospholipids (gel or liquid crystalline state), length and degree of unsaturation of the fatty acyl chains and presence of ergosterol on the antibiotic binding. Information about AmB binding, obtained at the molecular level (amount of bound drug and changes in its configuration, nature of the pores formed) and at the functional level (characteristics of the induced permeability pathways), was reviewed by Hartsel et al.[13] It should be noted that earlier studies were generally done on small unilamellar vesicles (SUV) which have special properties, not necessarily found in cellular membranes. Only recently, a more realistic model of large unilamellar vesicles (LUV) has been studied. It has been shown that pure phospholipid vesicles in liquid crystalline state, generally found in biological membranes, bind AmB poorly, and that binding is strong and complexes of stoichiometry 1:1 are formed with AmB in the presence of ergosterol in the membrane. It should be stressed that in cholesterol-containing membranes, which mimic mammalian cell membranes, the interaction appeared to be totally different, and any direct interaction between cholesterol and AmB could not be proven. Concerning filipin, recent fluorescence[14] and CD[7,15] studies have shown that incorporation of the antibiotic in sterol-free membranes is quite efficient, at variance with former results reported in literature.

Studies on mammalian cells corroborate the results obtained with cholesterol-containing liquid crystalline vesicles: with hepatocytes[16] or lymphocytes,[12] a 1:1 AmB-cholesterol stoichiometry could not be observed. Saturation of binding occurred for AmB concentrations corresponding to an AmB:cholesterol ratio of about 1:30. However, with erythrocyte ghosts at 15°C, CD spectra indicate that AmB is complexed with membrane cholesterol; the complex formation is saturable but not co-operative.[17] At 37°C, new spectra are observed, and their existence is conditioned by the presence of membrane protein: the binding is cooperative but not saturable. These results indicate that there may be two different modes of AmB complex formation with cholesterol-containing membranes, depending on the molecular organization of the membrane.

Only a few binding studies with fungal cells, particularly of their membranes, are available. It should be kept in mind that with these cells, the wall surrounding them should be considered as a barrier which prevents oligomers of AmB or its aggregates from reaching the membrane. It is indeed known that the fungal wall acts as a barrier for molecules of molecular weight higher than approximately 2000 Da. Actually, it was not observed in the dose-response curves of AmB-induced release of K^+ from *Candida albicans* (unpublished data). With *Saccharomyces cerevisiae*, it was shown that nystatin uptake decreases as the ergosterol concentration of the membrane increases, which is not an anticipated observation.[18] Prasad and Bolard have compared, by CD, the interaction of AmB with isolated plasma membranes of ergosterol-less mutants and wild type cells of *C. albicans* (unpublished data). Presence of ergosterol in the membrane did not appear to be a prerequisite for binding. The slight decrease in binding to ergosterol-less mutants could be attributed to changes in the degree of unsaturation of the phospholipid fatty acyl chains or the amount of phosphatidylserine in the membrane. However, in a study of another ergosterol-containing organism, *Leishmania mexicana*, it was shown by CD that the weaker binding of AmB, after heat transformation of the cells, correlated well with the lesser content of ergosterol in the membranes.[19] The specific enrichment of *S. cerevesiae* in phosphatidylcholine (PC), phosphatidylethanolamine (PE) or phosphatidylserine (PS) was shown to selectively protect the cells from nystatin action.[20,21]

B. CONFORMATION OF THE BOUND DRUG

As seen by CD, in the presence of gel state vesicles at a low antibiotic:lipid ratio ($R<0.03$), the conformation of AmB is totally different from that observed in the presence of liquid crystalline vesicles; spectra present four positive bands between 420 and 350 nm instead of a dichroic doublet centered around 330 nm and negative bands around 400 nm.[22] It has been proposed that, in the presence of gel state vesicles,[23] penetration of monomeric AmB into the lipid bilayer is highly unfavorable. AmB would therefore be restricted to the membrane interfacial region. Above $R = 0.03$, AmB is however able to disrupt individual lipid bilayers.

The study of the interaction of aromatic polyene antibiotics with small unilamellar vesicles revealed a much lower complexity as compared to that of AmB.[24] The number of polyene conformations corresponding to bound species is limited to two while with AmB, depending on the presence or absence of sterol in the membrane, the nature of the phospholipids, the time elapsed after the beginning of the interaction and temperature, nine different species were described. In the following sections are discussed the models which have been proposed for explaining the mechanism of action of AmB.

IV. MEMBRANE LIPIDS IN THE PRESENCE OF THE BOUND POLYENE ANTIBIOTIC

DSC measurements were performed by Janoff et al.[25] and Hamilton et al.[26] on particular AmB-phospholipid associations (liposomal formulations). 5 to 50 mol % of AmB were incorporated in a 7:3 dimyristoyl phosphatidyl choline (DMPC):dimyristoyl phosphatidyl glycerol (DMPG) mixture during the formation of liposomes. In both studies the preparations apparently differed only in the dispersion mode, either by bath sonication[24] or by simple vortexing.[25] By increasing the AmB content, in the first case, the main transition progressively disappeared, which corresponded (as seen by freeze-etch micrography) to the replacement of liposomal structures by ribbon-like structures. In contrast, in the second case, the transition peak persisted with a shoulder towards T>Tm. By the second procedure, complementary DSC studies were made on mixtures of AmB, with pure phospholipids having C_{14} saturated chains (DMPC) or C_{18} unsaturated mixed chains oleoyl stearoyl phosphatidyl choline (OSPC). In both cases, there was no AmB effect on the onset temperatures of endotherms. Besides, extension of the endothermic regions towards T > Tm was more pronounced with OSPC than with DMPC which reflected a higher thermal stability of the AmB association with OSPC. However, it was not apparent as to which factor, unsaturation or length of fatty acyl chains, was responsible for that. In this perspective, a study by one of the authors[27] was undertaken to examine if the length of the hydrophobic region of the phospholipid (the fatty acyl chains) and the polyene antibiotic (the polyenic chain) plays a role in the thermal stability of the association. Interactions between the pentaene filipin III or the tetraene nystatin A_1 and phosphatidylcholines with C_{12} (DLPC), C_{14} (DMPC) and C_{16} (DPPC) saturated chains were systematically studied. As seen in Figure 4, the onset temperatures of the endothermic regions were not shifted in the presence of the antibiotics, confirming the immiscibility of the polyene antibiotics with the phospholipid gel phase. However, the endotherms extended to T > Tm.

Complementary studies by CD show that within the temperature range of endotherms, the antibiotic remains bound to the membrane and that it is released at the completion temperature. Briefly, it could be said that the average transition temperature of phospholipids was shifted upward in the presence of polyene antibiotics, i.e., polyene antibiotics have a "solidifying" effect on the bilayers. While, in DMPC bilayers, this shift was also the largest in the presence of filipin III, in DLPC bilayers, it was the largest in the presence of nystatin A_1. Interestingly, within the temperature range of the corresponding endotherms, the association stoichiometries of DMPC-filipin III and DLPC-nystatin A_1 mixtures, separately determined by CD, remained constant: this showed that those mixtures were exceptionally thermally stable. Therefore, the hydrophobic association of the phospholipid and the antibiotic appears to be the factor which favors stability.[28,29]

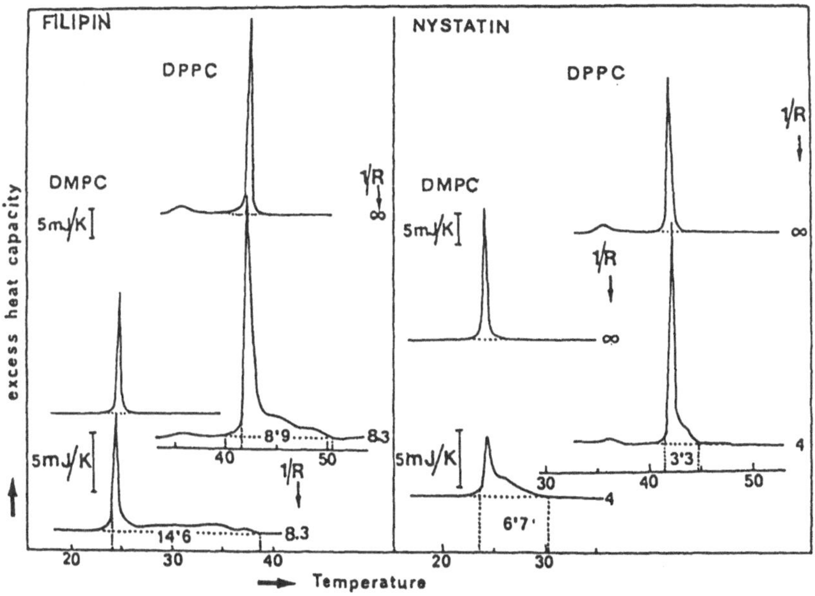

Figure 4. Thermograms of mixtures of DMPC and DPPC with filipin III (phospholipid/antibiotic = 1/R = 8.3) or nystatin A_1 (1/R = 4); lipid concentrations: 1mM.

^{2}H-NMR studies were performed with mixtures of AmB or filipin III with DMPC. Concerning AmB, comparison was made between cholesterol-free and cholesterol-containing [4'-$^2H_2$] DMPC bilayers in the presence of 5 to 30% of the drug. In the absence of cholesterol,[30] the spectra remained unchanged in the presence of AmB, at a temperature below Tm. However, above Tm, the spectra exhibited a broad sub-spectrum superimposed on a narrow doublet, characteristic of the pure phospholipid in the fluid phase with a maximum width of 120 KHz as in the gel phase (Figure 5). This reflects a partial AmB-induced motional restriction of the fatty acyl chains. From a dynamic point of view, it means that the corresponding phospholipids exchange with the bulk on a time scale greater than the time characteristic of ^{2}H-NMR ($>10^{-5}$ s). On the other hand, the amount of this broad component did not vary with temperature up to 40°C which could be due to the fact that the gel-like DMPC-AmB associations are thermally stable. In the presence of 30 mol % cholesterol,[31] the quadrupolar splitting exhibited no discontinuity throughout Tm. In compliance with the known regulatory effect of cholesterol on the membrane ordering and in the presence of AmB, a change in the spectra was observed only at the level of the center of the bilayer.

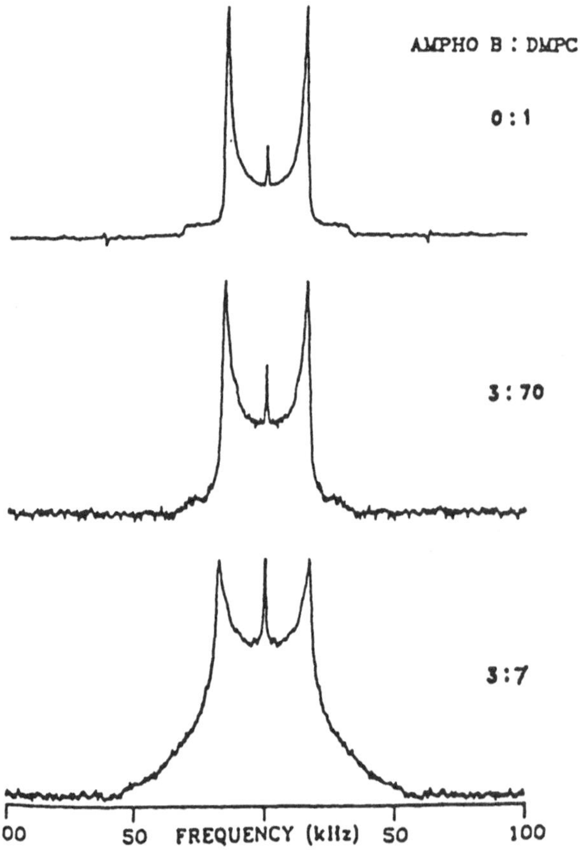

Figure 5. Effect of AmB on the ^{2}H-NMR spectra of [4′-$^{2}H_2$] DMPC membrane at 25°C. From Dufourc, E.J., Smith, I.C.P. and Jarrell, H.C., *Biochim. Biophys. Acta*, 778, 435, 1984. With permission.

The modification of the ^{2}H-NMR spectra of DMPC fatty acyl chains, by incorporating filipin III during the MLC formation, was investigated in presence or absence of cholesterol. In the absence of cholesterol, 12 mole % of filipin III induced a significant change in the spectra of perdeuterated DMPC. A broad component having the features of a gel phase spectrum superimposed on the fluid pure DMPC spectrum appeared between Tm and Tm+11°C. The corresponding immobilized phospholipid domains were interpreted as filipin-DMPC complexes. In the presence of 30 mol % of cholesterol, DMPC spectra did not exhibit any indication of a gel-like phase.[32] The authors concluded that the effect of filipin III was masked by cholesterol. However, it appears more suitable to examine this observation in the light of results obtained by CD. The changes in filipin CD spectra obtained by adding cholesterol-containing LUV with different antibiotic/lipid ratios showed that filipin binds competitively to cholesterol and phospholipid.[15] The features of filipin III CD spectra when the quantity of drug is more than that of cholesterol are the same as those obtained in the presence of cholesterol-free DMPC bilayers.[33] This suggests that a filipin-phospholipid association can occur when the sterol sites of fixation are occupied.

Two other NMR spectroscopic techniques were used to study $CDCl_3$:DMSO-d6 dipalmitoyl-PC (DPPC)/AmB solutions. By [^{1}H]-NMR, line broadenings were observed on lipid and AmB resonance and, by [^{31}P]-NMR, the titration of AmB by DPPC led to an increase in the line width and a small chemical shift of the [^{31}P] signal.[34] These findings support the formation of a complex between the drug and the oxygen atoms bound to the phospholipid phosphorus. The spatial proximity (3.6 to 4 Å) of the heptaenic chain of AmB and the lipid chain, estimated from Nuclear Overhauser Enhancement (NOE), seems to confirm this conclusion. Infra-red (IR) spectroscopy was also used to monitor the state of the membrane at the interface.[35] The effect of adding AmB to cholesterol containing DPPC bilayers on C=O bands of the acyl chains was examined as a function of temperature. In pure labeled 2-[1-^{13}C]-DPPC, the sn1 and sn2 chain C=O modes are distinct and show an abrupt broadening at the gel-to-fluid transition temperature, Tm. This broadening reflects an increase in motional freedom. Incorporation of cholesterol cancels this discontinuity, bands get broadened whatever the temperature. On spectra of DPPC: cholesterol:AmB layers (1:1:2), the line-broadening effect of cholesterol was amplified. This finding, which reflects a change in the charge distribution in thc region surrounding the carbonyl groups, was attributed to an AmB-induced change in water penetration.

The action of AmB (permeabilization to K^+) and its derivatives on erythrocytes (cholesterol-containing cells) or *S. cerevisiae* (ergosterol-containing cells) were comparatively examined by taking into account the cell concentration.[3] The dose-response curves behave differently. Irrespective of the erythrocyte concentration, the antibiotic activity starts from the same threshold concentration. In contrast, this activity can start well below this threshold if the fungal cell concentration is low. With human erythrocytes[36] and BALB/c B lymphoid cell line, A20,[37] vacidin A was shown to exhibit much higher protonophoric activity than AmB. The esterification of the carboxyl group eliminated the ability of the antibiotic to increase H^+ conductance. Unsubstituted aromatic heptaenes, at high concentration, increased the conductive flux of Cl^- in a concentration dependent manner.

It was shown that AmB generates, on cholesterol-containing EPL-LUV and SUV, highly selective permeability for K^+, with respect to the Cl^- anion model vesicles.[5] In this model the permeability to K^+ was studied on ergosterol and cholesterol-containing EPC-LUV.[3] The lipid concentration dependence of the dose-response curves paralleled that observed for the

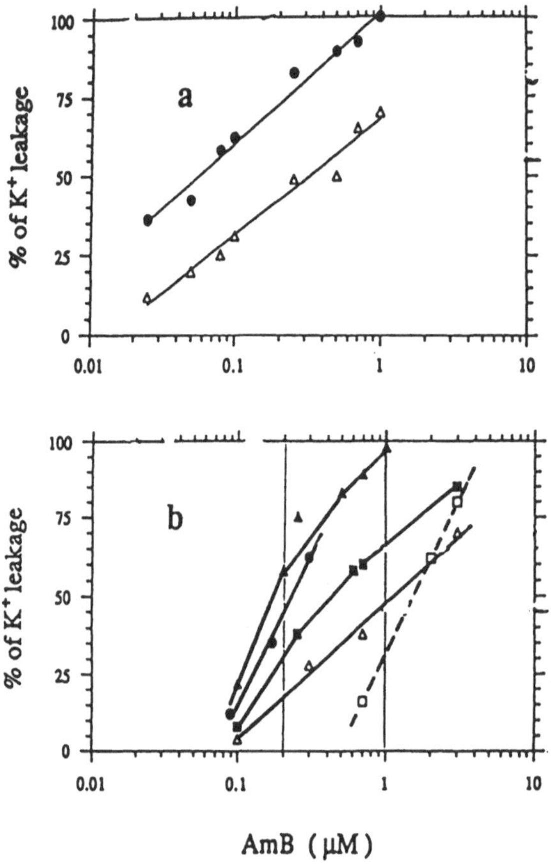

Figure 6. AmB-induced K^+ leakage from sterol-containing LUV (a): 10 mole% ergosterol EPC-LUV with EPC = 1.8 (Δ) and 0.18 mM (●); (b): 10 mole% cholesterol EPC-LUV with EPC = 2.2 (Δ) and 0.11 mM (●); 20 mole% cholesterol EPC-LUV with EPC = 2.2 (■) and 0.22 mM (Δ). Experiments were made at 23°C except (-------) at 37°C. From Bolard, J., Legrand, P., Heitz, F. and Cybulska, B. *Biochemistry*, 30, 5707, 1991. With permission.

whole cells. In the presence of cholesterol, the K^+ leakage occurs only when the drug concentration exceeds that of its self-association whereas in the presence of ergosterol, leakage depends only on the AmB-lipid molar ratio (Figure 6).

The permeability to alkaline ions is discriminated by the ^{31}P-NMR method either as a progressive leakage of the entire LUV population or as an all-or-none ("channel type") one. This may be due to the difference in the nature of the sterol present in the membrane[38] and also due to the chemical composition of the polar head group of the antibiotic.[39] With EPC-LUV, three cases can be distinguished: i) AmB develops a "channel"-type permeability irrespective of the nature of sterol. However, with the same dose, the permeabilization is ten times slower in the presence of cholesterol than that in the presence of ergosterol. ii) When the carboxyl group is substituted, the permeabilization is globally reduced and exhibits a mechanism which depends on the nature of sterol. In the presence of cholesterol (or a sterol whose side-chain is similar), the mechanism becomes a slow progressive one. Those antibiotics are, therefore, selectively toxic to ergosterol-containing membranes. iii) A substitution of the amino group leads to a big drop in the permeabilization efficiency irrespective of the sterol nature. These qualitative differences are probably due to different lifetimes of the ionic paths formed.

The pattern of the Na^+ leakage of EPC-LUV parallels the differences observed on the K^+ leakage of biological membranes.[40] This would suggest that EPC membranes have properties similar to biological membranes. We shall, however, see in the following section that it is not the same when the lipid chains of the phospholipid are saturated.

On EPC-LUV, AmB induces the permeability of Ca^{2+} which again behaves differently depending upon the nature of the sterol present in the membrane.[9] The AmB-induced Ca^{2+} flux increases linearly with the AmB-lipid molar ratio. While these straight lines merge in the presence of ergosterol, they diverge in the presence of cholesterol at the same AmB threshold concentration. Moreover, the same proportion of AmB is about ten times less efficient on cholesterol-containing membranes than on ergosterol-containing membranes.[9]

Another difference between cholesterol and ergosterol-containing LUV membranes is observed in "channel" formation. Depending on concentration and the time elapsed after mixing of AmB and LUV, two types of "channels" can be distinguished: "non-aqueous channels" which allow the passage of urea and small ions and some larger "aqueous channels", which allow the passage of glucose.[41] The first type of channels are formed at lower AmB concentration in ergosterol-containing membranes than in cholesterol-containing membranes. The nature of the phospholipid affects the channel formation. When the membranes were composed of fluid DMPC and ergosterol, "non-aqueous channels" were immediately formed which turned into "aqueous channels" in a few milliseconds. In contrast, with EPC-LUV, a fast ionic permeabilization occurs for an AmB molar ratio of 0.001 in the presence of ergosterol.[4] However, in the presence of cholesterol, this permeabilization is slow and requires more antibiotic.[41] These results agree well with those obtained by ^{31}P-NMR.

It should be emphasized that in all the aforementioned experiments, membranes were in fluid phase. Experiments on ergosterol-containing DMPC-LUV showed that the urea permeability was decreased by five-folds when the membranes were in gel phase.[41]

What happens in sterol-free membranes? Studies performed with such membranes emphasize the role of the radius of curvature of vesicles, as well as the importance of the fatty acid composition of bilayers. When the AmB concentration is a few mole %, with respect to the phospholipid, its permeabilization to alkaline ion is negligible in EPC-LUV.[5,41] However, it is significant in fluid DMPC-LUV.[41,42] In contrast, geometrical constraints imposed on SUV make them more permeable for the same concentration of AmB, even when the phospholipid is unsaturated.[5,42] The assumption that the presence of sterol is essential for permeabilization does not hold true in case of filipin. Indeed, this drug is often considered as a membrane-disrupting agent owing to the bulky complexes that it forms with membranous cholesterol. By using the carboxyfluorescein spectrofluorimetric method, it was demonstrated that filipin induces the same permeability for Na^+, whether or not they contain cholesterol, in EPC-SUV as shown in Figure 3B.[7]

V. MECHANISM OF ACTION

A. ASSEMBLY OF AMPHOTERICIN B MOLECULES IN THE MEMBRANE

There is no evidence about the nature of the transmembrane permeability pathways induced by polyene antibiotics. Information can only be obtained from the functional studies and the binding characteristics of the membranes. The classic single and double sided sterol/pore model is well known. An AmB channel would resemble a barrel and most likely the number of "staves" would be 8-12 AmB molecules. Furthermore, cholesterol and ergosterol fit quite well between the individual staves. Depending on the length of the phospholipid molecules constituting the membrane, one or two aligned barrels are necessary to span the membrane.

However, this simple model does not seem to be sufficient to reconcile all the experimental data.[43,44] Harstel et al. have proposed a synthesis based on a "raft" structure of self-associated AmB. They proposed that AmB forms dynamic amorphous structures, in competition with relatively well defined structures. The former ones would act as defect inducers. Its basis would be aggregates of AmB in solution, or rafts, with a very polar and a very non-polar surface. These rafts would initially bind to the lipid/water interface and then insert simultaneously with other rafts to form organized channels of varying sizes and organizations. Experimental evidence for an AmB aggregate adsorption orientation process, resembling the scenario outlined above, has been obtained from sterol containing monolayer studies.[45] After initial adsorption of AmB rafts, sterols could bind specifically to AmB molecules and "pull" them into the membrane. Non-specific or less specific enhancement of AmB activity by sterol could arise from spacing out of the lipid head group allowing easier adsorption of rafts.

B. ALTERATION OF THE OVERALL MEMBRANE ORGANISATION

The hypothesis of action at the level of the overall membrane organization traces back to twenty years from studies on model membranes.

It has again become significant, owing to the aforementioned recent results on filipin III and SUV[7] and on AmB and SUV[5,6] or LUV.[41,42] The organization of model membranes in the presence of the drug can be depicted from the aforedescribed DSC and ^{2}H-NMR results as follows:

i. the drug is embedded in an aggregated form in fluid membranes, within a more or less broad temperature range, corresponding to the temperature limits of endotherms,
ii. between those temperature limits, the drug exerts an ordering action on phospholipids throughout the entire length of fatty acyl chains by creating gel-like domains corresponding to antibiotic-phospholipid associations.

It seems reasonable to visualize the phospholipid-antibiotic aggregates with the rigid polyene chain of the antibiotic extended along the fatty acyl chains in order to optimize van der Waals interaction while the hydrophilic portions of the macrolide ring would be aggregated together. Within the temperature range where such aggregates are thermally stable, the lateral organization of membrane would look like a mosaic of fluid and gel domains. This observation claims in favor of a mechanism of permeabilization, based on the drug-induced lateral heterogeneity of the membrane. Thus, recent theoretical studies on the effects of the incorporation of diverse membrane-perturbing molecules seem relevant.[46] The authors take the model of the bilayers as a triangular lattice where one phospholipid interacts with the six nearest neighbors and perturbants are considered as interstitial molecules. The different acyl chain conformations are represented by ten states with a more or less great number of *gauche* rotations. They examined as to how such perturbants modulate the lateral density fluctuations around Tm and its impact on the local partitioning of the perturbant. The following results emerged:

i. the sharp peak of the specific heat characteristic of the transition of the pure phospholipid broadens while its height decreases,
ii. the local concentration of the perturbant becomes higher at the interfaces between gel and fluid domains over a broad temperature range around Tm,
iii. the maximum lateral compressibility of the bilayers at Tm becomes expanded on an increased temperature range.[47]

Based on a similar interpretation, the permeabilizing action of filipin III on EPC-SUV is assigned to the formation of gel-like domains (phospholipid-filipin aggregates) into the bulk and fluid at room temperature.[7] However, one must emphasize that such a mechanism can occur only when the molar ratio of the drug with respect to the phospholipid is at least a few mol %.

VI. CONCLUSION

There is good evidence that the antifungal activity of AmB does not solely result from the formation of ergosterol pores through the membranes of the fungi. In particular, it seems reasonable to assume that water-soluble oligomers of AmB can also induce cation permeability without direct participation of the ergosterol molecules, in a manner similar to that observed in cholesterol-containing membranes. Moreover, the sensitivity of fungal cells to AmB depends not only on the amount of ergosterol present in the membranes but also on the phospholipid composition. This dependence, especially the role played by the saturation of the fatty acyl chains, should be more operative as it is an important parameter of lipid peroxidation. Finally, the possibility that AmB may be internalized in *Candida* has not been explored, although it could have a negligible role in the activity of the drug. It should be noted that in the past years much more effort has been devoted to the elucidation of the mechanism of toxicity of AmB than to its antifungal activity. The reason is that, till now, it has not been possible to develop derivatives or lipid formulations of AmB itself (with the exception of aromatic heptaenes), whereas it was possible to find conditions under which AmB was less toxic. Consequently, efforts have focused on the latter aspect, neglecting the former which would certainly deserve more interest. For this reason it seems that methyl ester derivatives of aromatic polyene antibiotics, which are less toxic than their parent compounds while keeping higher antifungal activity, deserve special attention.

VII. REFERENCES

1. **Bolard, J.,** Mechanism of action of an anti-*Candida* drug: amphotericin B and its derivatives, in *Candida albicans, Cellular and Molecular Biology*, Prasad, R., Ed., Springer-Verlag, 1991, 215.
2. **Bunow, M.R. and Levin, I.W.,** Vibrational Raman spectra of lipid systems containing amphotericin B, *Biochim. Biophys. Acta*, 464, 202, 1977.

3. **Bolard, J., Legrand, P., Heitz, F. and Cybulska, B.,** One-sided action of amphotericin B on cholesterol-containing membranes is determined by its self-association in the medium, *Biochemistry*, 30, 5707, 1991.
4. **Cohen, B.E.,** Concentration and time-dependence of amphotericin B-induced permeability changes across ergosterol-containing liposomes, *Biochim. Biophys. Acta*, 857, 117, 1986.
5. **Hartsel, S.C., Benz, S.K., Peterson, R.P. and Whyte, B.S.,** Potassium-selective amphotericin B channels are predominant in vesicles regardless of sidedness, *Biochemistry*, 30, 77, 1991.
6. **Whyte, B.S., Peterson, R.P. and Hartsel, S.C.,** Amphotericin B and nystatin show different activities on sterol-free vesicles, *Biochem. Biophys. Res. Comm.*, 164, 609, 1989.
7. **Milhaud, J.,** Permeabilizing action of filipin III on model membranes through a filipin-phospholipid binding, *Biochim. Biophys. Acta*, 1105, 307, 1992.
8. **Bramhall, J.,** Electrostatic forces control the penetration of membranes by charged solutes, *Biochim. Biophys. Acta*, 778, 393, 1984.
9. **Ramos, H., Attias de Murciano, A., Cohen, B.E. and Bolard, J.,** The polyene antibiotic amphotericin B acts as a Ca^{2+} ionophore in sterol-containing liposomes, *Biochim. Biophys. Acta*, 982, 303, 1989.
10. **Hervé, M., Cybulska, B. and Gary-Bobo, C.M.,** Cation permeability induced by valinomycin, gramicidin D and amphotericin B in large lipidic unilamellar vesicles studied by ^{31}P-NMR, *Eur. Biophys. J.*, 12, 121, 1985.
11. **Bolard, J.,** How do the polyene antibiotics affect the cellular membrane properties?, *Biochim. Biophys. Acta*, 864, 257, 1986.
12. **Henry-Toulmé, N., Séman, M. and Bolard, J.,** Interaction of amphotericin B and its N-fructosyl derivative with murine thymocytes: a comparative study using fluorescent membrane probe, *Biochim. Biophys. Acta*, 982, 245, 1989.
13. **Hartsel, S.C., Hatch, C. and Ayenew, W.,** How does amphotericin B work? Studies on model membrane systems, *J. Liposome Res.*, 3, 377, 1993.
14. **Castanho, M.A. and Prieto, M.J.E.,** Fluorescence study of the macrolide pentaene antibiotic filipin in aqueous solution and in a model system of the membranes, *Eur. J. Biochem.*, 207, 125, 1992.
15. **Milhaud, J., Mazerski, J. and Bolard, J.,** Competition between cholesterol and phospholipid for binding to filipin, *Biochim. Biophys. Acta*, 987, 193, 1989.
16. **Binet, A. and Bolard, J.,** Recovery of hepatocytes from attack by the pore former amphotericin B, *Biochem. J.*, 253, 435, 1988.
17. **Szponarski, W. and Bolard, J.,** Temperature-dependent modes for the binding of the polyene antibiotic amphotericin B to human erythrocyte membranes. A circular dichroism study, *Biochim. Biophys. Acta*, 897, 229, 1987.

18. **Beezer, A.E. and Sharma, P.B.,** On the uptake of nystatin by *Saccharomyces cerevisiae*, *Microbios*, 31, 7, 1981.
19. **Ramos, H., Milhaud, J., Cohen, E. and Bolard, J.,** The effect of amphotericin B on *Leishmania mexicana* promastigotes and heat-transformed forms, *Antimicrob. Agents Chemother.*, 34, 1584, 1990.
20. **Rao, T.V.G., Das, S. and Prasad, R.,** Effect of phospholipid enrichment on nystatin action: differences in antibiotic sensitivity between *in vivo* and *in vitro* conditions, *Microbios*, 42, 145, 1985.
21. **Rao, T.V.G., Trivedi, A. and Prasad, R.,** Phospholipid enrichment of *Saccharomyces cerevisiae* and its effect on polyene sensitivity, *Canad. J. Microbiol.*, 31, 322, 1985.
22. **Madden, T.D., Janoff, A.S. and Cullis, P.R.,** Incorporation of amphotericin B into large unilamellar vesicles composed of phosphatidylcholine and phosphatidylglycerol, *Chem. Phys. Lip.*, 52, 189, 1990.
23. **Perkins, W.R., Minchey, S.R., Boni, L.T., Swenson, C.E., Popescu, M.C., Pasternack, R.F. and Janoff, A.S.,** Amphotericin B-phospholipid interaction responsible for reduced mammalian cell toxicity, *Biochim. Biophys. Acta*, 1107, 271, 1992.
24. **Mazerski, J., Bolard, J. and Borowski, E.,** Circular dichroism study of the interaction between aromatic heptaene antibiotics and small unilamellar vesicles, *Biochem. Biophys. Res. Comm.*, 116, 520, 1983.
25. **Janoff, A.S., Boni, L.T., Popescu, M.C., Minchey, S.R., Cullis, P.R., Madden, T.D., Taraschi, T., Grunner, S.M., Shyamsunder, E., Tate, M.W., Mandelsohn, R. and Boner, D.,** Unusual lipid structures reduce the toxicity of amphotericin B, *Proc. Natl. Acad. Sci. USA,*, 85, 6122, 1988.
26. **Hamilton, K.S., Barber, K.R., Davis, J.H., Neil, K. and Grant, C.W.M.,** Phase behaviour of amphotericin B multilamellar vesicles, *Biochim. Biophys. Acta*, 1062, 220, 1991.
27. **Milhaud, J.,** Unpublished results,
28. **Ganis, P., Avitabile, G., Mechlinski, W. and Schaffner, C.P.,** Polyene macrolide antibiotic amphotericin B. Crystal structure of the N-iodoacetyl derivative, *J. Amer. Chem. Soc.*, 93, 4560, 1971.
29. **Lewis, B.A. and Engelman, D.M.,** Lipid bilayer thickness varies linearly with acyl chain length in fluid phosphatidylcholine, *J. Biol. Chem.*, 166, 211, 1983.
30. **Dufourc, E.J., Smith, I.C.P. and Jarrell, H.C.,** Interaction of amphotericin B with membrane lipids as viewed by ^{2}H-NMR, *Biochim. Biophys. Acta*, 778, 435, 1984.
31. **Dufourc, E.J., Smith, I.C.P. and Jarrell, H.C.,** Amphotericin B and model membranes. The effect of amphotericin B on cholesterol-containing systems as viewed by ^{2}H-NMR, *Biochim. Biophys. Acta*, 776, 317, 1984.

32. **Dufourc, E.J. and Smith, I.C.P.**, ^{2}H-NMR evidence for antibiotic-induced cholesterol immobilization in biological model membranes, *Biochemistry*, 24, 2420, 1985.
33. **Milhaud, J., Mazerski, J., Bolard, J. and Dufourc, E.J.**, Interaction of filipin with dimyristoylphosphatidylcholine membranes studied by ^{2}H-NMR, circular dichroism, electronic absorption and fluorescence, *Eur. Biophys. J.*, 17, 151, 1989.
34. **Balakrishnan, A. and Easwaran, K.R.K.**, Lipid-amphotericin B complex structure in solution: a possible first step in the aggregation process in cell membranes, *Biochemistry*, 32, 4139, 1993.
35. **Green, P.M., Mason, J.T., O'Leary, T.J. and Levin, I.W.**, Effect of hydration, cholesterol, amphotericin B and cyclosporin A on the lipid bilayer interface region: an infra-red spectroscopic study using 2-[1-^{13}C]-dipalmitoyl phosphatidylcholine, *J. Phys. Chem.*, 91, 5099, 1987.
36. **Cybulska, B., Bolard, J., Seksek, O., Czerwinski, A. and Borowski, E.**, Identification of the structural factors of amphotericin B and other polyene macrolide antibiotics influencing selectivity of the permeability pathways induced in the red cell membrane, *Biochim. Biophys. Acta* (in press).
37. **Cybulska, B., Seksek, O., Henry-Toulmé, N., Czerwinski, A. and Bolard, J.**, Polyene macrolide antibiotics: indirect stimulation of the Na^{+}/H^{+} exchanger of BALB/c B lymphoid cell line, A20, *Biochem. Pharmacol.*, 44, 539, 1992.
38. **Hervé, M., Debouzy, J.C., Borowski, E., Cybulska, B. and Gary-Bobo, C.M.**, The role of the carboxyl and amino groups of polyene macrolides in their interactions with sterols and their selective toxicity – a ^{31}P-NMR study, *Biochim. Biophys. Acta*, 980, 261, 1989.
39. **Cybulska, B., Hervé, M., Borowski, E. and Gary-Bobo, C.M.**, Effect of the polar head structure of polyene macrolide antifungal antibiotics on the mode of permeabilization of ergosterol and cholesterol-containing lipidic vesicles studied by ^{31}P-NMR, *Mol. Pharmacol.*, 29, 293, 1986.
40. **Cheron, M., Cybulska, B., Mazerski, J., Grzbowska, J., Czerwinski, A. and Borowski, E.**, Quantitative structure-activity relationships in amphotericin B derivatives, *Biochem. Pharmacol.*, 37, 827, 1988.
41. **Cohen, B.E.**, A sequential mechanism for the formation of aqueous channels by amphotericin B in liposomes. The effect of sterols and phospholipid composition, *Biochim. Biophys. Acta*, 1108, 49, 1992.
42. **Milhaud, J., Hartmann, M.A. and Bolard, J.**, Interaction of the polyene antibiotic amphotericin B with model membranes; differences between small and large unilamellar vesicles, *Biochimie*, 71, 49, 1989.
43. **Hartsel, S.C., Benz, S.K., Ayenew, W. and Bolard, J.**, Na^{+}, K^{+}, Cl^{-} selectivity of permeability pathways induced through sterol-containing

membrane vesicles by amphotericin B and other polyene antibiotics, *Eur. Biophys. J.*, 23, 125, 1994.

44. **Legrand, P., Romero, E., Cohen, E. and Bolard, J.**, Effects of aggregation and solvent on the activity of amphotericin B on human erythrocytes, *Antimicrob. Agents Chemother.*, 36, 2518, 1992.
45. **Saint-Pierre-Chazalet, M., Thomas, C., Dupeyrat, M. and Gary-Bobo, C.M.**, Amphotericin B-sterol complex formation and competition with egg phosphatidylcholine: a monolayer study, *Biochim. Biophys. Acta*, 944, 477, 1988.
46. **Jorgensen, K., Ipsen, J.H., Mouritsen, O.G., Bennett, D. and Zuckermann, M.J.**, The effects of density fluctuations on the partitioning of foreign molecules into lipid bilayers: application to anaesthetics and insecticides, *Biochim. Biophys. Acta*, 1067, 241, 1991.
47. **Cruzeiro-Hansson, L. and Mouritsen, O.G.**, Passive ion permeability of lipid membranes modelled via lipid-domain interfacial area, *Biochim. Biophys. Acta*, 944, 63, 1988.

INDEX

—A—

—B—

—C—

—D—

—E—

—F—

—G—

—M—

—N—

—Q—

—R—

—S—

—T—

—U—

—V—

—W—

—Y—

—Z—